Rosina Pastore

Vedānta, Bhakti, and Their Early Modern Sources

Welten Süd- und Zentralasiens / Worlds of South and Inner Asia / Mondes de l’Asie du Sud et de l’Asie Centrale

Im Auftrag der Schweizerischen Asiengesellschaft / On behalf of the Swiss Asia Society / Au nom de la Société Suisse-Asie

Edited by
Blain H. Auer
Karénina Kollmar-Paulenz
Angelika Malinar
Nicolas Martin
Ingo Strauch

Volume 14

Rosina Pastore

Vedānta, Bhakti, and Their Early Modern Sources

Philosophical and Religious Dimensions of Brajvāsīdās's Prabodhacandrodaya Nāṭaka

DE GRUYTER

Swiss National
Science Foundation

ISBN 978-3-11-106318-8
e-ISBN (PDF) 978-3-11-106383-6
e-ISBN (EPUB) 978-3-11-106431-4
ISSN 1661-755X
DOI https://doi.org/10.1515/9783111063836

Library of Congress Control Number: 2023944391

Bibliographic information published by the Deutsche Nationalbibliothek
The Deutsche Nationalbibliothek lists this publication in the Deutsche Nationalbibliografie; detailed bibliographic data are available on the internet at http://dnb.dnb.de.

Cover image: Dinodia Photo
Typesetting: Integra Software Services Pvt. Ltd.
Printing and binding: CPI books GmbH, Leck

www.degruyter.com

MIX
Papier | Fördert gute Waldnutzung
FSC www.fsc.org FSC® C083411

Acknowledgements

I am deeply grateful to the numerous individuals who have supported me throughout the six years leading to the completion of my doctoral thesis, which serves as the foundation for this book. I am fortunate to have worked with Professor Maya Burger at the University of Lausanne (UNIL) as her assistant and PhD candidate. Her unwavering belief in my abilities, coupled with the intellectual nourishment derived from participating in her classes and engaging in countless discussions, has played an immense role in shaping both my academic growth and personal development. Perhaps the extent of its impact exceeds even our own imaginations. Furthermore, I would like to express my sincere appreciation to my thesis co-supervisor Professor Stefania Cavaliere, whose steadfast motivation and valuable suggestions have been indispensable throughout my journey, beginning from my undergraduate and master's years in Naples.

I am also indebted to the members of my jury, Dr Basile Leclère and Dr Anand Venkatkrishnan, whose unique approaches and encouragement have continuously pushed me to strive for precision in my work and academic stance. The members of the Department of Slavic and South Asian Languages and Civilisations at UNIL have also been instrumental in my progress, actively contributing to my daily efforts to completing this work. Their support and friendship have been invaluable. During my year in India at Santiniketan, made possible by a mobility scholarship funded by UNIL's Research Committee and the Centre hospitalier universitaire vaudois in Lausanne (2019–2020), I was warmly welcomed and supported by Professor Asha Mukherjee and her family, Professor Rameshwar Mishra of the Hindi Department and Professor Raj Kumar of the Benares Hindu University. All have been fundamental to the advancement of this research. I also wish to thank Jon Wilcox, who copyedited this book, harmonised the English, and provided valuable comments. The publication of this volume is funded by the Swiss National Science Foundation.

To my friends in Switzerland and Italy, I am immensely grateful for their unwavering companionship and support throughout the highs and lows of my journey as a PhD student. Finally, I want to acknowledge my family – my dad, mum, brother, and sister. Their affection and support have held both myself and this book together. It is to them that I dedicate this work.

https://doi.org/10.1515/9783111063836-202

Contents

Part II: **Annotated Translation of the *Prabodhacandrodaya Nāṭaka***

List of Tables and Figures

https://doi.org/10.1515/9783111063836-204

Note to the Reader

The Indic languages featured in this book are presented in a transliterated form using the Latin alphabet with diacritical marks, following the system outlined by Ronald S. McGregor in the *Oxford Hindi-English Dictionary*. Sanskrit, Brajbhāṣā, and Avadhī terms keep the final short -a, which is not retained for modern Hindi terms and expressions nor the names of authors. Hence, we have "Brajvāsīdās" (or "Brajvāsī") instead of "Brajavāsīdāsa" and "Tulsīdās" (or "Tulsī") instead of "Tulasīdāsa". The titles of lesser-known Brajbhāṣā and Avadhī works retain the final short -a, while titles of well-known works are presented in their most recognised format. For example, *Rāmcaritmānas* instead of *Rāmacaritamānasa* and *Sūrsāgar* instead of *Sūrasāgara*.

Terms of Perso-Arabic origin found in Brajbhāṣā texts are transliterated according to the method employed for Brajbhāṣā itself. Conversely, titles of Persian works are transliterated following the conventions outlined in John Platts's *A Dictionary of Urdu, Classical Hindi, and English*. Proper names of renowned historical figures of Perso-Arabic origin are transliterated without diacritical marks, such as "Dara Shukoh". Geographical place names are written with diacritical marks only when they appear so in literary works. For example, "Ayodhyā" is used when referring to the capital of King Rāma's reign in the *Rāmcaritmānas*.

Adjectives of Indic origin combined with English suffixes are presented in accordance with (British) English orthography and without diacritical marks, following the spelling of the *Oxford English Dictionary*. As such, terms such as Advaitic, Puranic, shastric, Vedantic, Upanishadic, and yogic are used instead of śāstric, Upaniṣadic, and so on. Nouns of Indic origin currently used in the English language also adhere to this rule, such as "yogi". All other Indic terms are transliterated with diacritics. English translations or explanations of these terms are provided, which may vary depending on the context, and are used interchangeably with the original terminology. Dates following Vikram era dating are prefaced with "VS", while other cases follow the Gregorian calendar.

The written representation of Indian poetry exhibits a remarkable variability, showcasing a range of influences, historical contexts, and linguistic factors. Consequently, it becomes essential to establish clear definitions of certain terms used in this book. A poetic line (or "line") denotes a single metrical line in a poetic composition. A "verse" is a grouping of poetic lines, which may consist of one or more lines, according to the organisation of that specific poetic composition. Within a poetic line, a "pāda" refers to a distinct section, typically separated by a single daṇḍa (comparable to a full stop), with the other half concluding with a double daṇḍa. Where quoted verse runs on in the text or notes, the end of the first pāda is indicated by a single vertical slash (|), while the end of the second is

https://doi.org/10.1515/9783111063836-205

signalled by double slashes (||), which may or may not coincide with the end of a poetic line. As such, vertical slashes are being used here in place of daṇḍas. It is important to note that a poetic line can comprise two or more pādas, depending on the particular metre employed. All translations provided in this book are my own unless otherwise specified.

List of Abbreviations

Texts

BhG	*Bhagavadgītā*
BhP	*Bhāgavata Purāṇa*
BS	*Brahmasūtras*
GH	*Gulzār-i Ḥāl* by Banvālīdās
KPC	*Prabodhacandrodaya* by Kṛṣṇamiśra
NS	*Nāṭyaśāstra* by Bharata
PcN	*Prabodhacandrodaya Nāṭaka* by Brajvāsīdās
RcM	*Rāmcaritmānas* by Tulsīdās
YS	*Yogasūtras* by Patañjali

Upaniṣads

BĀU	*Bṛhadāraṇyaka Upaniṣad*
CU	*Chāndogya Upaniṣad*
MU	*Muṇḍaka Upaniṣad*
ŚvU	*Śvetāśvatara Upaniṣad*

Dictionaries

BHK	Kalika Prasad (1956): *Bṛhat Hindī Koś*. Varanasi: Prakashak Gyanmandal.
HSS	Shyamsundar Das (1965–1975): *Hindī Śabdsāgar*. Kashi: Nagari Pracarini Sabha.
MW	Monier Monier-Williams (1899): *A Sanskrit-English Dictionary*. Oxford: Clarendon Press.
OHD	Ronald S. McGregor (1993): *The Oxford Hindi-English Dictionary*. Oxford; Delhi: Oxford University Press.
Platts	John T. Platts (1884): *A Dictionary of Urdu, Classical Hindi, and English*. London: W. H. Allen & Co.
SK	Dindayal Gupta / Premnarayan Tandan (1974): *Brajbhāṣā Sūr-koś*. Lucknow: Lucknow Vishvavidyalaya.
TS	Bholanath Tiwari (1954): *Tulsī-śabdsāgar*. Ilahabad: Hindustani Ekedamy.

Other

MSH	Modern Standard Hindi
PC	Prabodhacandrodaya (the story, not necessarily a specific text)
Skt.	Sanskrit
v./vv.	verse/verses
VS	Vikram Saṃvat

https://doi.org/10.1515/9783111063836-206

Introduction

This book delves into the realm of philosophy in early modern India, with the specific focus of exploring the positioning of Brajvāsīdās (b. 1730s–?), an author who wrote in Brajbhāṣā, in philosophical matters through his first composition, the *Prabodhacandrodaya Nāṭaka* (The Drama of the Rise of the Wisdom Moon, c. 1760 [VS 1817]; hereafter referred to as PcN). The PcN, which remains unedited and untranslated to this day, belongs to a tradition of allegorical dramas, a retelling of a story best known in its Sanskrit form as the *Prabodhacandrodaya*, composed by Kṛṣṇamiśra in the eleventh century (hereafter referred to as KPC). Within Brajvāsīdās's text, philosophical and religious speculations abound, challenging the conventional sources commonly employed to study Vedānta and bhakti in the early modern era. This book aims to demonstrate how an allegorical drama can contribute to tracing the history of philosophical ideas during this period.

By examining my own translation of the PcN, upon which this critical study is based, it becomes evident that Brajvāsīdās's imagery draws from, on the one hand, the Upaniṣads and the teachings of Vedānta philosophers and, on the other, the literary universe of bhāṣā (regional or local languages). Thus, this study seeks to illustrate that relying solely on Sanskrit shastras, intended as systematic and argumentative theorisations of a subject, when analysing a vernacular philosophical text partially obscures the author's creativity. The following two sections will further elucidate the interpretive frameworks and research questions guiding this work, as well as the methodologies employed. An overview of the book's chapters concludes the introduction. However, before proceeding, it is essential to clarify the approach taken towards the PcN and the Prabodhacandrodaya tradition.

In this study, the PcN is not considered a version of the original Sanskrit *Prabodhacandrodaya* but rather a telling or retelling. The terms "telling" or "retelling" are preferred over "version" or "adaptation" as they emphasise an aspect that will be explored in the subsequent pages, one that should not be overlooked: orality. Consequently, references will be made not only to the Prabodhacandrodaya tradition/paramparā (hereafter PC tradition/paramparā) but also to the Prabodhacandrodaya story (PC kathā). Similar to the Rāma story (Rāma kathā), this book demonstrates that "Prabodhacandrodaya" was not solely Kṛṣṇamiśra's *Prabodhacandrodaya* as a written text but also a tale that existed independently of a standardised version. People were aware of the story, which was recounted and retold orally according to their preferences, sometimes blending with other narratives. Furthermore, Paula Richman, in the introduction to her edited book

https://doi.org/10.1515/9783111063836-001

Many Rāmāyaṇas, astutely observes that (re)tellings of the Rāma story "vary according to historical period, regional literary tradition, religious affiliation, genre, intended audience, social location, gender and political context".[1] Building on this insight, this book investigates the multifaceted layers surrounding the PcN's creation, considering how Brajvāsī's work uniquely positions itself within the larger literary, philosophical, and religious landscape.

It is important to note that the approach taken in this study does not seek to untangle Vedānta and bhakti (commonly translated as "devotion"[2]). Instead, through a meticulous investigation of these intertwined dimensions, the book strives to present a compelling case for "Greater Vedānta(s)" and create a space within the realm of Vedānta philosophies for texts like the PcN. This prompts the fundamental question: How can we situate the texts that retell the PC story within the broader histories of Indian philosophy, specifically within the domain of Vedānta philosophies?

Greater Vedānta(s): Exploring Multilingual and Multi-Genre Philosophy

The historical narrative of Vedānta philosophies has predominantly been read through literature composed in Sanskrit. More specifically, it has been primarily examined through commentaries and sub-commentaries on the Upaniṣads, Bādarāyaṇa's *Brahmasūtras* (BS), and the *Bhagavadgītā* (BhG),[3] as well as in independent treatises, compendiums, and primers. Even notable exceptions, such as Surendranath Dasgupta's monumental *History of Indian Philosophy*, which includes chapters on the philosophical viewpoints found in non-technical texts,[4] have maintained a strong association between philosophy and Sanskrit.[5]

However, the mere juxtaposition of shastras with various other genres of texts in addressing Indian philosophy, as done by Dasgupta, does not seem to have exerted significant influence. A quick look at the table of contents of publications

1 Richman 1991: 16.

2 On the meaning of bhakti in the PcN, see footnote 31 of Chapter 1.

3 These three canonical texts are often collectively referred to as the prasthānatrayī, meaning "three sources". They are deemed to have epistemic authority, especially for the Vedānta schools. See Potter 1981: 5.

4 Dasgupta 1922. This multivolume work includes chapters on "The Philosophy of the Yoga-Vāsiṣṭha" and "Speculations in the Medical Schools".

5 In the preface, Dasgupta refers to the object of his study as "Sanskritic ways of thought" (1922, vol. 1: x).

focusing on the most extensively studied school of Vedānta, Advaita Vedānta,[6] reveals the limited attention given to non-shastric texts. Dasgupta's unease with the unsystematic nature of these texts seems to have had a lasting impact on the study of Indian philosophy, overshadowing his recognition of their importance to understanding Indian thought.[7]

Nonetheless, in recent times, scholars have sought to challenge the established assumptions about what qualifies as philosophy. For instance, Jessica Frazier raises questions about the prevailing notion of separate domains for philosophy and religion in the South Asian context. Instead, she argues for a "constitutive" relationship between the two and highlights the phenomenon of the "narrativisation of philosophy", which refers to the inclusion of philosophical discourses within texts considered primarily literary or theological.[8] Frazier emphasises that "popular religious poetry, drama and myth became the new clothing for philosophical ideas" in the medieval period.[9] This prompts us to ponder: What insights might we glean from philosophy in sources beyond commentaries and treatises?

This preoccupation, among others, has informed recent scholarly endeavours, notably special issues of the *International Journal of Hindu Studies* (edited by Michael Allen and Anand Venkatkrishnan) and the *Journal of Indian Philosophy* (edited by Jonathan Peterson), the latter of which includes articles by Allen and Venkatkrishnan.[10] The two issues are titled *New Directions in the Study of Advaita Vedānta* and *Pluralising the Non-Dual: Multilingual Perspectives on Advaita Vedānta, 1560–1847*. Several essays in these issues not only delve into philosophy beyond the confines of the shastras but explore philosophical ideas expressed in languages other than Sanskrit. Allen and Venkatkrishnan explain the need for such studies in the field of Advaita Vedānta because "scholars have yet to provide even a rudimentary, let alone comprehensive account of the history of Advaita Vedānta in the centuries leading up to the colonial period".[11] Exploring texts from this period is, they argue, crucial to enhancing our understanding of how the school was received and interpreted in later times. Thus, they propose a critical inquiry into overlooked periods, authors, texts, and languages within the realm of post-Śaṅkara (i.e. post-eighth century) Advaita Vedānta. In a complementary vein, Peterson

---

6 See e.g. Potter 1981; Narain 2003; Deutsch/Dalvi 2004. Chaudhuri 2012 also covers other Vedantic traditions.

7 With the sole exception of the *Bhagavadgītā*. Dasgupta 1922, vol. 2: 7.

8 Frazier 2018: 60.

9 Frazier 2018: 66.

10 Special issues: Allen/Venkatkrishnan 2017; Peterson 2020a. Articles: Allen 2020; Venkatkrishnan 2020. See the bibliography for details.

11 Allen/Venkatkrishnan 2017: 272.

emphasises that the exploration of multilingual sources has the potential to "fill out, and perhaps complicate, our understanding of the social and intellectual histories of Advaita Vedānta".[12]

The PcN is a drama composed in Brajbhāṣā during the early modern period in North India, evidently not a shastra. Consequently, it aligns with the genres of texts, periods, and languages examined in the aforementioned studies. However, it is essential to acknowledge certain caveats when adopting such a perspective. First, this book does not primarily contribute to "Greater Advaita Vedānta". Moreover, it does not focus solely on Advaita Vedānta itself. In my view, not only the study of Advaita Vedānta but also all the schools falling under the umbrella of Vedānta, as traditions engaged in Upanishadic exegesis, can benefit from insights derived from non-classical sources and languages.[13] Thus, I have chosen to omit the term "Advaita" in the heading of this section and instead opted for a neutral and pluralistic understanding of "Vedānta(s)".

Michael Allen has advocated for the concept of "Greater Advaita Vedānta".[14] This notion contrasts with the canon of "Classical Advaita Vedānta", which encompasses thinkers like Śaṅkara (eighth century), his students such as Padmapāda (700–750 CE), Citsukha (thirteenth century), and later thinkers like Madhusūdana Sarasvatī (sixteenth century). As previously mentioned, it includes Advaita Vedānta "also in narratives and dramas, 'syncretic' works blending Vedānta with other traditions, and vernacular works".[15] Allen proposes two fundamental criteria for identifying an Advaita thinker in the premodern era: explicit endorsement of the world's unreality and the doctrine of the individual self (jīva) being identical with brahman.[16] This leads us to question whether the PcN presents these views and, if so, how Brajvāsīdās aligns with Advaita philosophy.

---

12 Peterson 2020b: 2. See also Pahlarai 2013, which specifically examines the epistemological aspect of the *Vṛttiprabhākar* by Niścaldās (1791–1863), an independent treatise from the Advaita Vedānta perspective.

13 Publications on these traditions also tend to prioritise Sanskrit sources. See e.g. Gupta 2007; Okita 2014, 2020. See also Maharaj 2020, an edited volume that encompasses early modern thinkers like Jīva Gosvāmī (sixteenth century) within the category of "Classical Vedānta" and includes Ramakrishna Paramahansa (1836–1886), Aurobindo Ghose (1872–1950), and the French intellectual Romain Rolland (1866–1944) in Modern Vedānta. Although there is not enough space to expand this discussion here, it is worth noting that labelling early modern thinkers like Gosvāmī as "Classical Vedānta" might obscure the fact that their historical contexts and perspectives may have differed from those of Śaṅkara or Vedāntadeśika.

14 Allen 2013.

15 Allen 2017: 275–297.

16 Allen 2020: 58.

While we will return to these specific issues later in this study, it is important to consider that categorisations such as "Greater Advaita Vedānta" or "Greater Vedānta" should not create further divisions within the intellectual history of Vedānta. Instead, terms like "Greater (Advaita) Vedānta", with necessary parentheses, can be useful in signalling a distinct approach to studying Vedānta philosophies. Such terminology highlights a diverse body of sources that has been often overlooked by scholarship on these traditions. However, it is crucial to clarify that this book does not interpret the PcN as inherently opposing traditional texts and philosophers. Instead, its objective is to demonstrate that Brajvāsīdās's work draws inspiration from traditional sources while also incorporating other influences, albeit not exclusively and not without critical examination.

Anand Venkatkrishnan has argued for a "bottom-up" view of philosophy, wherein localised and non-scholastic forms of knowledge permeate a wide range of Sanskrit texts. He examines the commentary composed in Marathi by the poet Eknāth in the sixteenth century on the Sanskrit *Hastāmalaka Stotra*. The story revolves around Hastāmalaka, a mute boy who encounters Śaṅkara, the philosopher. When questioned about his identity and provenance, Hastāmalaka initially takes a Mahānubhāva anti-Brahminical stance. However, the conclusion of the commentary takes a Brahminical turn as Śaṅkara initiates the boy as a renunciate, followed by an exposition on the Daśanāmī order. Although traditionally associated with Śaṅkara as its founder, research has revealed that the Daśanāmī order emerged between the sixteenth and seventeenth centuries.[17] The presence of an unusual Daśanāmī denomination in Eknāth's commentary suggests that he might have been acquainted with this knowledge from local sources or direct contact with such groups rather than relying solely on Sanskrit sources.[18] This leads Venkatkrishnan to propose that Eknāth potentially engaged with sources beyond Sanskrit, although in the poet's Vedantic works he demonstrates familiarity with classical sources like the Upaniṣads and "Greater Vedānta" sources such as the *Bhāgavata Purāṇa* (BhP; ninth–tenth century) and the *Yogavāsiṣṭha* (twelfth–fourteenth century).[19]

Considering these studies, it becomes crucial to not merely view the PcN in relation to the KPC but to investigate its interactions with other texts and contexts closer in time, language, and inspiration. Moreover, the endeavour undertaken in this book to reflect on early modern Vedānta in bhāṣās departs from the inclination to exclusively compare Brajvāsīdās with figures like Śaṅkara or the founders

---

17 See Clark 2006.

18 Venkatkrishnan 2020: 16–17.

19 Venkatkrishnan 2020: 19.

of the tradition to which Brajvāsīdās belonged: specifically, Vallabha (1478–1530) and his son Viṭṭhalnāth (1516–1572). This approach stems from the belief that addressing the dearth of knowledge concerning early modern philosophical activity and, more specifically, enriching and complicating the history of Vedānta philosophies requires exploration beyond commentaries and treatises.[20] Moreover, it necessitates discarding the bias that knowledge must conform to a single model of systematicity to be considered efficient and valid – to be actual "Knowledge" – and instead examines genres where philosophical communication intersects with poetry and, in the case examined here, drama.[21]

In the context of the language of knowledge, Sheldon Pollock raises the notion of "the clear and untranscendable limits of vernacular textualization in the early modern period".[22] He argues that, aside from alaṅkāraśāstra (science of poetics) texts, there are close to no works in bhāṣā addressing philosophical schools of thought. He notes that only vaidika (works associated with the Vedas) are found.[23] This prompts the question of whether these vaidika works can be regarded as representative of how philosophy and philosophical knowledge were conceived and transmitted in local languages.[24]

---

20 See Ganeri 2017: intr.

21 In a volume on Advaita Vedānta edited by Debi Prasad Chattopadhyaya as part of a multivolume work on the *History of Science, Philosophy and Culture in Indian Civilization*, there is a section on "Advaita in the Vernacular". The essays in this section explore Advaita in various languages such as Tamil, Telugu, Malayalam, Kannada, Marathi, Gujarati, Hindi, Punjabi, Bengali, and Oriya. While the endeavour is laudable, when reading the chapter on Advaita in Hindi by Surendra K. Srivastava, one encounters several conceptual difficulties. The chapter primarily focuses on Niścaldās and his independent works on Advaita. The last part of the chapter discusses translations, commentaries in modern Hindi, and secondary literature from the twentieth century. Thus, early modern philosophy is reduced to Niścaldās and a single section on the "Advaita of the Saint Tradition". In this section, various authors and contexts that are in fact remarkably different from one another – Kabīr, Tulsīdās, Nānak, and the Maratha bhaktas – are combined under the vague banner of Advaita. Srivastava's essay predominantly presents a unified movement with anti-caste and anti-Vedic claims, subsuming it even further under the label of Advaita. This perspective does not align with the goals of the present book, which aims to provide a historical and philosophical analysis of the PcN without imposing a preconceived structure on the text. See Srivastava 2010: 582–599.

22 Pollock 2011: 28.

23 Pollock 2011: 41n26.

24 There remains a significant body of texts dealing with philosophical matters to be explored. For instance, Tyler W. Williams has uncovered references within the Niraṅjanī and Dādū sampradāyas that indicate the existence of "vernacular religious scholarship" in the early modern period, which suggests a rich corpus of untapped material. See Williams 2019a: 10–12. Even more than Vedānta, scholars have often assumed that Nyāya (Reason/Logic) remained untouched by

Furthermore, philosophy cannot be said to be divorced from poetry. In addition to composing scholastic texts in Sanskrit, many thinkers also expressed their ideas through poetry.[25] The BhP – along with the prasthānatrayī (Upaniṣads, BS, BhG), the primary scriptural basis for the Vedānta schools that flourished in Braj during the sixteenth century – exemplifies this connection, as it is itself a remarkable piece of poetry.[26] Could it have served as an inspiration, demonstrating alternative ways to explore profound philosophical matters?

Friedhelm Hardy provides valuable insights into a religious tradition that does not consider the BhP as one of its primary references. He demonstrates how the celebrated Viśiṣṭādvaita philosopher Vedāntadeśika (1269–1370), also known as Veṅkaṭanātha, incorporated quotations from both the poetry of the Āḻvārs and the Pāñcarātra scriptures in his *Dehalīśastuti*. Although Hardy acknowledges that the content of the poem may not precisely align with Vedāntadeśika's "proper" philosophy, he recognises that Vedāntadeśika may have had purposes beyond adhering to systematic notions. He suggests that the hymn, dedicated to a particular form of Viṣṇu, could provide a space for the convergence of conflicting traditions within Viśiṣṭādvaita (eleventh century) – namely, the Vaṭakalai and the Teṉkaḷai – on a central theme like prapatti (surrender to Viṣṇu).[27] Significantly, Hardy concludes that Vedāntadeśika's identity as both a kavi (poet) and a tārkika (philosopher) is not diminished or mutually exclusive and that exploring the technical and poetical output of a single author can significantly contribute to the study of Indian thought.[28]

However, notwithstanding Hardy's viewpoint, eclectic material, particularly allegorical drama, continues to face scepticism. Vedāntadeśika's work provides us with a further example. As a prolific author, he also crafted a retelling of the PC story from the Viśiṣṭādvaita perspective, titled *Saṅkalpasūryodaya* (The Rising of the Sun of [Viṣṇu's] Will). Steven Hopkins, while introducing the works attributed to this influential philosopher, highlights the *Saṅkalpasūryodaya* and acknowledges that:

---

bhāṣās prior to the colonial era. See Pollock 2011: 23–24. However, the PcN in fact addresses Nyāya through a treatment of six categories of debate (vāda). See Pastore 2023.

25 For example, the sixteenth-century Gauḍīya theologian Rūpa Gosvāmī also composed dramas. See Wulff 1984.

26 See Theodor 2016: 1–8. On the difficulties of translating the BhP, see Gupta 2018: 175–189. Among the traditions, chief were those founded by Vallabha and the Bengali Caitanya (1486–1533). The Gauḍīyas, for example, saw the BhP as the perfect commentary to the BS. See Gupta 2007: 105.

27 See Hardy 1979: 315–316. The arcāvatara (manifestation of Viṣṇu in a worshippable form) of Dehalīśa (Lord of the Porch) is one of the central deities of Tamil bhakti. See Shulman 2016: 115–116.

28 Hardy 1979: 317–318.

> *Saṃkalpasūryodaya* is a fine example of the limitations of works written to evoke *śānta* – "transcendental peace" – as their main aesthetic experience (*rasa*). Though *śāntarasa* is at the heart of some extraordinarily powerful poetry in Sanskrit, including (according to tradition) portions of the *Mahābhārata* and the Buddhist *kāvyas* of Aśvaghoṣa, most often its poets err on the side of self-consciousness. What to Aśvaghoṣa was merely the mixing of "bitter herbs with honey" becomes in many works of this genre a shoehorning of doctrinal views into verse, straining the already limited scope of philosophical allegory. Like other examples of the philosophical drama, Deśika's play sometimes, in M. R. Rajagopala Iyengar's phrase, "smells of the lamp."[29]

Although Hopkins's observations are not explicitly directed towards Vedānta but rather the conveyance of śāntarasa (sentiment of peace), they suggest a dismissive attitude towards sources that differ from what one considers canonical. Such biases hinder the careful expansion of self-imposed boundaries within scholarship. Interestingly, a few lines later, the scholar acknowledges that "the posturing and heavy allegories of Deśika's play do not get in the way of its appreciation by Vaṭakalai scholars and devotees. Instead, the play is often used as a veritable manual in the proper interpretation of Viśiṣṭādvaita philosophy".[30] While we have confirmation of the *Saṅkalpasūryodaya*'s value within its tradition, we lack similar information about the PcN. Nonetheless, we can endeavour to understand how it conveyed philosophical matters, which specific topics Brajvāsīdās sought to communicate, and which audience he aimed to address.

Returning to Frazier's words quoted at the beginning of this section, it is important to state that adopting a new literary genre did not merely provide a fresh appearance to the same body of thoughts. Rather, each author offered their own interpretation of philosophical doctrines. Additionally, the new garb or outward form influenced the modality through which these ideas were presented. This implies that the conventional approach to studying poetical works dealing with philosophy may not always yield fruitful results.

Angelika Malinar's essays on philosophy in the *Mahābhārata* propose alternative ways of reading philosophy in texts beyond the canon.[31] Malinar points out that scholars often identify "philosophical texts" within the epic and study them in isolation, detached from their context, as a separate category. Another approach she identifies is an attempt to identify a "philosophy of the epic", a coherent framework of thought that presupposes "a unity of meaning along with a unity of composition".[32] Furthermore, Malinar highlights that scholarship has as-

---

29 Hopkins 2002: 52.

30 Hopkins 2002: 53.

31 Malinar 2017a, 2017b.

32 Malinar 2017a: 589–591.

signed various positions to this "epic philosophy" within the history of Indian philosophy. It has been seen as a proto-philosophical form, mixed or transitional.[33] According to Malinar, Western academia's perception of the term "philosophy" carries biases that have hindered a serious consideration of texts like those found in the *Mahābhārata*. While these texts may not possess original philosophical value per se, they still hold fundamental importance as documents contributing to the overall history of Indian philosophy.[34] The unsystematic nature of texts like those in the corpus examined poses a challenge, as readers expect philosophical doctrines to be presented in specific chapters or in a more argumentative manner.[35] However, what is discovered in the epic does not always conform to a shastra intended as a "system" of philosophy. Often, it is expressed in narrative form and with affirmative statements. Malinar suggests that we consider that some teachings could have been included not solely for their philosophical content but for didactic or encyclopedic purposes or to enhance significance or appeal.[36]

These suggestions inform my approach to reading the Vedānta philosophical discourses in the PcN, focusing on personified concepts such as paramātama, māyā, jīvātama, and mana, the meanings of which I shall discuss later. The book acknowledges that their presentation in the PcN, similar to the *Mahābhārata*, does not follow a systematic exposition on a given topic but unfolds as a story. Consequently, the PcN sometimes presents different and contrasting viewpoints that I do not attempt to reconcile philosophically at all costs. Instead, I aim to consider other factors, such as their contexts of enunciation, their interrelationships, their placement within the development of the PcN story, and the overall function and purpose of the PC story for Brajvāsīdās.

In his monograph on Navya Nyāya (New Reason), Jonardon Ganeri proposes contextualism as an effective method for exploring Indian intellectual cultures. Building upon Quentin Skinner's theory of "illocutionary intervention", Ganeri expands its application through the concept of "intertextual intervention".[37] However, he cautions against applying the same expectations as Skinner's theory to the Indian context. Skinner's approach depended on examining the biographical,

33 Malinar 2017a: 592–595.

34 Malinar 2017a: 595–598.

35 Anxiety around this expectation seems evident in Dasgupta, for example, who, in the preface to his history of Indian philosophy, states that he covered the Brāhmaṇas (prose commentaries attached to the Vedas) concisely since he "did not think it worthwhile to elaborate the broken shreds of thought of this epoch". Dasgupta 1922, vol. 1: x.

36 Malinar 2017a: 600.

37 Ganeri 2011: 7.

political, and social circumstances surrounding a text.[38] In the realm of intellectual literature in Sanskrit, however, there is often an abundance of texts and a scarcity of information about the authors, resulting in a lack of contextual details.[39] Ganeri suggests an alternative approach by investigating "intertextual contexts", which involves identifying the kind of authorial intervention one seeks to make within one's framework of affiliation, which in the case of Sanskrit scholars is to a tradition of shastra.[40]

Regarding early modern Advaita Vedānta between 1500 and 1800, Christopher Minkowski also raises the question of approaching the tradition from a historical perspective rather than viewing it as a static phenomenon immune to change and variation. Minkowski proposes tracing interconnected authors, exploring institutional ties, examining geographical locations and movements, and discerning literary trends of the respective periods. These methods help construct a social history of Advaita and alleviate the lack of knowledge about these centuries.[41] That said, Minkowski acknowledges that studying the school during the early modern period poses unique challenges, as many thinkers and authors deviate creatively from conventional routes and established demarcations of the past.[42]

In the case of Brajvāsīdās, we also need to ask what it *means* to put his PcN into context. While it seems certain that, from a literary point of view, he wanted to make his intervention within the PC tradition, we lack substantial information about him beyond his Vallabhite affiliation.[43] Only recently have Vallabha's and his sons' works garnered renewed scholarly attention.[44] At the same time, it is important to reiterate that the PcN does not belong to a Sanskrit shastra tradition. Consequently, understanding Brajvāsīdās's overall intellectual intervention regarding philosophical and religious perspectives poses a challenge. Unfortunately, studies on eighteenth-century North Indian literature do not offer much assistance, as the handful of authors whose works have been partially explored seem to have different concerns to Brajvāsīdās or have received attention for other aspects of their poetry.[45]

Yet we may discern an intertextual context. Translating the PcN made me aware of its connection with the *Rāmcaritmānas* (The Lake of Rāma's Deeds, 1574;

38 Skinner 2002.

39 This is also true of literary works in regional or local languages. See Hawley 1988.

40 Ganeri 2011: 63–68. On contextualism, see also Minkowski/O'Hanlon/Venkatkrishnan 2015.

41 Minkowski 2011: 205–206.

42 Minkowski 2011: 212.

43 See subsection 1.1.1.

44 See e.g. Smith 1998, 2004, 2011, 2016; Buchta 2016.

45 For example, see Bangha 2014 and his work on Ānandghan, among which, Bangha 2007 and 2013. See also other works on court poets such as Cattoni 2020; Rajpurohit 2018; Busch 2018.

hereafter RcM), the major Avadhī work of the poet and bhakta Tulsīdās (1532–1623). He composed several works, not always clearly dated, displaying his poetical dexterity: the *Gītāvalī* (Collection of Songs [about Rāma], 1590?), the *Jānakī Maṅgal* (dedicated to the marriage of Rāma and Sītā), the *Pārvatī Maṅgal* (dedicated to the marriage of Śiva and Pārvatī, 1586), the *Vinaya Patrikā* (A Letter of Petition [to Rāma], before 1609), and the *Kavitāvalī* (Collection of Quatrains [about Rāma]).[46] Among his works, the RcM holds immense importance in the religious life of North India.[47] Tulsīdās's interpretation of the story of Prince Rāma of Ayodhyā is not only a fine work of poetry but also a central component of the bhakti traditions. By the eighteenth century, it had gained a distinguished reputation, as evident from the appraisal provided by court poet Bhikhārīdās in his *Kāvyanirṇaya* (Appraisal of Poetry, 1746), which gives an overview of bhāṣā poetry and language and categorises various bhāṣā poets, including Tulsīdās.[48]

The central argument of this book is that the RcM not only serves as a literary model for the PcN but also acts as a source for the Vedānta and bhakti perspectives within it. Thus, I have identified the RcM as the contextual framework for Brajvāsī's text. It is my contention that exploring the relationship between these two texts enriches our understanding of early modern Vedānta and bhakti expressed through the medium of bhāṣās. Why did Brajvāsīdās choose Tulsīdās, who placed utmost importance on Rāma, rather than a Vallabhite poet when his preferred deity, as a Vallabhite himself, was Kṛṣṇa? How can we theorise and demonstrate the philosophical and religious inspiration Brajvāsīdās found in the RcM for his PcN? These are questions that will guide my exploration.

Adaptive Reuse: Discussing a Method

To shed light on the Vedānta and bhakti presented in the PcN, this book adopts the conceptual approach of "adaptive reuse". Although relatively recent in the field of literary studies on ancient and medieval Indic literary production,[49] the

---

46 It is worth mentioning that Tulsī also composed the *Kṛṣṇagītāvalī* (Collection of Songs on Kṛṣṇa) in Brajbhāṣā. For a description of the contents of the aforementioned texts and of others attributed to the poet, see McGregor 1984: 109–117.

47 So much so that in the colonial period, George Grierson said in a meeting of the Royal Asiatic Society that "Over the whole of the Gangetic Valley his great work [i.e. the RcM] is better known than the Bible is in England." Grierson 1903: 459.

48 See McGregor 1984: 189.

49 The concept of adaptive reuse originates from the field of architecture and is often associated with the themes such as sustainability, conservation, and the environment. See Cummer/DiStefano 2021; Tam/Hao 2019; Plevoets/Van Cleempoel 2011.

broader domain of South Asian studies, as seen in Julia Hegewald and Subrata Mitra's edited volume, recognises re-use as a means of understanding the relationship between the past and present, particularly in the interplay of art and politics.[50] It is important to note here that "re-use", written with a hyphen, is not synonymous with "reuse" or "adaptive reuse". Elisa Freschi and Philipp Maas have elucidated the distinction between "adaptive reuse" and "simple re-use" concepts in the context of South Asian intellectual traditions, specifically literary production. Through the lens of adaptive reuse, one can make sense of authorship, creativity, and how an author acts as both an innovator and perpetuator of earlier forms and contents while establishing a relationship with the past.[51]

Simple re-use occurs when an author resumes the usage of an item without significantly altering its original purpose, essentially using it again. The immediate availability of the item determines its re-use, and typically, "the re-user does not want the re-used element to be specifically recognised as having been re-used".[52] On the other hand, in an ideal case of adaptive reuse, the reused item and its source are clearly identifiable. The element may have been repurposed, and the reuser intends for the audience to recognise the connection. Adaptive reuse is motivated by reasons such as prestige and credibility conveyed through the identified reused object. Additionally, Freschi and Maas propose a spectrum between "simple re-use" and "adaptive reuse", allowing for degrees of variation between the two. Instances of reuse can manifest as quotations, whether literal or not, and references, which raise questions such as: Are they explicitly attributed to an author? In Sanskrit texts, particles generally indicate the end of a quotation, but what conventions exist for bhāṣā texts?[53]

By integrating the RcM as an essential component of the context for interpreting the Vedānta and bhakti in the PcN, this book aims to clarify whether and how

---

50 Hegewald/Mitra 2012.

51 Freschi/Maas 2017: 11. In the context of discussing recurring verses between the Pāli Canon and the *Mahābhārata*, Bailey 2011 offers an essay that emphasises the significance of context and content in comparison, although not from the perspective of adaptive reuse.

52 Freschi/Maas 2017: 14. They give as an example of re-use utilising a pillar from an old building to fulfil the same function as a support in a new building. Adaptive reuse, on the other hand, would be using the pillar not just as a building block but also as a means of communication, for example conveying continuity with the past.

53 In these contexts, quotations can be of broadly two kinds: those that reproduce the source verbatim (literal) and those that modify the source to some extent (non-literal). The modified text can arise from either the presence of different recensions of the same text or intentional alterations made by the author. See Freschi 2015 for the particles and expressions employed in Sanskrit and for a broader discussion of the typology of quotations in Sanskrit texts. See the discussions in section 3.3 and Chapter 4 in this book for the cases involving the PcN and the RcM.

Brajvāsīdās adaptively reused the RcM. It argues that while Brajvāsī does not explicitly mention Tulsīdās, his engagement with the RcM goes beyond simple reuse. In fact, more than that, this book will demonstrate how simple re-use fits into a broader framework of adaptive reuse in the PcN, incorporating the elements of the latter proposed by Freschi and Maas: (1) the involvement of at least one consciously acting agent (2) aiming to achieve a specific purpose by (3) resuming the usage of (4) a clearly identifiable object after an interruption in its use.[54]

The methods employed to illustrate the close relationship between the PcN and the RcM include philology, where the book demonstrates how Brajvāsīdās reworks the RcM based on syntax, prosody, and vocabulary. Additionally, a comparative analysis of the philosophical and religious viewpoints of the two works is undertaken. From this perspective, the book employs the terminology of both intertextuality and adaptive reuse[55] but distinguishes between the two. Adaptive reuse is utilised as a heuristic tool to identify the particular relationship between the PcN and the RcM. While the PcN holds intertextual connections with various texts in Sanskrit and in bhāṣā, the crucial distinction lies in Brajvāsīdās's adaptive reuse of the RcM, which serves as a foundation of building blocks and conceptual units for his own text. The study will demonstrate that the RcM acts as a filter through which Brajvāsīdās consistently interprets Vedānta and bhakti to the extent that Tulsī's epic influences Brajvāsīdās's reading of other texts that share intertextual relationships with the PcN. By emphasising the author's creative activity and highlighting the practical process of composition, adaptive reuse brings attention to the author's agency in the textual construction.[56]

---

54 Freschi/Maas 2017: 13.

55 Intertextuality was first introduced as a hermeneutical tool by Julia Kristeva in the latter half of the twentieth century based on her interpretation of Mikhail Bakhtin's work. See Kristeva 1986: 34–61. The concept has been extensively discussed and elucidated by scholars throughout the years. Notable works include Allen 2011; Alfaro 1996.

56 Kristeva's development of intertextuality appears to shift the focus away from the author and their intentions, emphasising the rupture with the past, which may not be readily apparent in the case of the PcN. See Kristeva 1986: 37, 39. For the distinction between adaptive reuse and intertextuality, see Freschi/Maas 2017: 20–21. While their exploration of adaptive reuse aligns closely with other scholars' efforts to place limits on the potentially infinite and somewhat depleting dynamics of intertextuality, there are still discernible differences. See Alfaro 1996: 277–284. It is worth mentioning Gérard Genette's concept of "hypertextuality", which refers to the relationship between a text B and an earlier text A. However, the basic forms of hypertextuality proposed by Genette, such as the transformation and imitation of text A by a later text B, are grounded in Western literary genres like parody and travesty, making them challenging to apply to the text under analysis in this book. For more details, see Genette 1982: 7–19.

Outline of Chapters

This book addresses the questions raised in this introduction in two parts. Part I presents a critical study consisting of four chapters that offer a commentary and analysis of the subject of this book, each concluding with a summary of the chapter's findings. Part II comprises the first annotated English translation of the complete playscript of the PcN.

Chapter 1, "Brajvāsīdās's *Prabodhacandrodaya Nāṭaka*", begins with an overview of the state of the art of Brajvāsīdās and his works (sec. 1.1), including a subsection on his status as a Vallabhite author. It then focuses on the PcN itself (1.2), first presenting information on the editions and manuscripts of the PcN used in this study and discussing their relationship and the reception history of the PcN (1.2.1), and second, delving into the structure, narrative levels, plot, and opening of the drama (1.2.2). Moving beyond a descriptive analysis, these sections start developing the argument that the PcN embodies a broad Vaiṣṇava viewpoint, framed by Brajvāsīdās as a bhakti text with the potential to appeal to multiple communities. The chapter further examines the oral and performative dimensions of the text (1.2.3), followed by an investigation of its language and rich prosody (1.2.4).

Chapter 2, "The *Prabodhacandrodaya Nāṭaka* and the Prabodhacandrodaya Tradition", sheds light on the PC tradition. Its three analytical sections provide contextual background for the PcN by focusing on three earlier works within the paramparā that are explicitly related to the PcN: Kṛṣṇamiśra's Sanskrit *Prabodhacandrodaya* (2.1), Banvālīdās's Persian *Gulzār-i Ḥāl* (2.2), and Nanddās's Brajbhāṣā *Prabodhacandrodaya* (2.3). The chapter examines the intricate relationships between these texts and explores open questions that could contribute to a deeper understanding of the tradition with further investigation.

Chapter 3, "Tulsīdās's *Rāmcaritmānas* and the Prabodhacandrodaya Story", sets the stage for the discussion in Chapter 4. The goal of Chapter 3 is to investigate the reasons behind Brajvāsīdās's decision to draw on Tulsī's poem. It provides an overview of the influence of the RcM on subsequent works, establishing the connection between the RcM and the PC story (3.1). This association is traced through nineteenth-century retellings of the RcM and a close reading of the RcM itself. The central argument proposed is that the RcM assigns significant importance to the concept of viveka (discrimination or discernment) within its bhakti framework, particularly embodied by Bharata, Rāma's brother (3.2). As viveka plays a crucial role in the PcN, the chapter concludes by returning to Brajvāsīdās's text: it examines instances of adaptive reuse and re-use of the RcM in the PcN that can be seen as purely literary, without involving philosophical or religious discussions (3.3).

Chapter 4, "The *Prabodhacandrodaya Nāṭaka*'s Vedānta and Bhakti", focuses on the philosophical and religious dimensions of the PcN. This chapter highlights the distinctive role of the RcM among the PcN's sources, as Brajvāsīdās creatively reuses and adapts its text in multiple instances to construct his own text and Vedānta discourse. It is divided into two sections, with the first section analysing the metaphysical and ontological structure of the PcN, examining the personified concepts of paramātama puruṣa, māyā, the phenomenal world, and jīvātama puruṣa/mana, as well as their interrelationships (4.1). The second section explores the path to liberation within the PcN, establishing a connection between the metaphysical and soteriological dimensions (4.2). It refers back to the discussion in Chapter 3 regarding Bharata as an ideal "bhakta yogi" endowed with viveka in the RcM, highlighting its similarities to the figures of bhakta yogis or viveki depicted in the PcN. Additionally, it focuses on how Brajvāsīdās integrates yogic bhakti with the knowledge gained through reflection on the scriptures.

Following Chapter 4, the conclusion summarises the key points discussed in the previous chapters and revisits the questions outlined in the introduction. After the conclusion, the annotated translation of the playscript of the *Prabodhacandrodaya Nāṭaka* is presented, preceded by a note explaining the translation strategy employed to render the PcN in English and detailing the editorial interventions made in the text. The translation enables readers to become familiar with – and enjoy – the PcN, while the footnotes provide additional commentary on the specific aspects of the text not extensively covered in the critical study.

Part I: **Critical Study of the *Prabodhacandrodaya Nāṭaka***

1 Brajvāsīdās's *Prabodhacandrodaya Nāṭaka*

This chapter serves as an introduction to Brajvāsīdās and his *Prabodhacandrodaya Nāṭaka*. First, it provides an overview of Brajvāsīdās by examining the number of works attributed to him in the histories of Indian literature and how they relate to his affiliation with the Vallabhite tradition. The subsequent section focuses on the play itself, dealing first with the editions and manuscripts of the PcN and shedding light on its material and publication history and the availability of textual sources. By examining the editions, we gain insights into the landscape of Hindi publishing from the late nineteenth to the early twentieth century and where the PcN fits into that context. Additionally, the circulation of manuscripts of the play within a broad Vaiṣṇava context during the early modern period can be connected with the intended audience that Brajvāsīdās himself addresses in the PcN's opening verses. The exploration further unfolds with an exposition of the plot of the PcN, which aids the reader in better understanding the subsequent sections of the chapter. It then addresses the oral and performative dimensions of the text, as well as its language and prosody.

1.1 Brajvāsīdās's Life and Works

Like many early modern authors, the details of Brajvāsīdās's life remain largely unknown. According to Kishorilal Gupta, Brajvāsī was a brahmin from Mathura who belonged to the Vallabha sampradāya (religious tradition).[1] Although numerous historians and Hindi literature critics mention him, they provide limited additional information.[2] In addition to the PcN, Brajvāsīdās is attributed to two other works, one of which is the *Brajavilāsa*, composed in 1770. This text narrates the story of Kṛṣṇa in Braj, depicting his līlās (divine pastimes) with the cowherds and gopīs (cowherd girls), including his departure to Dvārakā. Ronald S. McGregor also attributes to Brajvāsīdās the *Purātankathā* (The Ancient Tale), which recounts the story of Rāma as narrated by Yaśodā to Kṛṣṇa.[3]

Building upon Krishnaprasad Gaur's research, Peter Friedlander attributes three additional compositions to Brajvāsīdās, all centred around Kṛṣṇa. These include the *Aghāsuravadhalīlā*, which portrays the slaying of the wicked Aghāsura; the *Mānacaritralīlā*, which describes the proud behaviour of the gopīs; and the

1 Gupta 1967: 471.

2 See Brajratnadas 1944: 47; Garcin de Tassy 1839–1947: 131; Sitaram 1926: 74; Shukla 1942: 251.

3 McGregor 1984: 159.

https://doi.org/10.1515/9783111063836-002

Mākhanacoralīlā, which captures the playful acts of Kṛṣṇa as the butter-thief.[4] It is possible that the *Purātankathā* and these three līlās are not separate works but rather sections extracted from the *Brajavilāsa* narrative. The current form of the text, as found in lithographic editions, presents such distinct units.[5] Furthermore, according to Gupta, these episodes from the text also circulated independently in manuscript form as "instalments".[6]

Despite the scarcity of information about Brajvāsīdās, he was not entirely unknown. He is mentioned in two nineteenth-century texts: Horace Hayman Wilson's *Religious Sects of the Hindus* and Frederic S. Growse's *Mathura: A District Memoir*. Discussing the textual sources of the Vallabha tradition, Wilson writes:

> Amongst the votaries in general, various works upon the history of Krishna are current, but the most popular are the Vishnu Padas, stanzas in Bhasha, in praise of Vishnu, attributed to Vallabha himself; the Braj Bilas, a Bhakha poem of some length, descriptive of Krishna's life, during his residence at Brindavan, by Braj Vasi Das.[7]

Similarly, Growse describes the *Brajavilāsa* as "the popular Hindi authority for what concerns Rādhā's life and loves".[8]

The familiarity and appreciation of Brajvāsīdās's works among the general public are evident through the choices made by the renowned publishing house Naval Kishore Press to publish it in 1875. Ulrike Stark's examination sheds light on the circumstances surrounding the Lucknow publisher's foray into Hindi publishing in the 1860s, focusing on titles that would resonate with the readership. Based on potential market demand, this strategic approach involved simultaneously publishing classics and new books. Religious literature in Brajbhāṣā and Avadhī, such as Tulsīdās's RcM and Sūrdās's (c. 1483–1563) *Sūrsāgar*, was among the genres published.[9] Notably, even within the initially limited selection of reli-

4 Friedlander 1996: 236. Unfortunately, I was unable to consult Gaur's catalogue (1964).

5 See e.g. Brajvāsīdās 1873: 5–7.

6 Gupta 1967: 470. Although the term "instalments" suggests that Brajvāsīdās intentionally released these episodes in serial form, in fact it seems that he composed *Brajavilāsa* in its entirety and select līlās were subsequently copied and circulated independent of the full work. I was able to consult an (almost complete) illustrated paper manuscript of the *Brajavilāsa* held at the Allahabad Museum (MS 455). This manuscript consists of 428 folios, with the first folio featuring an illustration titled *Gokula ko le cale*, depicting Vasudeva rescuing the new-born Kṛṣṇa from the prison, where he and his wife Devakī were held captive by the evil Kaṃsa, and bringing him to Gokula.

7 Wilson 1958: 75.

8 Growse 1883: 75, unnumbered note.

9 See Stark 2007: 385–391.

gious texts, the *Brajavilāsa* found its place, already recognised as a "pre-print classic", according to Stark.[10]

The PcN can be found among the early publications in the nāṭaka (drama) genre, published in 1875.[11] Stark notes that the publication of dramas began relatively late in the publishing venture, coinciding with what Stark calls the formative years of Hindi drama, closely associated with the renowned figure of Bhārtendu Hariścandra (1850–1885). This thinker and author from Benares made establishing a new theatre for Hindus part of his mission. To accomplish this, he reworked several older plays, among them the third act of the PC story. Vasudha Dalmia highlights that Hariścandra was familiar with Brajvāsīdās's version of the play:

> With the verse translation into Brajbhasha in 1760 by Brajvasidas, who was, like Harishchandra, a member of the Vallabha *sampradaya*, the play had become securely ensconced within the Vallabha tradition. Harishchandra was familiar with Brajvasidas's version as he makes explicit mention of it in his essay "Natak".[12]

In "Nāṭak", Hariścandra in fact criticises dramas composed in Brajbhāṣā and other regional languages because, in his opinion, they do not fully adhere to the conventions of Sanskrit drama and are merely nāṭakas in name only.[13] It is within this context that he references the PcN. By contrast, he adapts the third act of the Sanskrit KPC somewhat literally into his own work, *Pākhaṇḍ Viḍamban* (A Satire of the Heretics).[14]

In this regard, the PcN, along with all literature composed in the eighteenth century, has been subjected to historiography that has often linked political instability resulting from the gradual disintegration of the Mughal dynasty and the ascendency of the East India Company to a perceived "decline" in all other domains.[15] However, it is crucial to re-evaluate the literature produced during this century,

---

10 The *Brajavilāsa* was first published by Lalluji Lal's Sanskrit Press in 1817 and later by Naval Kishore in 1866. See Stark 2007: 392–393. Naval Kishore also published a condensed version of the text in 1898, titled *Brajavilāsa Sārāvalī*. This abridged version, composed by Govarddhandās, was intended "for the benefit of all those loving Kṛṣṇa". As well as mentioning Brajvāsīdās, Govarddhandās pays homage to Vallabha and Viṭṭhalnāth. For more details, see Govarddhandās 1898: 1.

11 Stark 2007: 421.

12 Dalmia 2006: 51.

13 See Hariścandra 1987: 575.

14 See Pastore (forthcoming b) for a discussion of the position of Hariścandra's drama in the PC tradition.

15 For example, McGregor 1984: 132 states: "Despite its achievements, the period was on the whole one of limited development and abating originality."

freeing it from value judgements while acknowledging the era's precarious and shifting political conditions.[16]

Throughout the eighteenth century, the prevailing trend of composing commentaries or adaptations in regional languages continued from previous centuries. However, this practice encompassed not only Sanskrit sources but also compositions in other local languages.[17] An example of this trend can be found in the works of the court poet Sūrati Miśra (c. 1710–1740), who wrote commentaries on Bihārīlāl's *Satsaī* (seventeenth century) as well as Keśavdās's manuals, the *Rasikpriyā* (1591) and the *Kavipriyā* (1601).[18] As mentioned earlier, assessing the extent and nuances of these literary practices is challenging due to the substantial lack of study on the corpus produced in North India during this period.[19] The translation and examination of the PcN thus serve to bridge this gap in scholarship. But, before delving into the PcN itself, it is pertinent to briefly explore how Brajvāsīdās's literary output aligns with the expectations of a "Vallabhite author".

1.1.1 Brajvāsīdās, the Vallabhite Author?

Vasudha Dalmia suggests that the shared Vallabhite affiliation between Hariścandra and Brajvāsī could have influenced the former's recognition of the latter's work. The PC story was not new to the Vallabhite sampradāya, as it had been retold in the sixteenth century by one of the esteemed eight seals (aṣṭa chāpa) poets of the tradition, Nanddās (1530–c. 1585).[20] However, Chapter 4 of this book will demonstrate that the PcN does not convey an explicitly Vallabhite connotation. Does this raise doubts about Brajvāsī's affiliation with Vallabha's Puṣṭimārga

16 Many later studies that touch upon the PcN continue to adopt a biased approach. Their comparison of bhāṣā nāṭakas with Sanskrit dramas is only partially useful, as it measures the early modern compositions against a standard that we have limited knowledge of their adherence to. A relevant example is Ojha 1995.

17 Adaptations of the BhP, particularly Book X, increased significantly during the eighteenth and nineteenth centuries. See McGregor 1984: 156; Horstmann 2018: 123–174.

18 See McGregor 1984: 187.

19 One notable exception is Cattoni 2020. See also Williams 2020a.

20 The aṣṭa chāpa refers to the eight most renowned vernacular poets of the Vallabhite tradition, who lived between the times of Vallabha and his son Viṭṭhalnāth (fifteenth–sixteenth century). To explore their works, see McGregor 1984: 83–88. The connections between the PcN and Nanddās's PC are discussed in section 2.3, while the influence of Vallabhite philosophy and religious perspectives on the PcN is examined Chapter 4.

(path of nourishment)? After all, scholarship has revealed that the Vallabhites were keen on assimilating poets, including Sūrdās, into their tradition.[21]

When considering Brajvāsī's overall literary output, we find that the PcN and the *Brajavilāsa*, written a decade apart, are two distinct compositions. While the former does not overtly pay homage to Vallabha, the latter does.[22] The PcN discusses the concept that Kṛṣṇa and Rāma are two saguṇa forms (rūpa) of brahman, allowing the bhakta to choose the form they prefer.[23] On the other hand, the *Brajavilāsa* unquestionably elevates Kṛṣṇa as supreme, making his story its subject matter. Moreover, the PcN delves into philosophical inquiries, whereas the *Brajavilāsa* serves as a hymn to bhakti without addressing metaphysical matters. Despite the disparities between the two texts, I argue that they can both be regarded as part of Brajvāsī's wider intellectual project.

To begin with, numerous early modern authors composed both works that could be categorised as "sectarian", reflecting specific religious and philosophical inclinations, and works that did not lean towards any particular school of Vaiṣṇava Vedānta. Nanddās, for instance, authored religious texts yet also wrote and adapted works on literary theory.[24] Similarly, Tulsīdās is credited with composing the *Śrī Kṛṣṇagītāvalī*, dedicated to the Braj deity and composed in Brajbhāṣā.[25] Furthermore, the fact that Vallabha and his immediate successors formulated their philosophical views in opposition to Advaita Vedānta did not entail a complete rejection of Advaita interpretation. Instead, it involved a reinterpretation of Advaita ideas and values. Chapter 4 will delve into the role of viveka in the PcN and how this Vedantic concept, present as early as Śaṅkara's commentary on the BS in the eight century, is framed within the work's bhakti context.[26]

Although the significance and function of viveka in the PcN will be connected mainly to the RcM's elaboration of it, it is important to highlight that viveka held value in the Vallabhite tradition as well. Vallabha composed sixteen treatises in Sanskrit, known as the *Ṣoḍaśagranthas*, with the eighth one titled *Vivekadhairyāśrayanirūpaṇa* (An Examination of Discrimination, Resolve, and Refuge). Vallabha esteemed viveka (discrimination, discernment) as an antidote to avidyā (misconception),[27] a means for the jīva (individual self) to attain a correct understanding

---

21 See Hawley 2005: 181–193.

22 Brajvāsīdās 1873: 14, unnumbered verse.

23 See Chapter 4.

24 For Nanddās's religiously oriented works, see McGregor 1984: 98–99. For his works on rhetoric, see McGregor 1984: 125–126.

25 McGregor 1976.

26 See subsection 4.2.2.

27 See Narain 2004: 306.

of its existence and the world, recognising that everything is Kṛṣṇa's manifestation. It also entails the awareness that everything occurs by God's will and favour.[28] This treatise garnered attention and was the subject of commentaries from adherents of the Vallabhite tradition, including members of Vallabha's family such as Gokulnāth (1552–1641) – Vallabha's grandson and Viṭṭhalnāth's son – who was more renowned for his hagiographies (vārtā) in Brajbhāṣā.[29] Following him, Kalyāṇarāy (1568–?), Viṭṭhalnāth's grandchild, authored a commentary, while subsequent commentaries were written by Gokulotsav in 1630 and Cācā Gopeśaṇ in 1660.[30]

Considering these precedents, it can be conceived that viveka must have been a concept widely discussed in Vallabha's Puṣṭimārga. One could view the PcN and the *Brajavilāsa* as complementary works. The PcN elucidates the method by which an individual can understand their true nature and cultivate detachment from the phenomenal world through discrimination. By doing so, they become prepared to practise unrestricted bhakti, which the text understands as forming or sharing an exclusive bond with a deity.[31] On the other hand, once such awareness is attained, the *Brajavilāsa* immerses the individual in the depths of Kṛṣṇa's story, akin to plunging one into an ocean of divine narratives.[32]

1.2 The *Prabodhacandrodaya Nāṭaka*

In composing the *Brajavilāsa*, Brajvāsī skilfully crafted his own unique identity by explicitly positioning his text as a direct descendant of Sūrdās's songs rather than

---

28 See Barz 1976: 62–63, 71. As far as I am aware, no comprehensive study of this specific treatise exists. For an English translation of the *Ṣoḍaśagranthas*, see Redington 2000.

29 Specifically, the *Caurāsī Vaiṣṇavan kī Vārtā* and the *Do Sau Bāvan kī Vārtā*. The former was originally composed orally and later transcribed by Gokulnāth's nephew, Harirāy (1591–1716). For more details, see Barz 1994. For the *Do Sau Bāvan kī Vārtā*, see Parikh 2009.

30 See Potter 1970–2019.

31 I translate the word bhakti as "bond with god" or "bond" and bhakta as "bound to God". The term "bond" captures the specific connection that exists between individuals while remaining broad enough to encompass various types of connections. Brajvāsīdās's concept of bhakti involves shifting from a bond with the world to a bond with a god, which is why this translation seems fitting.

32 Vallabha viewed liberation as nirodha, which he defined as detachment from the world and attachment to Kṛṣṇa. In this interpretation, Vallabha redefined a term that carries significant meaning in Buddhist, yogic, and Tantric traditions. For more information, see Smith 1998: 495–503. It is worth noting that Brajvāsī does not employ the term nirodha in his PcN. Instead, he adapts the definition from YS 1.2 and uses the term nigraha. See subsection 4.2.1.

solely as an heir to the BhP.[33] In contrast, in the PcN, composed a decade before the *Brajavilāsa*, Brajvāsī seems less confident in explicitly acknowledging the RcM as his inspiration. While this literary practice will be explored in relation to other authors who adapted the RcM in Chapter 3, the subsequent subsections of this discussion delve into the PcN's editions and manuscripts, circulation, and reception. Following that, the structure, narrative levels, and plot of the PcN will be introduced, accompanied by an analysis of its oral and performative features, as well as an examination of the language and prosody that make it distinctive.

1.2.1 Editions and Manuscripts

The primary source used for this study is a lithographic copy of the PcN published by Naval Kishore Press in 1875 in Lucknow, 115 years after Brajvāsīdās composed his text in 1760. While it cannot be considered a canonical critical edition, as no information is available regarding the editorial process or the manuscripts used to establish the text, it is a valuable resource. I was also able to trace two other lithographic copies of the PcN: one published by Gyanbhaskar Press in Barabanki in 1902[34] and the other by the Benares Light Press in 1874. Unfortunately, the former, held at the National Library in Kolkata, is in a fragile condition, making it impossible to study without compromising its integrity. The latter, held at the Gopal Narayan Public Library in the Dulhin Bazar district of Bihar in a locality called Bharatpura, provides some intriguing insights.[35]

Interestingly, the 1874 Benares Light Press edition appears to be connected to the 1875 Naval Kishore Press edition. Gopinath Pathak, the owner of the Benares

33 Brajvāsīdās 1873: 5–6: Śrīśukadeva kahī harilīlā | sunī parikṣita saba guṇaśīlā || sūradāsa soi hari rasasāgara | gāyo bahuvidhi parama ujāgara || phaila rahyo so tribhuvana māhīṁ | gāvata sunata suyaśa haraṣāhīṁ || vividhi prakāra carita hari kere | tāmahi varaṇe sūra ghanere || so vaha prīti rīti sukhadāī | mere mana atiśaya kari bhāī || so to kathā amita vistārā | mopai pāyo jāta na pārā || tāmeṁ brajavilāsa sukhadāī | so kachu kahihauṁ kara caupāī || (The venerable Śukadeva narrated Hari's playful activities; Parikṣita, virtuous, listened to all. That *Ocean [of] Hari's Nectar* – Sūrdās sang [it] in many ways, extremely distinguished. It spread in the three worlds; its reputation delights [people through] singing and listening. Sūr has described Hari's manifold deeds variedly. That way [of] love, bestowing happiness, was exceptional to my mind. Therefore, that renowned tale [with] infinite expansion is unparalleled for me. [Kṛṣṇa's] sports [in] Braj, joyful, [are described] in it; I will narrate something [of] it in caupāī.)

34 McGregor 1984: 159. Brajvāsīdās 1875 is a copy held by the Bodleian Library and has been fully digitised: http://dbooks.bodleian.ox.ac.uk/books/PDFs/N13445609.pdf.

35 Located approximately sixty kilometres from Patna, the library was established in the latter half of the nineteenth century by Gopal Narayan Singh, a descendant of a former nobleman in the area. For more information, see Gopal Narayan Library (n.d.).

Light Press, established the press in 1860 and had ties to the royal court of Benares. Pathak collaborated with court poets such as Narayan Kavi, Munshi Harbans/Haribamsh Lal, and Babu Avinashi Lal. Together, they not only published religious classics in bhāṣā but also court poetry and translations from Sanskrit literature.[36] The cover page of the Benares Light Press edition of the PcN states that Chedilal Vaishya published the text after correcting it based on the instructions of the aforementioned Munshi Harbans/Haribamsh Lal and Babu Abinashi Lal.[37]

Haribamsh Lal also collaborated with Naval Kishore Press. In 1875, the Lucknow publishing house released an edition of the *Bhagavadgītā* with Lal's Hindi commentary, which had been published by the Benares Light Press the year before.[38] It is plausible to assume a similar trajectory for the PcN's Lucknow edition since it was published a year earlier in Benares by Gopinath Pathak's press. Both editions of the PcN were composed for the entertainment (manorañjana) of spiritually advanced individuals (sants). On the final page of its list of errata (śuddhipatra), the 1874 Benares edition includes two dohās (rhyming couplets consisting of twenty-four syllabic instants in each line), the second of which states that the PcN explores the oneness/unity (ekatā) of brahman and jīva, offers the joy (sukha) of knowledge (jñāna) and bhakti, and that its essence (sāra) is a reflection (anumāna) on reality (tatva). While this provides a fitting summary of the PcN's philosophical and religious perspective, there are further connections between the two editions.

The list of errata in the Benares edition provides additional links between the two editions. The list spans a little over five pages, each containing four columns. The columns include the page number, line number, incorrect word (aśuddha), and correct word (śuddha). The errors encompass various cases, primarily involving the misprinting of vowels, either missing or needing to be changed from short to long or vice versa, as well as consonants.[39] There are a few instances of additions or omissions of terms from the aśuddha to the śuddha column, and substitutions of terms are rarer still.

These changes raise questions about the modifications made between the two editions and the actions taken by the editors, potentially including the aforementioned Haribamsh Lal, during the republication of the text. Most of the corrections, such as shortening long vowels or lengthening short vowels, seemed to be made for the sake of prosody and were adopted by the Naval Kishore team.

---

36 See Orsini 2004: 117–118.

37 The passage reads: lāla chedīlāl vaiśya ne munśī haribaṁś lāl aur bābū abināśī lāl kī ājñānusār śodh kar prakāśit kiyā.

38 Stark 2004: 263, 275n28.

39 For example, from "svātī" to "svāti" or "kṛkṣma" to "kṛṣṇa".

However, there are a few cases where this did not occur. While it is not possible to conduct a complete comparison of the two texts here,[40] it is useful to examine a few instances that shed light on the ties between the two editions and the intellectual process behind creating a new edition of the PcN.

On the fourth page of the Benares Light Press list of errata, there is a correction that the Naval Kishore edition did not adopt. The correction changes the verb, which in the sentence is translatable as "touch", from "chūvaata" to "chuata", but the Naval Kishore text reads "chūvata".[41] This choice appears to have been made based on prosody, as the amended "chuata" does not fit the metre's scheme. The metre, known as a somarājī, consists of a fixed sequence of three syllables at the end of each pāda: the first long and the last two short.[42] Keeping "chūvaata" would result in four syllables, while "chuata" has three, but its first syllable is short.[43] For this reason, "chūvata" seems to have been the better option in this case.

On the same page of the Benares errata, the unattested "svābhi" is changed to "svāmi". The Naval Kishore edition accepts "svāmi" but lengthens the final syllable to "svāmī" for metrical reasons. This adjustment is necessary to maintain the thirteen mātrās (syllabic instants) required for the 13 + 11 scheme of the dohā.[44] By doing so, the reading of the Naval Kishore edition improves upon the Benares Light Press edition.

However, there are a few cases where the corrections made by the Benares Light Press were valid but not applied in the Naval Kishore edition. In one instance, the verse follows a caupāī metre, consisting of sixteen syllabic instants in each pāda. The Benares errata corrects "ati ādara" (extreme respect) to "ādara" (respect), omitting the "ati". The Naval Kishore edition does not incorporate this change, and it remains unclear why such a decision was made based on the information available to us.[45] Retaining the "ati" compromises the caupāī scheme, as

40 Not least because it remains uncertain whether the Naval Kishore team had access to *only* the Benares Light Press copy while preparing their own publication.

41 PcN 3.88: kaho sevarā yoṁ nahiṁ pīvata | hama to hāḍa hātha nahiṁ chūvata || ([Buddhist Monk:] I am not going to drink; I do not touch bones [with my] hand.) English versions, unless alternative translations are being discussed or unless otherwise stated, are found in the annotated translation of the play in Part II.

42 Throughout this subsection, various metrical schemes such as dohā, caupāī, and others are discussed in detail.

43 See "Prosody" in subsection 1.2.4.

44 PcN 3.87: aba e svāmī saṅgurū kāpālī taji koha | śiṣya kījiye āpano karikai choha || ([Jaina Ascetic:] Now, O master, true teacher, Skullman, do not be angry. | Make me your disciple, have mercy ||.)

45 PcN 3.22: ati ādara karikai āniya dhāmā | bhāṁti bhāṁti dījai biśrāmā || ([Jaina Ascetic:] Bring him respectfully to your house, make him rest in every way.)

the prefix contributes two mātrās, resulting in a total of eighteen syllabic instants. As a result, the pāda becomes hypermetrical.

In another case, we can deduce why the Benares Light Press amendment was not retained. Pathak's edition modifies "mṛdula paṭa" to "komala paṭa", introducing a complete word change. Both adjectives, mṛdula and komala, convey a similar meaning: "soft, delicate".[46] However, the Naval Kishore edition preserves "mṛdula", and the only plausible explanation again relates to prosody.[47] The verse follows a caupaī metre with fifteen syllabic instants in each pāda. "Komala" accounts for four mātrās, thus altering the total number of mātrās and making the verse hypermetrical. On the other hand, "mṛdula" counts as three mātrās and, therefore, is more suitable in this context.

The readings from the list of errata of the Benares Light Press, primarily concerning the prosodical aspect of the text, have not posed any problems for my translation based on the Naval Kishore edition. Unfortunately, the Benares text does not offer helpful clues for other editorial issues that I have encountered and attempted to resolve, regardless of how small they may be.[48] However, there was one instance where a Benares amendment enhanced my understanding of a verse. The dohā in question appears in a dialogue between Peace and her friend Compassion at the beginning of Act III. Compassion tries to console Peace, who is desperate because she cannot find her mother, Faith. Verse 3.10 reads as follows:

yatana kariya dhīraja dhariya milahi kuṁvari tuva māya
jahāṁ tahāṁ khojiya caliya janihu dhīra akulāya

Make an effort [to] be calm, you will meet your mother, O young girl.
Wherever you go, search [for your] mother, don't be focused and agitated.

Evidently, "janihu dhīra akulāya" (don't be focused and agitated) makes little sense. The Benares errata proposes a modification to the sequence "janihu dhīra": by omitting the enclitic emphatic particle -hu and changing "dhīra" (stable, focused) to its antonym, "adhīra" (unstable, unfocused). This adjustment maintains the same number of syllabic instants (six) between "janihu dhīra" and "jani adhīra" but alters the meaning. The reading with "jani adhīra" aligns well with the sense of the sentence, as Compassion is urging Peace not (jani) to be unstable (adhīra) and agitated (akulāya), conveying a more coherent message. In my translation, I have re-

46 See HSS 4008.

47 PcN 2.87: kahāṁ mṛdula paṭa sukhada śarīra | kahahu kahāṁ balakala ke cīra || ([Materialist:] [If] soft garments giving pleasure to the body [exist, then], say, why [wear the] clothes of tree bark?).

48 Refer to the Note to the Translation in Part II.

versed the negation and rendered it "be focused and patient". The few instances examined above clearly demonstrate that the Naval Kishore editors were not merely duplicating the existing content found in their possession, specifically the Benares copy. Rather, they consistently reworked the texts available to them in various ways. Moreover, these examples highlight the notion that the PcN enjoyed significant familiarity and circulation among the printing houses of Uttar Pradesh during their early years.

Another avenue for assessing the reception and circulation of a text is through its manuscripts. I have come across three manuscripts of the PcN. One is kept in Datiya, Madhya Pradesh,[49] while the other two are preserved at the Vrindavan Research Institute in Vrindavan. These latter two manuscripts, MS 17098 and MS 14008, provide limited colophon details but still offer valuable information. Both manuscripts are written on paper, in the pothī (stack of folios) format, and are covered with a red cloth known as bastā. They feature paratextual elements in red ink, such as introductory formulas and daṇḍas, while the main text is in black ink. MS 14008 is missing its first two folios and is characterised by bold and exquisite handwriting. In contrast, the writing in MS 17098 appears thinner and slightly less precise, perhaps suggesting a hurried copying process. Both manuscripts include folio numbering, verse numbering, and prosodic notations, with approximately nine or ten lines per folio. Furthermore, the contents of the written texts seem to align with each other.

Regarding differences, the text in MS 17098 is enclosed within a rectangular frame outlined with red ink. I suspect that these margins may have been drawn prior to the writing process because, as observed in the first folio, the text occasionally extends slightly beyond the boundaries of the vertical border lines. Additionally, the last line of the same folio is entirely written over the lower horizontal border line of the frame. While MS 17098 lacks a specific date, the colophon in the final folio of MS 14008 indicates that it was copied in the "bright half of the month of Āṣāṛh (June/July), VS 1908", corresponding to 1851 in the Gregorian calendar. This implies that it was copied almost a century after Brajvāsī composed the text in 1760 (VS 1817).

---

49 The MS 159 is currently held in the archaeological museum (Jilā Purātattva Saṅgrahālaya), and unfortunately, I was unable to access and examine it. The information regarding this manuscript has been obtained from National Mission for Manuscripts (n.d). However, it is important to note that these records often lack completeness or precision. While many titles such as *Prabodhacandrodaya* or *Prabodhacandrodaya Nāṭaka* are listed, there is no additional information available regarding the author, which poses challenges for further research. According to the listings, two manuscripts of the PcN were reported to be held by the Gopal Narayan Library. However, during my visit to the collection, I could only locate the lithographed copy published by the Benares Light Press and not the original manuscripts. The process was further complicated by the fact that the library's catalogues are not centralised, handwritten, and inconsistent in their notations.

Examining the paratextual elements of both manuscripts, it becomes evident that these copies were intended for public use rather than personal consumption. The presence of folio and verse numbering, indications of the metres, and the relatively organised layout, while not unique features, serve as aids to readers working with a lengthy text like the PcN. The opening of the PcN sets the tone for its intended and prospective audience: the community of devotees or truth-knowers (satsaṅga), among whom Brajvāsī heard the PC story from his teacher.[50] While a more extensive discussion of the PcN's introduction will follow in the next subsection, a final observation about its circulation can be gleaned from one of the manuscripts. The colophon of MS 17098 is concise but contains a line that provides a clue about the intellectual and institutional context in which the manuscript was copied. The anonymous scribe writes in red ink:

śrīrādhāvallabho jayati śrīhitaharivaṃśo jayati
atha prabodhancandrodaya liṣyate

Victory to Śrī Rādhāvallabha! Victory to Śrī Hitaharivaṃśa!
Now the *Prabodhacandrodaya* is written down.

Apart from "likhyate" being written as "liṣyate" according to bhāṣā orthographic conventions, this brief indication of obeisance (technically termed maṅgalācaraṇa) mentions the revered deity (rādhāvallabha) and the founder (hitaharivaṃśa) of one of the religious traditions of the Braj region. Hitaharivaṃśa (1502–1552) established his sampradāya in Vrindavan in 1535. Consequently, he initiated one of the Vaiṣṇava bhakti traditions that centred their religious life and practice around Vrindavan and Kṛṣṇa.[51] Despite being established during the same period and in the same place, Hitaharivaṃśa's tradition exhibited several differences from those of Vallabha and Caitanya (1486–1533). From a theological point of view, the Rādhāvallabhas emphasised the ontological superiority of Rādhā over Kṛṣṇa, as indicated by the word rādhāvallabha itself, referring to Kṛṣṇa as the "Dearest of Rādhā". They did not seek to establish their authority by claiming continuity with the traditions of Viṣṇusvāmī and Madhva as others did.[52] Significantly, they did not consider Sanskrit texts the most authoritative sources for their bhakti. Prioritising the loveplay of Kṛṣṇa and Rādhā, understood as a married couple, they regarded classical texts as secondary in importance compared to the lyrics composed by Hitaharivaṃśa himself. The congregational singing of these poems (samāj gayān) remains a

50 PcN 1.1–1.8.
51 See Entwistle 1987.
52 See Hawley 2011; Horstmann 2009, 2011.

crucial aspect of Rādhāvallabha worship to this day.[53] Consequently, the tradition did not view Vedantic reflection as fundamental and rejected celibate vows.[54]

The question arises: What would a Rādhāvallabha seeker find in a work like the PcN? The views presented in the PcN and those of the Rādhāvallabha tradition appear contradictory. The PcN narrative does not directly involve Rādhā or Kṛṣṇa but is instead infused with Vedantic imagery. Chapter 4 will demonstrate that the PcN's broader objective is to combine bhakti – a yogic bhakti – with viveka as the faculty capable of extracting knowledge from the Upaniṣads. Bhakti and viveka are also closely linked to the cultivation of vairāgya (detachment) from the material world. Given the information available at present, it is challenging to provide a definitive answer regarding how the PcN might fit within the conceptual framework of Hitaharivaṃśa's tradition. One possibility is that the manuscript was not intended for use within the Rādhāvallabha context.[55] However, no evident indications in the text suggest an alternative context, audience, or sponsor for the manuscript.

At this stage, it is worth considering the distinctive Vaiṣṇava perspectives presented in the PcN, both broad and specific. Brajvāsī's drama opens with an invocation to the bhaktas of Hari and consistently emphasises the significance of the satsaṅga. It promotes an inclusive understanding of Vaiṣṇava belief, acknowledging a Supreme Brahman (or Ultimate Principle) that is simultaneously both nirguṇa (unqualified or devoid of attributes) and saguṇa (qualified or endowed with attributes). Moreover, it upholds the individual's freedom to choose the identity of the saguṇa aspect, with Kṛṣṇa and Rāma the primary alternatives. Since the PcN does not explicitly advocate a Vallabhite viewpoint, these elements could potentially find acceptance among Vaiṣṇavas in eighteenth-century North India, including Rādhāvallabhas.[56] Additionally, Brajvāsī subtly displays his preference for Kṛṣṇa through references scattered throughout the text. For instance, he praises (through the words of the First Actor) Vrindavan as the capital (rājadhānī) where the divine couple of Kṛṣṇa and Rādhā reside.[57] These elements further support the notion that the PcN could be embraced within the Rādhāvallabha tradition.

---

53 See Beck 2012; Williams 2020b.

54 See Snell 1991: ch. 1.

55 Tyler Williams has observed that manuscripts of various works on topics such as Advaita Vedānta, yoga, and commentaries on sacred scriptures were often transmitted outside the circles of a single tradition. Williams 2019b: 288.

56 For a detailed discussion on these points, please refer to Chapter 4.

57 See PcN 1.38 where Kṛṣṇa and Rādhā are described as a "matchless couple" by the First Actor, with Kṛṣṇa's capital in Vṛndāvana. For the plot of the play, see subsection 1.2.2. On the characters of the drama, see the lists of dramatis personae and additional characters in Part II.

Furthermore, even though Vedantic reflection may not have been the primary focus of the Rādhāvallabha sampradāya's activities and philosophy, there are traces of engagement with classical Vedantic texts and the PC story.[58] Priyādās Śāstrī, a disciple of Gosvāmī Sanehīlāl residing in Patna during the late nineteenth century, composed various Sanskrit works. His *Nija Mata Darpaṇa*, a kind of doxography, analyses five Vedantic doctrines and concludes with a description of the Rādhāvallabha tradition. He also wrote a commentary on the *Īśa Upaniṣad*, interpreting it as the story of Kṛṣṇa and Rādhā, as well as the *Śrī Vyāsanandan Bhāṣya*, a commentary on the first three sutras of the first chapter of the BS.[59] Another notable figure in this context is Rādhāmohan Dās, who, according to Goswami Lalitacharan, was an esteemed author of the Rādhāvallabha tradition living in the early nineteenth century. Dās's writings reveal his familiarity not only with texts for bhakti but also with the content of the Upaniṣads and their varied interpretations (vedāntaśāstra).[60] Additionally, Brajlāl Gosvāmī (1658–?), who composed works in both Brajbhāṣā and Sanskrit, authored the *Premacandrodaya Nāṭaka* in 1688, written in Sanskrit. The title of this drama implies a connection with the PC tradition.[61] The increased engagement with Vedantic texts starting from the early nineteenth century, along with the presence of a drama inspired by the PC story, provide substantial circumstantial evidence of an earlier interest that might have prompted the copying of the PcN for internal use within the Rādhāvallabha context.

1.2.2 Structure, Narrative Levels, Plot, and Opening

The previous subsection addressed two main aspects: first, it focused on the primary source of the present book, the Naval Kishore Press edition of 1875, and second, it explored the life of Brajvāsī's PcN beyond its original period and milieu of composition, tracing its publication history up until the late nineteenth century. Examining paratextual material from two manuscripts provided insights into the early modern circulation of the text. In this subsection, we will discuss the structure, narrative levels, and plot of the PcN, followed by an analysis of its opening verses.

---

58 According to Alan Entwistle, later commentators of Hitaharivaṃśa's lyrics focused primarily on theology and doctrine. For details, see Entwistle 1987: 155.

59 Lalitacharan 1957: 592–599.

60 Lalitacharan 1957: 587–588.

61 It also points to the substitution of prabodha with prema (love), which reflects the Rādhāvallabha tradition's inclination towards the love-play of Rādhā and Kṛṣṇa. It is worth noting that in *Premacandrodaya Nāṭaka* the protagonist is not Viveka (Discrimination) but Premacandra (Love Moon). See Lalitacharan 1957: 583–585.

This analysis will shed light on how Brajvāsī crafted his nāṭaka to resonate with multiple communities of Vaiṣṇava bhaktas in the eighteenth century and beyond.

Structure and Narrative Levels

The cover page of the 1875 edition presents the story as "viveka aura moha kā samvāda" (a conversation between Discrimination and Bewilderment). Interestingly, the two characters never actually meet, which prompts the question of what this conversation (samvāda) signifies. To my mind, the conversation can be seen as characteristic of the entire story. This is not only because the plot unfolds through dialogues but the PcN, as a whole, juxtaposes divergent teachings in alternating acts. Act I serves as an introduction, Acts II and III explore the nature of Bewilderment and the heretics, while Act IV elaborates on the tools at Discrimination's disposal. Act V portrays the rejection of Bewilderment and his followers, and Act VI continues the objective of the previous act, culminating in ultimate liberation, consisting in the triumph of the opinion (siddhānta) supported by Discrimination.[62]

The PcN possesses a three-level narrative structure that frames the work, which is established in its first act. This structure is outlined in Table 1:

Table 1: Narrative Levels established in Act I.

	Extradiegetic Level	Intradiegetic Level	Diegetic Level
Verses	1.1 to 1.23	1.24	1.25 to 1.177

The diegetic and intradiegetic levels of a narrative are that of the story world and the events that exist within it, while the extradiegetic level stands outside these. In verses 1.1 to 1.23, Brajvāsīdās first pays obeisance to the sants before discussing his spiritual predicament and how he became acquainted with the PC story, which he has decided to retell, giving the date of his composition as "the year [VS] 1817" (v. 1.23), that is, 1760. He attributes the invention of the tale to Kṛṣṇadās Bhaṭṭ (Kṛṣṇamiśra; v. 1.14), who composed it to impart Vedānta teachings to his disciple and strengthen his bhakti (devotion), jñāna (knowledge), and vairāgya (detachment). This extradiegetic narration makes use of the first person – although Brajvāsīdās does also refer to himself in the third person in verses 1.7,

62 Such a structure is also typical of doxographies. See Nicholson 2010: 9–14. On the particular nature of the dialogues (vādas) in Act VI, see Chapter 4, especially section 4.1. For more detail on the plot, see below.

1.23, and 6.167 – and stands outside the fictional universe occupied by the characters whose story it tells. This level of narration returns in Act VI as Brajvāsīdās closes the tale in the first person at verses 6.166 to 6.168. The intradiegetic level can be seen as beginning in verse 1.24, where Brajvāsīdās stops describing the circumstances of the composition of his drama and again mentions Kṛṣṇadās Bhaṭṭ. The latter assumes the narrative role from verse 1.25 and introduces King Kīratibrahma, a successful ruler who is nonetheless dissatisfied and yearning for spiritual insight.[63] The inclusion of Kṛṣṇadās Bhaṭṭ at this moment is crucial, as he will narrate the story of the play taking place at the court of the king to his disciple, who is not named. This intradiegetic level of narration is thus *inside* Brajvāsīdās's tale but *not* inside the play being performed in the court. It reappears at the beginnings of Act III, Act IV, and Act V, where Kṛṣṇadās Bhaṭṭ answers a question posed by his disciple (vv. 5.5–5.6). This second frame definitively concludes in Act VI, in verses 6.164 to 6.165, where we are informed by Brajvāsīdās that Kṛṣṇadās Bhaṭṭ's disciple "became discriminating, learned, [with] clear understanding, and incomparable".[64] Finally, the diegetic level consists of the narrated events that make up the play itself as opposed to any *telling* of it, whether that be Brajvāsīdās's extradiegetic narration or Kṛṣṇadās Bhaṭṭ's narration of how the play unfolds in the court of King Kīratibrahma. As such, this third level encompasses the rest of the text, the plot of the play itself, written in the third person by the author and comprising spoken lines by characters, stage directions, scene setting, and so on.

Plot

Having outlined the structure and narrative levels of the PcN, it is essential to provide a summary of its plot to gain a comprehensive understanding of the main events in the story. As illustrated above, the PcN is an allegorical drama in which a fictive version of a historical figure tells his disciple the tale of the *Prabodhacandrodaya* in the form of a play being performed by an entourage of actors in the court of King Kīratibrahma. Kṛṣṇadās's intradiegetic narration serves as a frame for the performance of the play and is itself bookended by Brajvāsī's extradiegetic narration, which acknowledges Kṛṣṇadās as the original inventor of the PC tale, refers to other adaptations or translations of it, and frames his own work as a retelling of it in the vernacular language to a new audience. Taken as a whole, the PcN can thus be thought of as a drama about a story being told about a

63 Likely based on the eleventh-century Candella king Kīrtivarman of Jejākabhukti, who also appears in the KPC. See footnote 42 of Chapter 2.

64 PcN 6.165: phīkī pari gai citta teṁ viṣai vāsanā rūpa | bhayo vivekī viṣada mati paṇḍita parama anūpa ||.

play being performed. The plot summaries below, reflecting how I have structured the translation in Part II, encompass all three of these narrative worlds.

ACT I

After six benedictory verses dedicated to Hari and the sants (the truth-knowers, spiritually advanced individuals),[65] Brajvāsīdās introduces himself and his composition, inspired by the tale devised by Kṛṣṇadās Bhaṭṭ (Kṛṣṇamiśra) for his disciple. He then introduces Kṛṣṇadās Bhaṭṭ who begins to narrate the story of King Kīratibrahma, who, despite defeating his enemies and bringing prosperity to his kingdom, remains dissatisfied. The king's discontent arises from his lack of self-knowledge and knowledge of the Absolute (Supreme Brahman). At this point, the First Actor (naṭa, who is also the stage-manager, sūtradhāra) and his entourage enter the court of King Kīratibrahma. He converses with his wife, the Actress (naṭī), and shares with her the divine order he has received from the brahman himself as a voice from the sky to perform the play *Prabodhacandrodaya* at the royal court. His wife questions whether a king deeply immersed in worldly affairs can truly attain the tranquillity of the brahman-gnosis.

Next, Desire (kāma) and his wife, Passion (rati), enter the scene. They discuss the ministers of Discrimination (viveka) and Bewilderment (moha). Desire explains the family's situation to Passion, describing how Supreme Self (paramātama puruṣa) and Material Nature/Illusion (prakṛti/māyā)[66] had a child named Individual Self (jīvātama puruṣa), whose own son, Mind (mana), had produced Bewilderment and Discrimination, among other children. Bewilderment was born from Mind's first wife, Engagement-with-Action (pravartta, Skt. pravṛtti), while Discrimination, the hero of the play, was born from Mind's second wife, Disengagement-with-Action (nivartta, Skt. nivṛtti). Bewilderment aims to take control of the whole universe with the assistance of his first minister, Desire, and other allies, such as Anger (krodha), Greed (lobha), and Intoxication (māda), causing harm to Discrimination's party. Mind remains in their company, oblivious to his own true nature, which brings immense suffering to his father, Individual Self. A cosmic war involving all of them is predicted.

---

65 Verses 1.1 to 1.6 constitute the benediction, which opens the play, prior to Brajvāsīdās's first-person narration. See the subsection "Opening" below. It was customary for dramas to open with the praise of a deity (nāndī). For instance, in the KPC, homage is paid first to the impersonal brahman and then to Śiva the yāmin (the Controller), highlighting his yogic prowess. See Kṛṣṇamiśra 2009: 4–5.

66 On the distinction between the translated names Material Nature and Illusion, see subsection 4.1.2 and footnote 17 of the annotated translation (Act I).

Desire tells Passion of his concern upon learning of a prophecy that states if Discrimination reunites with his long-forgotten second wife, Upaniṣad, they will give birth to twins: a daughter named Knowledge (vidyā) and a son named Wisdom Moon (prabodhacandra). These two offspring will bring about the destruction of Bewilderment and his wing of the family. Discrimination enters the scene alongside his first bride, Good Reason (sumati), and further explains the familial relationships among Supreme Self, Material Nature/Illusion, Individual Self, and Mind, the last of whom remains ignorant and attached to his twin children with Engagement-with-Action: his son I-maker (ahaṃkāra) and his daughter Misconception (avidyā). Discrimination reveals that the only way to defeat Bewilderment is through the birth of Knowledge and Wisdom Moon. Thus, he must rekindle his relationship with Upaniṣad. Queen Good Reason agrees that Discrimination and Upaniṣad must reunite to achieve this goal.

ACT II

Deceit (dambha), one of Bewilderment's ministers and I-maker's grandson, reveals Bewilderment's plan to conquer Vārāṇasī completely. As part of this plan, Bewilderment had sent Deceit and other allies to the holy city. The First Actor instructs the troupe to set up the scene of Vārāṇasī. I-maker, Bewilderment's brother (but treated as a minister), enters the scene disguised as an old priest and surveys the banks of the Ganges. He becomes infuriated by the discerning people in Vārāṇasī, that is, ascetics and yogis dedicated to their spiritual practices and vows, which are identified as acts of bhakti. His attention is drawn to an appealing hermitage, where he decides to check on the prominent mahant (abbot) residing there. The mahant turns out to be Deceit in disguise, who has settled there with his disciples to corrupt the bhakta yogis.

After their encounter, Bewilderment arrives. A Materialist (cārvāka) and his disciple also appear, and Bewilderment is pleased with the Materialist's speech about his doctrine. The Materialist has been sent by Kaliyuga, Bewilderment's ally, and informs Bewilderment that they have an advantage over Discrimination as people are increasingly turning towards unrighteousness. However, the Materialist also says that where Bond-with-Viṣṇu (viṣṇubhakti) resides, they have no hope of winning. Meanwhile, a messenger arrives with news that Discrimination's allies, Faith (śraddhā), her daughter Peace (śānti), and Duty (dharma), are attempting to persuade Upaniṣad to reconcile with Discrimination. Distressed at this news, Bewilderment summons Anger and Greed and orders them, along with their wives, Violence (hiṃsā) and Thirst (tṛṣṇā), to track down Peace and Faith and have them killed or captured. He then visits his lover, False Viewpoint (mithyādṛṣṭi), seeking her strategic expertise to help him hinder Faith's progress.

ACT III

Peace engages in a conversation with her friend Compassion (karuṇā), expressing her concern over the disappearance of her mother, Faith. No one has been able to locate her, so Peace decides to search for her among the heretics. Initially, a Jaina Ascetic (jatī) appears, followed by a Buddhist Monk (sevarā). The two engage in a heated argument about their respective doctrines, invoking their false Faiths. Then a Skullman (kāpālī) makes an entrance. Initially subjected to ridicule, he eventually manages to persuade the other two men to embrace his path of power-seeking, especially after his own Faith manifests and seduces them. The Jaina Ascetic reveals that they are actually servants of Bewilderment and must find Duty's Faith, a supporter of Discrimination. Through his divination abilities, the ascetic discovers that Faith is seeking refuge with Bond-with-Viṣṇu within the hearts of those who focus on Hari. Duty is with Faith, having fled out of fear of Desire. The Skullman decides to dispatch his daughter, Bhairava's-Science (bhairavī vidyā), to capture Faith and Duty.

ACT IV

Friendship (mayatrī) enters the scene and recounts the attack on Faith and Duty by Bhairava's-Science, highlighting their rescue by Bond-with-Viṣṇu. Bond-with-Viṣṇu, filled with anger, decides to provide full support to Discrimination. She conveys a message to Discrimination through Peace. Discrimination enters and summons Investigation-into-Reality (bastubicāra) to counter Desire, explaining the methods to be employed. Subsequently, Patience (chamā, Skt. kṣamā) is summoned to confront Anger, while Contentment (santoṣa) is called upon to combat Greed.

Following these arrangements, Discrimination prepares to depart for Vārāṇasī and encounters Dispassion (bairāga) and Relinquishment (tyāga), whom Bond-with-Viṣṇu sent. Finally, Discrimination embarks on his journey to the sacred city and stops at the Bindu Mādhava temple along the way. There, he offers prayers to Mādhava, seeking victory over Bewilderment.

ACT V

The war is over. Faith enters the Shrine of the Wheel, a pilgrimage site near the Ganges in Vārāṇasī, where she is residing with Peace, to inform Bond-with-Viṣṇu of Discrimination's victory. Faith shares with Peace and Bond-with-Viṣṇu the details of the battle, describing how Discrimination sent a delegation led by Science-of-Reasoning (nyāyaśāstra) to negotiate with Bewilderment. Despite the offer to leave Vārāṇasī, Bewilderment stubbornly refused. Faith enumerates Discrimination's allies, including the Traditional Texts (agama, Skt. āgama) and Six Philosophies (śāstra ṣata). Peace expresses curiosity about their alliance against Bewilderment, to which Faith explains that they all originate from the Vedas. When the Vedas are threatened, they unite to defend their common foundation. Although the various doctrines may seem distinct, ultimately, they all lead to the realisation of the single Supreme Brahman.

Faith proceeds to describe how Bewilderment and his allies were defeated, leaving Bewilderment himself in hiding along with the Supernatural Powers (siddhi). Meanwhile, Mind, aware of the demise of that branch of his family, becomes desperate and contemplates suicide. In response, Bond-with-Viṣṇu sends Sarasvatī, the goddess of learning, to help Mind understand the futility of suffering for his loved ones. Sarasvatī approaches Mind and imparts wisdom about the impermanence of the world and the transient nature of familial bonds. She teaches him to focus on brahman, who exists in both unqualified and qualified forms. This qualified form is particularly suited for the bhaktas, who must direct their devotion to it.

Gradually, Mind comprehends the teachings, and his long-forgotten son, Dispassion, appears. Dispassion further teaches Mind to detach himself from his ego and possessions. Sarasvatī instructs him to live with his second wife, Disengagement-with-Action, entrust his duties to his son Discrimination, and hold love for his father, Individual Self. Only by embracing this path will Supreme Self manifest within his heart.

ACT VI

Peace enters the scene and reports that Discrimination has told her to bring Upaniṣad to him. She encounters Faith and enquires about the relationships between Mind and Individual Self and between Supreme Self and Material Nature/Illusion. Peace also seeks an update on Discrimination's rule. Faith addresses all of her enquiries and informs Peace that Mind had temporarily succumbed to the influence of the Supernatural Powers sent by Bewilderment but that Science-of-Reasoning had restored him to the right path.

Individual Self enters and describes the state of bondage, expressing gratitude to Bond-with-Viṣṇu for finally allowing him to glimpse the other shore. Peace and Upaniṣad join Individual Self and Discrimination. Individual Self asks Upaniṣad to recount her experiences during her period of separation from Discrimination. Upaniṣad shares that she endured much suffering at the hands of Sacrificial Science (yajñavidyā), Hermeneutics (mīmāṃsā), and Science-of-Reasoning. None of them could grasp her teachings, and they ultimately sent her away. In desperation, she sought refuge with her daughter Gītā in a temple on a mountain inhabited by Vaiṣṇavas. Upaniṣad proceeds to explain the perspectives of each of the three doctrines, and Discrimination criticises them.

Upaniṣad then reveals to Individual Self that he is not distinct from the Supreme Self and God. Initially, Individual Self struggles to comprehend this truth. Still, with

Discrimination's guidance, he comes to grasp the meaning of the great sayings (mahāvākyas) of Upaniṣad and learns how to focus on his own self through negative reasoning (neti neti). Individual Self realises his identity as the Supreme Self, and, then Discrimination and Upaniṣad's daughter, Knowledge, emerges and eradicates all remnants of Misconception and Illusion's manifestations. Knowledge then directs her concentration towards the Supreme Self, accompanied by Mind. Soon after, Wisdom Moon appears, and Individual Self embraces him. All traces of ignorance dissipate, and Individual Self praises the power and benevolence of Bond-with-Viṣṇu. He vows never again to forget the single eternal brahman and pledges his love solely to the Absolute. Bond-with-Viṣṇu appears and offers Individual Self a boon. Initially answering that he is content and has no further wishes, he then changes his mind and requests her support for other individual selves lost in the ocean of existence.

We are returned to the king's court where the First Actor dances at the fulfilment of his divine duty to put the play on and Kīratibrahma is said to be contented with the performance. As Kṛṣṇadās Bhaṭṭ concludes his narration, his disciple falls at his feet, his attachment to the world fading, and is said to have become discerning with clear understanding. Brajvāsīdās then declares that those who hear or read this tale shall become free of misconception and ego and achieve a bond with God through discriminative wisdom. He ends the drama by declaring that the world will continue to suffer if this lesson goes unheeded and directly addresses the community of fellow bhaktas a final time, imploring them to forsake desire for sensory phenomena and devote themselves fully to Hari.

Opening

Despite the PcN's core story largely adhering to the narrative established by Kṛṣṇamiśra's PC of the eleventh century and preserved in numerous retellings of the PC story,[67] a careful analysis of the PcN reveals that Brajvāsīdās skilfully reimagined the tale from his unique perspective. This distinctive interpretation becomes evident right from the outset, as demonstrated by the opening verses of the PcN, which will be examined in this subsection.

The PcN opens with eight dohās that provide valuable insights into the bhakti and Vedānta worldviews espoused in the text. Brajvāsīdās begins the drama by exalting the haridāsa ananya, "the single-minded devotees of Hari".[68] These verses

67 For more on the KPC, see section 2.1. For a summary of its plot, see Kapstein 2009: xviii–xxvi.
68 PcN 1.1: "I bow to the beautiful lotus-feet [of] the single-minded devotees of Hari | by a merciful sidelong glance of theirs all the gods are pleased ||." Similarly, at the beginning of the RcM, Tulsīdās distinguishes between good and bad company (RcM 1.4–1.7). See also Lutgendorf 1991: 118–119.

employ the terms "hari" (Hari) and "prabhu" (Lord) to signify the Absolute.[69] While Hari specifically refers to Viṣṇu within the text, it is broad enough, much like prabhu, to accommodate individuals with diverse Vaiṣṇava beliefs. Similarly, the mention of the "saccidānanda" (existence-consciousness-bliss) nature of the Highest Principle reflects this inclusive approach.

Dohās 1.3–1.4 reveal the Vedānta aspect of the text. While they maintain a broad appeal by stating that the Lord is both nirguṇa and saguṇa, they also invoke scriptural authority by quoting the "neti neti" (Neither this, nor that) of the *Bṛhadāraṇyaka Upaniṣad* (BĀU) 2.3.6. However, the reference to brahman being both unqualified and qualified reality is not only an inclusive tool but the actual perspective espoused in the PcN on the Highest Reality.[70] Following this, dohās 1.5–1.6 refer to the sants and emphasise the significance of the community of the truth-knowers, the satsaṅga. This theme recurs throughout the first two levels of narration (Brajvāsīdās and Kṛṣṇadās Bhaṭṭ) of the PcN tale, which only begins properly with the arrival into Kīratibrahma's courtly hall of the First Actor at verse 1.26.

While dohās 1.1 to 1.6 encompass the benedictory verses, dohā 1.7 marks the moment when Brajvāsīdās personally enters the narrative and speaks in the first person (albeit he does refer to himself in the third person in verses 1.7 and 1.23). This section, until verse 1.24, further contributes to shaping the PcN as a bhakti story that might resonate with an eighteenth-century audience. Dohās 1.7 and 1.8 are vinaya verses (verses of humility), continuing until verse 1.11. Notably, the author's self-introduction as a bhakta and the extensive address to the community of bhaktas highlights the framing of the PcN as a bhakti narrative. We see this particularly in verse 1.21:

nahīṁ catura nahiṁ rasika vara nahiṁ kavi yukta udāra
pācholai harijana kahata lehaiṁ sādhu sudhāra

[Since I am] not an expert, nor the best connoisseur, nor a gifted and skilful poet,
at the end I will say to the devotees of Hari: "O virtuous ones, please improve [my composition]!"

I have translated the first pāda of this dohā in the first person because, like the preceding and subsequent verses, it is Brajvāsī speaking and providing insights into himself and the composition of his drama. However, the first line could also

69 PcN 1.2: "In whose hearts, pure abodes, there is always peace and a sentiment of joy, | there, Hari, who is existence, consciousness and bliss, dwells perpetually ||." PcN 1.5: "Bearing such a Lord in their hearts, they think about the world. | [Their] conduct [towards] the whole universe is similar to water and lotus leaves ||."

70 See Chapter 4, especially subsections 4.1.1 and 4.1.4.

be rendered as: "Not [to] the expert, neither [to] the best connoisseur, nor [to] the gifted and skilful poet, at the end [to] the devotees of Hari, I say: 'O virtuous ones, please improve [my composition]!'." The emphasis on the community of bhaktas as an incontestable source of authority is accomplished through the use of "catura" (expert), "rasika" (connoisseur), and the eloquence of the "kavi" (poet) as a contrast. This authority is literary – the devotees of Hari (harijanas) are asked to improve the composition – but what makes the community distinctive is their spiritual authority, to which the subsequent dohā alludes.[71]

Returning to the opening of the play, after the eight initial dohās, Brajvāsī adds, in verse 1.9:

> In the end, how could I – impure, wretched, vile, lacking the merit connected to the aims of human life, full of sin – speak? Dependent on the objects of the senses, how far can I still suffer the blows of the so-called skilful Desire, Anger, Greed, Bewilderment, and Pride? It is your power to save [those] wicked like me; you are the saviour [of] the vilest and the river I desire. Please bestow me the offering of agreeable, simple, wise truth-knowers. Having come for refuge, I will stay in the company of the truth-knowers.

This is a vinaya verse in that it surrenders to and recognises complete reliance on a higher power. Such verses are typical of someone positioning themselves as a dāsa (servant) of the Lord.[72] Stylistically, it resembles the tone of Tulsīdās's *Vinaya Patrikā*:

Whom the silent sages venerate after leaving all [worldly] company, by silent prayer, austerities, and rites,
[Someone] weak, extremely vile, and base like me, [by] what effort may attain him?

Hari [is] taintless, [but my] tainted heart makes me perceive duality.
Why would a gander come to a lake where crow, vulture, heron, and pig [dwell]?

71 See PcN 1.22, which reads: "Hari's devotees observe compassionately the mark [on] the body of those | whose forehead has been drawn [by] the creator, be it an auspicious or inauspicious sign ||." Similarly, in RcM 1.8.2, Tulsīdās asserts that knowledgeable rasikas, if lacking love for Hari, will be unable to appreciate his work. David Shulman, discussing Tamil legends, aptly observes that in these narratives, many authors undergo a transformation from being mere poets (kavi) to poet-devotees (bhakta-poets): "The *kāvya* poet is not allowed to survive in isolation from the overriding *bhakti*-ethos." According to Shulman, the poet in these instances becomes a vessel through which the divine expresses itself, and the overarching ethos of bhakti is prioritised over individual poetic craftsmanship. See Shulman 1993: 18. While the claim to "spontaneity" and lack of craftsmanship in the composition of the PcN is not extensively developed, the presence of a "voice from the sky" (gagana girā) instructing the First Actor to perform the PcN suggests that the play is sanctioned by the divine. This reinforces the divine inspiration behind the performance. See verses 1.32–1.45.

72 See Dimock 1966: 49–51.

Whose refuge learned people take [and] extinguish the fearsome threefold suffering,
Extreme Pride, Bewilderment, and Greed reached even there – Envy does not fade even in heaven.[73]

Tulsī's *Vinaya Patrikā* verses depict the author as a humble devotee, overcome by malevolent embodiments.[74] Aside from providing biographical insights into Brajvāsī, I perceive these PcN verses as both indicating and constituting the community to which Brajvāsī directs his work. While dohās 1.5 and 1.6 establish the image of the truth-knowers as the ideal company, in verse 1.9, Brajvāsī distances himself from them, thereby aligning himself with the bhaktas who may still perceive a need for spiritual progress ("How can I [. . .] speak?"). The path to such profound insights and understanding is illuminated through the PC story itself, whose origins he recounts in verses 1.13 to 1.15. By assuming the stance of dāsya bhāva (attitude of service), he establishes and reinforces the necessity of his PcN for the community of bhaktas. Verses 1.13 to 1.15 narrate the tale of Kṛṣṇadās Bhaṭṭ (Kṛṣṇamiśra), who composed the Sanskrit PC to impart salvific wisdom to a young disciple with a penchant for kāvya.[75] Brajvāsī prompts readers/listeners to first identify with him and then to identify with Kṛṣṇadās Bhaṭṭ's disciple, who displays devotion to his teacher but harbours resistance towards traditional learning.

After making a case for the imperative need to get acquainted with the enlightening PC story, the person entrusted with conveying it on behalf of Kṛṣṇadās Bhaṭṭ is none other than Brajvāsīdās himself, using Kṛṣṇadās Bhaṭṭ as his narrator. That Brajvāsīdās makes the Sanskrit teacher his double is further suggested by the name Kṛṣṇadās, meaning "servant of Kṛṣṇa", which carries the same meaning as Brajvāsīdās, "servant of he who dwells in Braj", that is, servant of Kṛṣṇa. Thus, the extradiegetic vinaya verses can be viewed as serving three purposes: encapsulating the PcN author's emotions, making the text pertinent to a bhakti-oriented audience, and establishing the authority of Brajvāsīdās's retelling.

---

73 Verses 185.2–185.4: sakala saṅga taji bhajata jāhi muni japa tapa jāga banāvata | mo sama manda mahākhala pāṁvara kauna jatana tehi pāvata || hari niramala malagrasita hṛdaya asamañjasa mohi janāvata | jehi sara kāka kaṅka baka sūkara kyoṁ marāla tahaṁ āvata || jākī sarana jāi kobida dāruna trayatāpa bujhāvata | tahūṁ gaye mada moha lobha ati saragahuṁ miṭata na sāvata ||. Tulsīdās 1962: 313.

74 The songs of the *Vinaya Patrikā* were composed in the first decades of the seventeenth century. It is believed to be one of the last works composed by the poet before his death in the 1620s. See Allchin 1966.

75 The term "kāvya" refers to belles-lettres or poetic literature, to which, canonically, drama belongs as a literary genre providing entertainment. For further discussion of these verses within the salvific context of the PcN, see section 4.2.

This shift in perspectives is also observed in some of Sūrdās's early vinaya verses. In many of these poems, he enumerates renowned bhaktas, evoking, through their names alone, the stories that bind them to God. By placing his own signature in the final verse of the poem, he carves out a place for himself among these luminaries. Simultaneously, the perceived distance he acknowledges between himself and these mythical figures – since they have attained salvation while he is yet to be saved – according to John S. Hawley, "moves him in the direction of his hearers". Therefore, speaking on his own behalf and asserting his entitlement to God's favour does not remain a private matter pertaining to an individual but rather a "public" one: "The signature required by a *pada* provides Sūr with a chance to inject into his poems an element of subjectivity or reflexivity – an 'I' – that is potentially inclusive to us all, all the people who are his audience."[76]

Within this "turbulence of identities",[77] Brajvāsīdās sets himself apart from the closest associates of God, thus positioning himself as an equal member within the broader community of bhaktas. Subsequently, he presents himself as their prospective "guide" on the path to liberation and true bhakti, much like Kṛṣṇadās with his disciple (śiṣya). In this manner, both he and the others will become deserving of the company of the sants and have the opportunity to experience the highest form of bhakti, surpassing the four traditional ends of human life (puruṣārthas).[78] Through the performance of dāsya bhāva in the vinaya verses, Brajvāsīdās imagines and defines the bhakti public for the PcN, where the story serves as a mise en scène for the triumph of viveka and bhakti. He fosters reflexivity within his audience through statements like "[those] wicked like me", prompting the audience to identify with him and actively participate in the fruition and dissemination of the PcN. Furthermore, he actively engages with the universe of Tulsī's and Sūr's vinayas, establishing a "rhetorical" circularity with these two authors and their works. The distinction from Tulsī's and Sūr's poems lies in the fact that the vinaya verses of the PcN *also* initiate a discourse on the PC literary tradition.[79] The opening of the PcN thus underscores the significance of the community of bhaktas, an

---

76 This and the previous quotation is from Hawley 1992: 234. For other vinaya verses by Sūrdās, see Bryant / Hawley 2015: 650–751.

77 The phrase Gary Tubb uses to refer the multiple levels of representation in Kavikarṇapūra's Sanskrit *Caitanyacandrodaya*, itself part of the PC tradition. See Tubb 2014: 710.

78 This point is reiterated several times in the early passages of the PcN. See e.g. verses 1.6 and 1.10–1.11.

79 See Novetzke 2007 for insights on bhakti as an embodied performance or phenomenon that contributes to the formation of publics of reception. The PC literary tradition will be the subject of Chapter 2.

element that resonates with all the Vaiṣṇava traditions that might have come across Brajvāsīdās's text.[80]

The author's intention to address a community with a strong didactic purpose is evident not only in the setting of the PcN but also in other sections of the composition. One such instance is the description of Discrimination's ministers, which combines elements of Pātañjala and other forms of yoga, expanding their ethical dimension as a set of guidelines for the ideal bhakta.[81] Another example is Good Reason's speech in response to Desire's accusations in Act I, where she extends her moral judgement of Bewilderment's general to "many men" (bahutaka puruṣa).[82]

These instances seem to confirm Tyler Williams's observation regarding the formation of reading publics, which are

> constituted not only simply by people who "read" the same texts, but by people who share the social contexts and affective regimes associated with reading [. . .]. In the case of early modern north India, the social, performative, and affective fingerprints of these readers have been left all over the books they once loved.[83]

The social aspect is also reflected in the fact that even in manuscript form, a text was intended to be complemented and explained by a teacher. This implies the presence of the spoken and oral sphere and possibly actual performances. This raises the question: Apart from the potential illocutionary aspect of Brajvāsīdās's utterances, did orality and performance play a role in the PcN? And if so, how?

1.2.3 Orality and Performance

Until now, I have primarily treated the PcN as a written text. However, considering how Brajvāsī shaped his public through his persona as a bhakta and cultivated the necessity for his PcN to be understood among the community, it is reasonable to interrogate the role of orality and performance in the PcN. After

---

80 Various bhakta authors, despite their distinct positions on the spectrum of bhakti, praise the value of the satsaṅga. For instance, Tulsīdās in his *Vinaya Patrikā* and Kabīr in his poems emphasise its significance. On the former, see Tulsīdās 1962: 118. On the latter, see Dharwadker 2003: 117 and 184 sākhī 34.

81 See Passion's description of Discrimination's ministers at verses 1.73 to 1.78, especially 1.74 (on Moral Principle and Observance). See also Friendship's mention of the "four sisters" (Friendship herself, Sympathy, Gladness, and Compassion) at verse 4.21. On the relationship between bhakti, bhakta, and yoga in the PcN, see Pastore (forthcoming a).

82 See PcN 1.126: "Many men [in] the world are stupid, ignorant, full of their own evil-doings; those who are virtuous and blissful do not look [for] nor see others."

83 Williams 2019b: 167.

all, the text is referred to as a nāṭaka, the principal dramatic genre according to classical sources on theatre (nātya), a kāvya meant "to be seen and heard".[84] Orality and performance extend beyond the written word; nevertheless, this does not imply the exclusion of the written aspect but instead emphasises its relationship with them.[85]

Although there is limited evidence of actual performances of the PcN or texts within its tradition,[86] the PC story was not always confined to a definitive book but resonated in the minds of early modern audiences, as the RcM attests.[87] The PcN, in fact, carries traces of the oral circulation and transmission of the PC story, as evident in verse 1.12:

satasaṅgati paranamya guru pada paṅkaja rākhi śira
prathama su kaha atiramya granthotpati jihi bidhi sunī

In the company of the truth-knowers I bowed, keeping my head on the lotus-feet of my teacher.
First of all, he described the extremely enjoyable origin of the book, [which I report] in the way I listened.

That Brajvāsī became acquainted with the PC story orally through the community of the truth-knowers is repeated in 1.16: "I heard in this way about the origin of the book [composed by Kṛṣṇadās] in the company of the truth-knowers." The verse also anticipates its effect (phala), by saying that: "[For one who] listens, understands, [and] reads [this work] with delight, the adversity [of] the world will vanish."[88] The presence of orality is further emphasised in verse 1.24:

kṛṣṇadāsa bhaṭa śiṣya soṁ kahata kathā parabodhi
naṭalīlā ke byāja kari paramatatvamaya śodhi

Kṛṣṇadās Bhaṭṭ narrated to his disciple a story, awakening [him],under the pretext of a play of actors, consisting of the Highest Reality, purifying.

---

84 As verse 1.11 of Bharata's *Nāṭyaśāstra* (NS) defines it. See also Ghosh 1951: 3.

85 A recent edited volume by Orsini / Butler Schofield 2015 delves into the "interdependencies of written text, sound, performer, audience and meaning" (quotation from p. 4).

86 However, Basile Leclère's study on a Jaina retelling of the PC story, titled *Moharājaparājaya* (The Defeat of King Bewilderment, twelfth century) by Yaśaḥpāla, offers several arguments that align with those presented in this subsection, supporting the idea of a mise en scène in medieval dramas. See Leclère 2013: ch. 6.

87 See subsection 3.2.1 for references to the PC story in the RcM.

88 In the original: [. . .] | sunai samujhai paḍhai ruci soṁ miṭai jagata vipatti ||.

This conceptual prominence of the spoken word and instruction is not uncommon among early modern poets.[89] However, Brajvāsī also alludes to granthas (books) and mentions in verse 1.18 that the PC kathā by Kṛṣṇadās was made into a Persian manuscript/book (bhāṣā yamana kitāba) by Balīrām (i.e. Banvālīdās). While the PC tradition and the sources of Brajvāsī's PcN will be explored in subsequent chapters, it appears that Brajvāsī combines knowledge from both oral and written sources in his retelling or, at the very least, he underscores this aspect of his literary practice.

While Keśavdās composed manuals for poets inspired by written Sanskrit precedents he encountered,[90] it appears to have been important to Brajvāsī that the composition of a nāṭaka involved an oral and aural dimension.[91] Although it is impossible to determine for certain whether Brajvāsī gives more importance to the oral account by his guru or the written text by Balīrām,[92] we can make some additional observations concerning performance.

While Brajvāsī does not explicitly mention in the verse regarding the phala that those who "see" the PcN shall benefit from it (preferring instead "listens" and "understands"), he does make it clear that even reading the text was not intended as a solitary endeavour. In fact, he mentions at the end of the play that King Kīratibrahma, upon witnessing the deeds performed by the actors, is content.[93] Brajvāsī further strengthens the connection between the PcN and performance in verse 1.27 by providing, through Kṛṣṇadās's narration, a detailed description of the company

---

89 Williams 2019b: 163n25 highlights how even literate authors like Tulsīdās and Nanddās have paid minimal attention to the material aspect of their writings.

90 See Busch 2015: 254.

91 The interaction between oral and written traditions in literature is complex and fluid. An example of this can be seen in the texts employed by the Marathi kirtankāras, which clearly serve the purpose of performance. See Novetzke 2015.

92 In this context, the insights of Velcheru Narayana Rao and David Shulman regarding cāṭu, free-standing verses in Sanskrit, Telugu, and other South Indian languages, are pertinent: "Whether it is a written poem in oral circulation or a verse composed orally is an issue of little significance in the oral-literary culture of premodern India. What is important is that a cāṭu is remembered and used in social communication among a community of people who constitute themselves as a group by sharing a certain body of knowledge and ideology. The cāṭu defines, expresses, and communicates such knowledge and ideology among specific communities." Rao/Shulman 1998: 6–7. For Brajvāsī, the PC is a story that deserves to be retold, as it has the capacity to convey a message to a community he defines in the opening of his drama. While his exact source is relevant to some extent, its significance is limited.

93 See PcN 6.162: bahuri sabhā ke maddhi naṭa kīnho nirtta sudeśa | bhayo prasanna caritra lakhi kīrata brahma nareśa || (The First Actor danced beautifully. King Kīratibrahma, having seen the performance, became content.)

of musicians, actors, and particularly actresses who arrive at King Kīratibrahma's court to perform the play:

> With much expertise, and as beautiful as a song [of] passion, someone brings the tāla and a mṛdaṅga, someone else brings a small ḍhola and the muhacaṅga; someone a flute, a vīṇā, and an upāṅga, someone else a large sārangī. Someone brings a sitāra, someone else a khanjarī and a kartāla. A big company of beautiful young women; looking at them [even] the bodiless Desire, thirsty, blushes. Extremely lovely, delicate, youthful, long-haired, all pretty-faced and exceptionally expert, [with] lively eyes [like] a wagtail [or] a fish. Well-dressed and decorated with ornaments, the whole assembly is incredibly charming. All versed [in] dancing and singing, similar to the women [of] the celestial singers. Looking at them, [even] Urvaśī would hesitate. All competent [in] the ways [of producing] amusement, they describe various stories. They assume different kinds [of] disguise, according to the stories [they narrate]. Clever, they bring [into the performance] many speculations; looking [at them], the heart rejoices. Such a unique assembly, they can take on distinct appearances. They all arrived in front of the king and pronounced auspicious words.

This verse vividly portrays a dynamic acting company accompanied by musical instruments. The author emphasises the competence and versatility of this professional crew, who can skilfully play multiple instruments and portray various characters through disguises and costumes. The description, which in the Brajbhāṣa text abounds with "someone . . . someone" (kou . . . kou), evokes the bustling atmosphere of a crowded assembly hall (sabhā), reminiscent of similar descriptions found in Keśavdās's *Kavipriyā*. In both works, the authors praise the literary, musical, and dancing abilities of a group of courtesans.[94]

Although the portrayal in the PcN may follow certain conventions, it still carries a distinct "local flavour" by mentioning the sitar (sitāra). This musical instrument has an ambiguous history: traditionally believed to have been invented by Amir Khusrau in the thirteenth century, it appears that it was actually introduced at the court of Muhammad Shah (1719–1748) in Delhi by the musician Khusrau Khan.[95] The earliest mention of the sitar in a text dates back to 1739,[96] and sources from the late eighteenth century onwards associate the instrument with both courtly and informal settings, as well as different playing styles.

Another way performance is incorporated into the PcN is through the stage directions that describe the scene's setting and the mentions of the backstage area, dancing and music sequences, entry and exit of actors, indications of moods

---

94 See Busch 2015: 252.

95 This question is summarised on the basis of earlier secondary literature by Allyn Miner, who provides further insights drawing on primary sources and oral accounts. See Miner 1997: ch. 1.

96 Specifically, Dargah Quli Khan's *Muraqqa'-i Dehlī*. See Miner 1997: ch. 2 for other sources.

and facial expressions, and inclusion of costumes.[97] Furthermore, the use of various metrical forms throughout the text suggests that it was composed with a form of performance in mind. Consider, for example, verse 1.36, spoken by the First Actor:

jāko śruti gāvai kahata na āvai neti neti kahi hāra rahe
go girā atītā carita punītā bhanata ananta na anta lahe

jākī yaha māyā jagata upāyā saba maiṁ saba te nyārā
jo aguṇa anūpā bahu guṇa rūpā adbhuta rūpa apārā

The revealed texts praise him, [yet], not being able to define [him], they say, defeated: "[He is] neither this, nor that."
Beyond the senses and speech, his pure deeds are said [to be] infinite, they do not have an end.
[Through] his Material Nature he manifested the world, [present] in all, [yet] separated from all.
Who [is] without qualities, [and yet] a unique form with many qualities, a wonderful form, infinite.

This text comes alive not through silent reading but oral recitation. Its poetic metre is the tribhaṅgī, a rhythmic pattern with three yatis (caesuras).[98] The yatis occurring after the tenth and eighteenth syllabic instants establish the verse's rhythm, creating rhyming pairs such as "gāvai" and "āvai", "atītā"[99] and "punītā", "māyā" and "upāyā", and "anūpā" and the first "rūpā". Additionally, the musical effect is heightened by alliterations, as seen in "atītā carita punītā" and "jākī yaha māyā jagata upāyā". In the second part of the fourth pāda, with "aguṇa anūpā bahu guṇa rūpā adbhuta rūpa apārā", the alliteration is on both vowels (a and u being the only ones used) and consonants, with repetitions of ga, ṇa, ra, ha, and pa.

The convergence of the written and oral dimensions is further exemplified in the closing verses of the final act, at 6.167 and 6.168:

bhakti hoya bhagavanta kī aura viveka prakāsa
bhakti binā tribhuvana dukhī kaha brajavāsīdāsa

jo cāhai ānanda sadā re brajabāsīdāsa
tau kīje hari bhakti ko choḍi viṣai kī āsa

97 See e.g. PcN 1.29–1.31; 1.58–1.61. See also the Note to the Translation in Part II.

98 See Sinha 1964: 56. For more on the prosody of the PcN, see the relevant part of subsection 1.2.4. For another meaning of yati, see footnote 39 of Chapter 2.

99 The text of the lithographic copy has atīta, which I have amended to atītā for metrical reasons.

There will be a bond with the Blessed Lord and the effulgence of discrimination.
Brajvāsīdās says: "Without a bond with God the three worlds are distressed.

O servants of he who dwells in Braj (Kṛṣṇa), if you wish for eternal bliss,
then forsake desire for the objects of the senses and share a bond with Hari."

Brajvāsī skilfully plays with the meaning of brajabāsīdāsa. In verse 6.167, I have interpreted "kaha brajabāsīdāsa" in the second line as the author's signature, "Brajvāsīdās says". This translation, in my view, harmonises the ending of the PcN with its beginning, emphasising the author's direct appeal to the community of bhaktas or harijana. In contrast, in verse 6.68, I have chosen to translate brajabāsīdāsa as "servants of he who dwells in Braj (Kṛṣṇa)", that is, the servants of the brajabāsī. Although the PcN is a lengthy narrative and not a song/short lyric like those of Sūrdās or Kabīr (?–1518), Brajvāsī's way of addressing the audience echoes the style popularised by these authors and others. He employs interjections like "re" to capture the audience's attention and signals a change in the sentence by using "tau".[100]

The question of how the PcN was performed can be understood by examining its potential relationship with actual eighteenth-century dramatic forms. Kathryn Hansen, in her research on the early appearance of Svāṁg (and later, Nauṭaṅkī) theatre, points out that texts such as *Āʾīn-i Akbarī* by Abul Fazl (1551–1602) and Maulana Ganimat's *Nairang-e Iśq* (1685), among others, hinted at performers who were likely not associated with devotional contexts.[101] The few lines mentioned by Hansen, which highlight the acting abilities of such individuals, indeed resonate with the description found in the PcN. She acknowledges that religious līlā performances, particularly the more informal ones held in villages, served as folk forms that paved the way for the secular Svāṁg and Nauṭaṅkī.[102] In her discussion, she considers the dramas of the PC tradition composed in Brajbhāṣā. Hansen draws on a study by Gopinath Tiwari, who asserts that the bhāṣā nāṭakas (vernacular dramas), such as those in the PC tradition, employ the terms svāṁga and khyāla to refer to theatrical forms. Tiwari notes that the PcN contains eight instances of the term "svāṁga" in that specific sense.[103] However, while svāṁga appears sixty-eight times in the PcN, its meaning is not that of a particular dramatic form but simply denotes a specific actor or actress, which I have translated as

100 The inclusion of the adverbial particle contributes to a spoken, colloquial tone in the sentence, as seen also in the arillas of the early seventeenth-century poet Bajid. Arillas are short poems containing proverbs, idioms, questions, and exclamations popular in the Dādūpanth tradition and prevalent in Rajasthan. For more details, see Bangha 2015: 380.

101 Hansen 1992: 61–62, 314n10.

102 Hansen 1992: 60–61.

103 Hansen 1992: 62; Tiwari 1959: ch. 3.

"the actor playing . . .". To take just three examples, verses 1.114, 6.152, and 6.155 read as follows:

yoṁ bāta kahī jaba kāma rāya
    tihi antara svāṁga viveka āya

When King Desire uttered such words, the actor playing Discrimination arrived.

aise devī upaniṣada aura viveka bhuāla
    gaye sabhā te nikasikai doū svāṁga utāla

The two actors playing Upaniṣad and Discrimination exited the hall swiftly.

tā antara paṭa graha te āyo
    svāṁga prabodha canda chavi chāyo

Then, having arrived from backstage, the lustre of the actor playing Wisdom Moon spread.

Hence, it cannot be concluded that the PcN demonstrates an awareness of a distinct form of folk theatre such as Svāṁg.[104] Nevertheless, Tiwari's hypotheses should not be disregarded, as he attempted to challenge the viewpoint of most historians of Hindi literature and theatre who argued that Brajbhāṣā nāṭakas were essentially long poems and not plays due to their deviation from the conventions of Sanskrit drama.[105] Tiwari contended that folk dramas served as a model for the Brajbhāṣā nāṭakas. However, as Hansen points out, and as my discussion of the term svāṁga in the PcN demonstrates, it is challenging to refute or support such an assumption without a more thorough examination of these Brajbhāṣā dramas.

Certain elements in the PcN could lend support to Tiwari's hypothesis. First, the limited number of scene indications and divisions in the PcN resembles traditional Nauṭaṅkī, where acts or scenes are not explicitly demarcated.[106] In the PcN, no explicit separation is indicated, for instance, between the prastāvanā (prologue) and the commencement of a specific act. Additionally, changes in setting and the entry or exit of characters are not always explicitly mentioned. For example, in verse 2.6, the First Actor instructs his troupe to recreate Kāśī in the assembly hall, but it is unclear at

104 It is also worth noting that the mention of khyāla (e.g. vv. 1.87, 2.122, and 3.98) in the PcN clearly refers to "thought, opinion, conception, idea", rather than the poetic style or folk theatre. See Platts 498.

105 See e.g. Gupta 1958: 41. See also section 1.1.

106 See Hansen 1983: 82.

what point he leaves the stage (or assumes another role) before the appearance of the actor playing I-maker a few lines later (from v. 2.14). Could these omissions indicate something that transpired beyond the confines of the written page? We don't know.

Second, despite upholding a bhakti worldview with a strong philosophical inclination, the PcN challenges a rigid demarcation between religious and secular theatre. Its narrative revolves around a king (Mind) who must learn the art of ruling, wherein dispassion or detachment (vairāgya) is not simply about disengagement from the social realm but cultivating the equanimity necessary for just governance over the mind, body, and kingdom through the exercise of reason/discernment (viveka). In this sense, the PcN shares a thematic affinity with the stories found in traditional Nauṭaṅkī, where kingship plays a pivotal role. Early dramas, in particular, portrayed kings who had to establish the legitimacy of their power through renunciation and detachment from material desires.[107]

By transcending the confines of the "religious" and "secular" theatre categories, there is undoubtedly more to discover.[108] This discussion holds particular significance for the analysis presented in Chapter 4, where it will be demonstrated how the PcN adaptively reuses the RcM. Could it be that Brajvāsī's utilisation of some of the RcM's rhymes indicates not only inspiration drawn from the latter text but also its oral rendition? The investigation conducted in the subsequent subsection reveals several Avadhī linguistic features within the language of the PcN, further bridging the gap between the two works. Additionally, the prosodic aspects offer valuable insights into the potential oral exposition of the PcN.

1.2.4 Language and Prosody

Although the metaphors and similes employed in the PcN pose challenges in terms of understanding and translation, the language itself is relatively straightforward, with certain syntactical relationships clarified by postpositions. Grammatically, the foundation is Brajbhāṣā, yet it incorporates elements from Avadhī and, to a lesser extent, Khaṛī Bolī.[109] A brief examination of these linguistic fea-

107 On the evolution of these dramas towards a critique of sovereignty and power, see Hansen 1992: 124–125, 263.

108 See Vatsyayan 2007: 162; Sharma 2020.

109 Brajbhāṣā and Avadhī are among the main vernacular languages of literary production employed predominantly in North India from the fifteenth to the late nineteenth century. Brajbhāṣā has generally been identified as a western dialect, specifically the language spoken in the Braj region, south/southwest of Delhi. Avadhī, as the name suggests, is identified as the local language spoken in the Avadh region, around the city of Benares or Vārāṇasī in the east. Traditionally,

tures follows. Special attention is given to the usage of the ergative particle "ne", which is uncommon in Brajbhāṣā poetry but appears in other literature in the Vallabha tradition, suggesting that Brajvāsī may have been familiar with those works. Additionally, the lexicon of the PcN is considered, as the presence of Sanskritic elements is expected given the genre of the text, while the inclusion of Persian vocabulary serves as a reminder of the political context of Mughal North India during the eighteenth century. Lastly, an investigation into the metrical aspects of the PcN will be undertaken.

Avadhī Elements

While the Brajbhāṣā employed in the PcN aligns with the descriptions found in numerous studies on early modern authors and their works,[110] it is important to take note of the Avadhī elements present in the text, as they hold significance for the adaptive reuse of the RcM in the PcN, which will be discussed in Chapter 4. The influence of Avadhī is primarily observed in the verbal system of the PcN, particularly in the forms of the perfect participle. In addition to the Brajbhāṣā forms, we encounter the following Avadhī forms:

- addition of -va- to roots ending with -ā, as seen, for example, in verse 2.109:

  āya tinhana bhūpahi śira nāvā
  prathama krodha bala balaki sunāvā[111]

- ending in -eū, as seen, for example, in verse 5.147:

  hvai nirāśa tuma teṁ so raheū
  koṭi bhāṁti saṅkaṭa bahu saheū[112]

Brajbhāṣā has been associated with poetry dedicated to Kṛṣṇa, such as the songs of the *Sūrsāgar* by the blind poet Sūrdās. Avadhī, on the other hand, is most notably the language used in the composition of the RcM, as well as in the Sufi premākhyāns. However, it has been demonstrated by researchers that Brajbhāṣā extended beyond religious poetry and became a medium for scholastic endeavours as well. Another language commonly associated with the Delhi region is Khaṛī Bolī, which gained prominence as a literary form from the eighteenth century onwards and is considered the basis of MSH. For an overview of the literary languages of early modern North India and their evolution, see McGregor 2003.

110 See e.g. McGregor 1968: 127–225; Snell 1991: 92–113; Pauwels 1996: 57–71.

111 "They arrived and bowed to the king. At first Anger boasted about his strength." In this example, the -va- is augmented to -vā- by the poet to fulfil the requirement of having two syllabic instants in order to achieve the desired count of sixteen in each pāda of the caupāī metre. See Kellogg 1989: 320.

112 [Sarasvatī:] "He lived disappointed by you; he suffered a million difficulties." See also Saxena 1938: 241–247.

- kaha used as the perfect participle of kah-, as in verse 2.43:

  ahaṃkāra alaga jasoṁ tinako kaha jhijhikāra
  re mūrakha mote nahīṁ baḍo mahanta tumhāra[113]

Furthermore, the PcN extensively employs the inceptive imperfect,[114] formed by combining the imperfect and perfect participles. The latter is conjugated, while the former retains the -ta ending to indicate the initiation of an action in the past. An example of this usage can be seen in verse 2.41:

beṭhana lagyo bahori ahaṁkara ḍhīga dambha ke
dambha śiṣya kari sora tākahaṁ puni ḍāṁṭata bhaye[115]

Another Avadhī verbal form employed in the PcN is the polite imperative, marked by the ending -iya. An example of this can be found in verse 3.10: "yatana kariya dhīraja dhariya milahi kuṁvari tuva māya | jahāṁ tahāṁ khojiya caliya janihu dhīra akulāya ||."[116] Lastly, the PcN also makes (albeit limited) use of the Avadhī polite possessive adjective and pronoun "rāvara",[117] which corresponds to the MSH "āpkā", as well as the nouns of agency formed with the suffix -vāra (pl. -vāre):

bolo hama bairāga aru tyāga duo asavāra
bhakti rāvare pāsa ko paṭhayo hamaiṁ guhāra (v. 4.97)[118]

kachu yaka loga baudha matavāre bani baiṭhe tāpāsā
lāgo kahana sevarā tinasoṁ kara pustaka parakāsā (v. 3.36a)[119]

113 [I-Maker *(in a fit of rage, answers as if they were alone)*:] "Fool, your mahant is no more important than I!" See Kellogg 1989: 320.
114 See Kellogg 1989: 319.
115 "He then sat next to Deceit. Deceit's Disciple, protesting, scolded [him] again."
116 See subsection 1.2.1 for a discussion of the syntax of this line and its amended translation.
117 See Saxena 1938: 197–198.
118 "[They replied:] We are Dispassion and Relinquishment, two riders. | Bond-with-Hari, calling us to help, sent us to you ||." "They replied" is technically closer to the beginning of the Brajbhāṣā line, but in my translation I have rendered this as solely Dispassion speaking for dramaturgical reasons.
119 "Some other actors, disguised as intoxicated Buddhists, sat around carrying out austerities. The monk started talking to them, showing them his book."

Kharī Bolī Elements

In addition to the Avadhī features observed in the verbal system of the PcN, there are also detectable elements of Kharī Bolī in its postposition system. The locative postposition "meṁ" is used alongside classical Brajbhāṣā forms such as "mahiṁ", "madhya", and "māhīṁ". Similarly, the Kharī Bolī postposition "ke pāsa" can be found alongside "ḍhīga" and "samīpa".[120]

The Ergative Particle "Ne"

The particle "ne" is primarily associated with Kharī Bolī and MSH, but it also finds a place in some Brajbhāṣā texts.[121] While the origin, first usage, and functions of the particle remain topics of debate,[122] a pattern can be observed in its usage in the PcN as a marker of the agent in sentences where the verb is in the perfect tense. Out of the eleven occurrences of "ne" in the PcN, five involve the verb kar-, three involve kah-, and the remaining three, paṭh-. An example for each of these verbs is given below:

kahyo saṭha preta mama guru taiṁ ne karyo (v. 3.71)
bahuri śāradhā ne kahyo e duhitā tehi ṭhāṁva (v. 6.44)
tabahīṁ paṭhyo bhakti ne svāṁga sarasvati āya (v. 5.89)[123]

Furthermore, "ne" is only used once each with the perfectives of the verbs sun-, lā-, de-, and mār-:

rājana rājā moha ne sunyo viveka vicāra (v. 2.1)
tāhī antara krodha ne hiṃsā laī bulāva (v. 2.119)

---

120 Additionally, "sātha" is often paired with the possessive "ke" as in Kharī Bolī. It is found together with the other Brajbhāṣā postpositions "saṅga" and "sahita".

121 Among such Brajbhāṣā texts, Andrea Drocco quotes prose works spanning from the seventeenth to the early nineteenth century, including Mahārājā Jasvant Siṃh's *Prabodha Nāṭaka* (a PC retelling), Gokulnāth's *Caurāsī Vaiṣṇavan kī Vārtā*, and Lallū Lāl's *Rājanīti*. See Drocco 2016: 213.

122 For a summary of diverging scholarly perspectives, see Reinöhl 2016: 59–63. See Dahl/Stroński 2016: 1–37 for additional insights.

123 3.71: "[He said:] Fool, you made of my teacher an evil spirit"; 6.44: "[Then Faith said:] 'O daughter, in that moment [. . .]'"; 5.89: "Sarasvatī, sent by Bond-with-Viṣṇu, arrived." In the first example, "He said" (kayho) in the original refers to Kapalika; in my translation, this is dropped, and the speaker identified, rendered as Skullman. The same applies to the second example, where the introductory clause "The Faith said" is dropped, and indeed they are several instances throughout my translation. See the Note to the Translation in Part II.

jāni parata kou preta ne tohi diyo upadeśa (v. 3.69)
jabahīṁ bastuvicāra ne māryo kāma pracāra (v. 5.63)[124]

In the PcN, the "ne" primarily accompanies singular common nouns, which can also be understood as proper nouns due to personification (as in v. 5.89). Only in one instance do we find the second-person singular pronoun "taiṁ" (v. 3.71). The usage of this particle in the PcN aligns with the tendencies observed in the *Caurāsī Vaiṣṇavan kī Vārtā,* transcribed by Harirāy sometime in the seventeenth or eighteenth century.[125] In the *Vārtā,* among the perfective clauses with verbs like kar- and kah- that feature "ne", the majority have the particle paired with proper or common nouns as agents.[126] The presence or absence of "ne" appears to depend on the agent's role. This raises the question of whether Brajvāsīdās was familiar with the *Vārtā* and whether it influenced this aspect of his composition of the PcN, given the importance of both works in the Vallabha sampradāya and the infrequent appearance of "ne" in earlier Brajbhāṣā works.

Vocabulary

While the PcN demonstrates flexibility in its verbal and syntactical systems, its vocabulary largely adheres to standard Brajbhāṣā. It incorporates both loanwords (tatsama) and loanwords adapted to the phonology of Brajbhāṣā (tadbhava) from Sanskrit in equal measure. Apart from the customary formulas found in the paratexts, the occurrence of purely Sanskritic elements is irregular and often adjusted for prosodic reasons. For example, consider verse 1.48:

kimi chūṭai ika bārahī inakī mati tādeva
    e bādhaka sukhasānta ke sata māraga kemeva[127]

The interrogative term "kimi", to be understood as the Sanskrit "kim" with the Brajbhāṣā exclusive emphatic particle -i, has been employed by other early modern authors.[128] However, the rhyming words "tādeva" and "kemeva" are not com-

---

124 2.1: [First Actor to Kīratibrahma:] "O king, Bewilderment got to know about Discrimination's plans"; 2.119: "At that moment Anger called Violence"; 3.69: [Buddhist Monk to Skullman:] "I must know: did an evil spirit give you this teaching?"; 5.63: [Faith:] "Thus Investigation-into-Reality killed the visible Desire."

125 See footnote 29 in this chapter.

126 Drocco 2016: 225 discusses the use of "ne" with the same verbs but paired with the first-person singular, honorifics, all other pronouns, and titles.

127 [Actress:] "How can [a king]'s thoughts escape [these attachments] just once? | They are obstacles to joy and peace; how, indeed, [can there be] the true path?".

128 The frequent use of "kimi" can also be observed in Tulsīdās's RcM.

monly found in Brajbhāṣā dictionaries. "Tād" corresponds to the Sanskrit "tad", with the first syllable lengthened for metrical reasons, while "kema" is a form of "kim".[129] Both are combined with the Sanskrit emphatic adverb "eva". The presence of these elements, which differ from nouns, in a single dohā can be seen as a deliberate choice to create a "Sanskritic" sound in the sentence.[130]

At the same time, Brajvāsīdās existed in a multilingual world, and his PcN's register incorporates Persian and Arabic vocabulary from various domains. Many terms relate to imperial power and the court. In addition to Sanskrit-origin words for "minister", such as saciva and mantrī, Brajvāsīdās also employs Persian-derived terms like divāna and saradāra for it or to mean "(army) chief/general".[131] There is also mention of the darabāra, meaning "court" or "courtly hall", and its nakība, a "herald" who announces the titles of the king upon his arrival or who served more generally as a "servant". The PcN also includes dagābāja, meaning "traitor" or "cheater", and jurrābāja, referring to a "white falcon". The term hukuma, meaning "order", in the sense of imperial commands, is also present.[132] A similar approach is observed in the military sphere, where the army is referred to as fauja. Specific army units mentioned include savāra (riders), piyādā (infantry), and jamāta (battalion).[133] The term bakhatara refers to the iron armour worn by combatants.[134] Another, broader category of terms of Perso-Arabic origin includes common words such as fakīra (beggar, ascetic), dimāga (brain, thinking), khabara (news, information), khyāla (thought), gulāma (servant), yāra (friend), tamāśā (show, spectacle), the rhyming batāśā (sugar-cake), and yaka, the term for the number "one".[135]

As we will see in Chapter 4, Brajvāsīdās discusses religious and metaphysical concepts in a manner consistent with the traditions of bhakti and Vedānta. Furthermore, through the Perso-Arabic vocabulary employed, we catch a glimpse of the everyday universe and sociohistorical environment accompanying the author's philosophical and religious concerns, providing a more concrete understanding of

129 HSS 1039 identifies it as a Gujarati form.

130 The frequent use of the Sanskrit possessive adjective "mama" alongside common Brajbhāṣā forms has the same effect. For the "Brajification" of the Upanishadic formulas "tat tvam asi" and "neti neti", see subsection 4.2.2.

131 See PcN 1.74–1.75 and 5.73.

132 See PcN 1.26 (courtly hall), 4.13 (white falcon), 4.36 (cheater), 4.89 (herald), and 4.99 (orders).

133 See PcN 4.91, 4.54, and 4.95 (infantry, army).

134 See PcN 2.4: [Deceit:] "Wear an armour, constrict the body [by] keeping firm the cloth around the hips!" The term bakhatara derives from the Persian baktar, meaning "iron armour, coat of mail, cuirass" according to Platts 160.

135 See e.g. PcN 1.30 (friends), 1.50 (spectacle, sugar-cake), 1.87 (thought), 2.16 (brain), 2.42 (one), 2.91 (servant), 2.100, and 3.45 (beggar).

his time.[136] The prosody of the PcN reveals another aspect of Brajvāsīdās's personality, wherein he exhibits, or aspires to exhibit, greater literary dexterity than he may openly acknowledge.

Prosody

One notable characteristic of the PcN, highlighted by McGregor, is the "great metrical virtuosity" displayed in Brajvāsīdās's work.[137] In fact, the first act of the PcN alone includes thirty-eight types of metre (vṛtta). These encompass both varṇa vṛtta, which is based on the order and number of syllables (varṇa), and mātrā vṛtta, which is based on the quantity of syllabic instants (mātrā) (See Table 3). According to Maheshwari Sinha, the prevalence of Sanskrit varṇa metres in early modern vernacular poetry is often due to poets having a Sanskrit text as a reference.[138] However, this does not appear to have been the case with the PcN, as Kṛṣṇamiśra's PC switches between verses (ślokas) and prose. This suggests that Brajvāsīdās took the initiative to introduce varṇa vṛttas (See Table 4).

As this book does not specifically focus on the metrical style of the PcN, it is sufficient to simply outline some of its central features in this subsection. A more thorough examination of its prosody would require a deeper study. Therefore, this subsection explores the most frequently employed metre in the PcN, the dohā, and briefly illustrates its various functions within the work. Furthermore, it demonstrates how Brajvāsīdās utilises dohās as a sort of "guiding thread" throughout his drama.[139] At the end of the discussion, Tables 3 and 4 list the metres found in Act I of the PcN.

The caupāī metre is frequently employed in the PcN, second only to the dohās.[140] Still, Brajvāsīdās does not seem to privilege the stylistic alternation of caupāī-dohā/soraṭha, which had been established since the times of the Sufi premākhyāns (lit. "love stories") and the RcM.[141] In the extended narrative sections, he extensively utilises other metres, such as the tomara chanda[142] and the bhujaṅ-

136 For insights into how the lexical choices of early modern authors did not always align with political or religious intentions, see Busch 2010.

137 McGregor 1984: 159.

138 Sinha 1964: 3.

139 In Chapter 4, the discussion explores how the PcN creatively incorporates elements from the RcM, considering its own metrical structure and adapting RcM verses accordingly.

140 I counted 414 dohās and 130 caupāīs.

141 See de Bruijn 2010: 125.

142 This metre consists of twelve mātrās in each of the four pādas, with each line ending in one long syllable followed by a short one. See Sinha 1964: 39. An example can be found in PcN 1.27, although occasionally both final syllables may be short, as in: bahu śiṣyatā ke sātha | sundara

gaprayāta chanda.[143] Additionally, the dohās serve not only to conclude preceding longer sequences[144] but also fulfil various purposes that do not necessarily indicate a couplet following the previous passage. They can also serve as an introduction to an explanation or perform a different function altogether, for example in verse 1.72:

A war, a disease, an enemy, a fire, a debt, a king, a great ascetic, a serpent –
consider them rightly, do not diminish [them], stay alert at all times.[145]

Here, the dohā serves as a warning from Passion to her husband Desire about Discrimination's powerful ministers. These ministers are extensively portrayed in the subsequent verses.

Furthermore, the dohās often play a role in advancing the narrative, and as a result, they can also contain scene indications. For instance, verses 1.30 and 1.31 read as follows:

First Actor *(after dancing for a moment, raising [his] arm)*: Friends, stop the ḍholas a little and all be quiet!

When they stopped singing, [only] the sound of the instruments could be heard.

Due to their synthetic and propelling nature, the dohās in the PcN serve as a means to follow the story to a certain extent solely by reading them. Given the length of the PcN, it is likely that these couplets acted as a guiding thread, encapsulating the main events of each act of the drama. To provide an example, I conclude this subsection by presenting in Table 2 a selection of dohās from the first act of the PcN, starting from the beginning and continuing until slightly after the entry of Desire and his wife Passion in verse 1.102. It is unnecessary to quote the first eight dohās of the PcN, as I have already demonstrated how they establish the bhakti and Vedantic mood of the play.

mano rati gātha || kou liye tāla mṛdaṅga | kou ḍholakī muhacaṅga || kou bena vīna upaṅga | kou bahu barī sāraṅga ||.

143 This metre belongs to the jagatī class, characterised by each pāda consisting of twelve syllables with varying moraic value. For details, see Sinha 1964: 27–28. An illustration of this metre can be found in PcN 1.42, which reads: phaṁda phāṁdatehī bhajā jāsu cāhā | rahyo bīcahī maiṁ kiyo jo umāhā || paryo dāra tāte ju santapta jñānī | tajyo chīra haṃsā piye cūka pānī ||.

144 An instance demonstrating this function can be seen in PcN 1.79.

145 rāri roga ripu agni ṛṇa nṛpati tapodhana byāla | itane ganiya na choṭa kari sajaga rahiya saba kāla ||.

Table 2: Synthetic Reading of the PcN through its Dohās.

Dohā	Story
1.11 and 1.12	Brajvāsīdās refers to the importance of the community of the truth-knowers and the origin of the PC story.
1.14	Brajvāsīdās recounts how the Kṛṣṇadās Bhaṭṭ composed a perfect book, consisting of Vedānta.
1.17–1.24	Discussing how he composed his own version, Brajvāsīdās reflects that he could not relate the story in the Sanskrit, Prakrit, or Persian languages and had decided instead to render it in a bhāṣā.
1.26	The First Actor and his company enter the assembly hall.
1.28–1.31	The actors sing the chorus and set up the stage. The First Actor and Actress dance and begin their dialogue.
1.40 and 1.43	The First Actor tells the story of King Kīratibrahma, who has prospered but seeks higher spiritual insight.
1.45	The First Actor exalts the community of the truth-knowers.
1.48 and 1.51	The Actress objects: How can a king attain peace and the true path?
1.53	The First Actor responds: When the light of knowledge shines, ignorance is dispelled for everyone.
1.55, 1.58, and 1.60	Desire and Passion arrive in the royal hall. They dance together.
1.67–1.69	Desire states that his power is effective even against Indra, the Moon-god, and Brahmā.
1.72–1.73 and 1.79	Passion advises her husband not to underestimate his enemies, says that a war with Discrimination is coming, and lists his eight powerful ministers.
1.89	Desire replies.
1.90–1.92	Passion enquires about Discrimination and Bewilderment's family and asks how they are related. Desire begins his reply.
1.94–1.95, 1.97,[a] 1.100, and 1.102	Desire's reply continues: Brahman and his union with Illusion/Material Nature; Mind, his wives, and the birth of Discrimination and Bewilderment; Mind's favour towards Bewilderment.

[a]This transition in dohā 1.97 may appear abrupt as it shifts from narrating the story of Supreme Self and Illusion to the birth of Bewilderment from Engagement-with-Action and Discrimination from Disengagement-with-Action. However, dohā 1.100 clarifies that Mind is the father of the two kings.

The purpose of this synthetic reading approach remains open to interpretation. Definite answers cannot be provided. While the PcN may have been primarily conceived as a written text, it is intimately connected to the oral and performative aspects, as discussed in the previous subsection. Therefore, it is plausible that the itinerary outlined by the dohās, serving as a shortcut, could have been beneficial for those who sought a quicker overview of the text or even for recitation purposes, without delving into the extended narrative and lengthier philosophical sections.

Table 3: Mātrā Vṛttas in Act I.

Denomination	Mātrās[a]	Frequency
Catuḥpadī	28 (a type of rolā)	1
Caupāī	16	23
Chappai	1 rolā + 2 ullālā[b]	3
Dohā	24 in a 13 + 11 scheme	77
Harigītikā[c]	26 to 30	6
Jhūlanā[d]	30	2
Kuṇḍaliyā	1 dohā + 1 rolā	6
Manamohana[e]	8	1
Marahaṭā[f]	29	1
Padmāvatī[g]	27 to 30	1
Rolā	28	7
Śobhana	24 to 26	1
Soraṭhā	24 in a 11 + 13 scheme	6
Sugati[h]	25 to 26	3
Tomara	12	3
Tribhaṅgī	32	1

[a]Meaning the number of mātrās or morae in each pāda.
[b]A couplet counting twenty-six mātrās in each pāda in a 15 + 13 scheme. See Sinha 1964; Prasad 1938.
[c]Often noted only as chanda, this denomination can also be found in the RcM. According to Lutgendorf 1991: 16, it is often set to music and sung. See also Prasad 1938: 69.
[d]Sinha 1964: 50 presents only one type of jhūlanā with twenty-six syllabic instants. However, Prasad's treatment allows for the possibility of jhūlanā with higher syllabic instants. See Prasad 1938: 78–79.
[e]This metre is not described in Sinha 1964 and does not correspond to the scheme presented in Prasad 1938: 47.
[f]See Prasad 1938: 71.
[g]This metre should not be confused with the metre of the same name consisting of thirty-two mātrās in Sinha 1964: 56–57 or Prasad 1938: 75.
[h]The verse with this denomination in Sinha 1964: 31 and Prasad 1938: 43 does not match what is found in the PcN. The closest form appears to be what Prasad refers to as sugītikā. See Prasad 1938: 66.

In essence, the dohās can be seen as another indication of the interplay between the written, oral, and performative dimensions of the PcN.

Table 4: Varṇa Vṛttas in Act I.

Denomination[a]	Syllables	Frequency
Amṛtagati	9 to 11	2
Ardhabhujaṅgī	6 to 7	1
Bhujaṅgaprayāta	12 in a 6 + 6 scheme	1
Brahmarūpaka	16	1
Cañcalā	16	1
Campaka[b]	8 to 10	1
Dodhaka	11 in a 6 + 5 scheme	1
Hākalikā	11	1
Kavitta	31 to 32	4
Kumāralalitā/Kumāralatā	7 to 8	2
Kusumavicitrā	10 to 12	1
Paṅkajavāṭikā	13	1
Savaiyā	22 to 26	3
Siṃhavilokita	14	1
Somarājī	12 to 15	1
Sugīti	18	1
Sundarī	12	3
Tāraka	13	1
Toṭaka	12 in a 4 + 4 + 4 scheme	1
Vidyunmālā[c]	8 in a 4 + 4 scheme	1

[a]Unfortunately, I was unable to locate information regarding two metres used in Act I: saṅgīti and rāsā.
[b]See Prasad 1938: 134, who refers to it as campakmālā.
[c]These metre is a variety of anuṣṭubh. See Prasad 1938: 125.

Chapter Summary

This chapter has provided an introduction to the PcN, starting with an exploration of Brajvāsīdās and his works through various sources, followed by an overview of his background as a Vallabhite author. The second part of the chapter focused on the PcN itself. The primary source for this study is a lithographic copy of the PcN published by Naval Kishore Press, and its relationship with another lithographic copy published a year earlier by the Benares Light Press has been highlighted. Additionally, manuscript sources have revealed that the PcN was copied in a non-Vallabhite milieu, specifically within the Rādhāvallabha tradition. The reasons behind the circulation of the PcN in a different tradition from its origin have been identified, including the broader Vaiṣṇava perspective promoted by Brajvāsī's drama, with particular emphasis on the significance of the community of the truth-knowers.

The subsequent analysis has demonstrated that Brajvāsīdās composed the PcN with a threefold narrative structure, aiming to make the story relevant for a public of bhaktas. In the opening dohās of the PcN, the author acknowledges the importance of scriptural authority, which is fundamental in Vedānta, but does not invoke a deity more specific than Hari. Instead, his attention is on seeking the blessings of the community of truth-knowers. Brajvāsī strengthens the bhakti aspect by presenting himself not so much as an accomplished poet but primarily as a bhakta with the attitude of a dāsa . He draws inspiration from the vinayas composed by Tulsīdās and Sūrdās, which provide a paradigm for him to identify with his audience and vice versa. Through this reflexivity, Brajvāsī creates a space for his composition within the community of bhaktas, offering a tale that uplifts them from worldly subjugation. The PC story is presented not as his own invention but as a product of the expertise of Kṛṣṇadās Bhaṭṭ, an experienced teacher who crafted it for a challenged student/disciple, to whom he narrates it as a play performed at the court of King Kīratibrahma. The implicit connection between the student and the wretched bhaktas serves the purpose of making Brajvāsī their guide through his own composition, as he is the only one capable of bridging the Persian medium to the vernacular language (bhāṣā). Therefore, this chapter has established that the opening of the PcN not only shapes a wide-ranging bhakti audience but also showcases the author's awareness of the vernacular literary precedents that enable him to fulfil his goal.

Furthermore, the discussion has revealed that Brajvāsī makes references to Kṛṣṇadās Bhaṭṭ's (i.e. Kṛṣṇamiśra's) PC, which he himself heard as a story from his guru among the satsaṅga. This highlights the oral and performative aspects of the text, further supported by its designation as a nāṭaka. The chapter argues that the connection between the PcN and performance can be inferred not only from Brajvāsī's preamble and paratextual formulas but also from the extensive description of the musicians and actresses in King Kīratibrahma's court. The text's aspiration for shared and collective enjoyment is also demonstrated through its use of prosody, particularly a preference for alliterations, and the final appeal to the residents of the Braj region with the poet's signature ("Brajvāsīdās says"). Additionally, the chapter establishes that although the term "svāṁga" in the PcN does not indicate a discrete dramatic form, the drama possesses certain formal elements and a plot that resonate with those found in early Nauṭaṅkī. These indications challenge the traditional assumption prevalent in the scholarship of Hindi literature and theatre that considers Brajbhāṣā dramas as nothing more than lengthy poems with limited dramaticity.

The final subsection of the chapter delves into several aspects of the language and prosody employed in the PcN. It highlights various Avadhī features present in the verbal system, further supporting the overarching argument of this book that the RcM served as one of the sources for the PcN. Furthermore, while the PcN predominantly employs a Sanskritic vocabulary for philosophical and religious matters, it also incorporates a group of Perso-Arabic terms. These terms serve as a reminder of the Persian language's status as the language of power and its presence in everyday life and interactions. Before exploring the connections between the PcN and the RcM in Chapter 3, the next chapter aims to provide additional insights into the literary tradition to which the PcN belongs, what I call the PC tradition.

2 The *Prabodhacandrodaya Nāṭaka* and the Prabodhacandrodaya Tradition

The introductory part of the PcN suggests that the relationship between the texts of the PC tradition is not a simple one-to-one translation or adaptation from one text to another. Instead, it indicates that each new codification of the PC story results from the convergence of two or more previous tellings, some of which may not even be in written form. While Chapters 3 and 4 demonstrate how the PcN responded to texts beyond the tradition itself, this chapter focuses on placing it within its own textual lineage. It discusses its relationship with three texts that preceded it chronologically: Kṛṣṇamiśra's Sanskrit *Prabodhacandrodaya* (sometime after 1060; KPC), Banvālīdās's Persian *Gulzār-i Ḥāl* (The Rose Garden of Ecstatic States; 1662/1663; GH), and briefly, Nanddās's Brajbhāṣā *Prabodhacandrodaya* (sixteenth century).[1]

As mentioned in the introductory chapter, this study does not approach the comparison of similarities and differences between an original work (mūla), in this case, the KPC, and later iterations of it, including the PcN. Saroj Agrawal's pioneering study on the PC tradition in Hindi, titled *Prabodhacandrodaya aur uskī Hindī Paramparā*, adopts this approach by categorising the texts within this segment of the tradition into distinct categories. First, there are texts based solely on the KPC, then those based on the KPC as well as other works, and finally, those based solely on other well-known works. Brajvāsī's PcN belongs to the second category as a versified translation that expands upon the source material.[2] While this classification appears reasonable, it is not entirely clear why Agrawal considers the PcN as a translation (anuvāda) rather than, for example, an adaptation (rūpāntara). On the one hand, she defines anuvāda as "works that have been written solely for the purpose of translation".[3] On the other hand, rūpāntara refers to "dramas that, although based on the plot of the [Sanskrit] *Prabodhacandrodaya*, also contain original parts in addition to the translation".[4] According to Agrawal, expanding upon the source material alone is insufficient to be classified as an ad-

1 Bhuvadev Dube's *Prabodhacandrodaya* (1893), which drew inspiration from the PcN, is discussed in Pastore (forthcoming b).

2 Agrawal 1962: 225–226.

3 Agrawal 1962: 202: "jin racnāoṁ ne keval anuvād ke uddeśya se likhī jākar, mūl se apne sambandh ko banāye rakkhā hai, unheṁ ham 'anuvāda' ke nām se abhihit karte haiṁ."

4 Agrawal 1962: 202: "rūpāntar kī śreṇī meṁ ve nāṭaka āte haiṁ jo likhe to gaye haiṁ 'Prabodhacandrodaya' kī kathāvastu ke ādhār par hī, parantu, jo anuvād ke sāth kuch maulik aṃś bhī rakhte haiṁ."

https://doi.org/10.1515/9783111063836-003

aptation.[5] At the time of her writing, the fact that early modern works were often not strictly translations, as implied by the term anuvāda, but rather adaptations or retellings had not yet received extensive scholarly attention.[6] Therefore, her study is valuable as it not only presents an anthology of extracts from the texts, some of which are from manuscript sources, but also has a flattening effect on the tradition, particularly the texts identified as anuvādas. Agrawal's study does not delve into the philosophical and religious concepts developed within these PC retellings, leaving a gap in the analysis.[7] This raises the question: What would be the significance of reading or studying a text that merely repeats a previous one?[8]

A new study on the PC tradition has emerged in Hindi in the form of a PhD dissertation titled "*Prabodha Candrodaya Nāṭak* ke Hindī Rūpāntar aur Antaḥdhārmik Saṃvād".[9] The author, Naveen Singh, undertakes a more ambitious project than Agrawal, as the first chapter not only explores the KPC but also extensively examines the Sanskrit adaptations.[10] The title of the dissertation refers to rūpāntara, which appears to correspond to the second category outlined by Agrawal. Singh argues that these works are not mere word-for-word translations (anuvā-

---

5 Agrawal describes two other categories with more originality: autonomous allegorical dramas (svatantra rūpak nāṭak) and works partially influenced by Kṛṣṇamiśra's PC (aṃśataḥ prabhāvit nāṭak). The former draws inspiration from Kṛṣṇamiśra's PC in terms of style but differs in characters, plot, and objective. The latter type of drama is only partially allegorical. See Agrawal 1962: 202.

6 Some Jaina works can be categorised as "iconic" or "literal" translations. See Cort 2015. For translation in the South Asian context, see Ramanujan 2004: 134–144.

7 Agrawal devotes an entire chapter to the KPC, delving into its "dharm-darśan". This suggests that there is little to say about the religious and philosophical aspects of the retellings because everything has already been said about the topic. See Agrawal 1962: ch. 3.

8 The neglect of such literature, often labelled as "mere translations", can be traced back to the colonial era. While bhakti literature in bhāṣā generally faced less colonial-bred prejudice than court poetry, the category of "translations" encompassed works from both traditions that were equally overlooked due to the perceived lack of originality in their Sanskrit originals. Sanskrit, being associated with the Vedas and epics, held greater importance, while bhāṣā works served as a reminder of Mughal rule, seen as foreign by the Hindi-sponsoring intelligentsia. Tulsīdās stands out as the only bhakti author of a retelling who was not portrayed in a negative light, making him an exception to the prevailing bias. For more insight into the construction and perception of the canon of Hindi literature during colonial times (from both colonial officials and Indian intellectuals), see Dalmia 2017. See also Mody 2012 and Busch 2011: ch. 6, the latter of which offers further reflections focusing on court literature.

9 Singh 2020. While my access to the work was limited to the table of contents, the bhūmikā (introduction), and the initial section of the chapter concerning the adaptations of the sixteenth and seventeenth centuries, it still provides valuable insights.

10 Singh 2020: 1–125.

das).[11] Additionally, he employs the terms punarracnā (reworking or recreation) and bhāvānuvāda (free translation), although it remains unclear whether there are distinct differences between these terms, and if so, what.[12]

The investigation proceeds chronologically, with three subsequent chapters focusing on different periods following the chapter dedicated to Sanskrit adaptations. The first period is the sixteenth and seventeenth centuries, followed by the eighteenth century, and finally, the nineteenth and twentieth centuries. In the fifth chapter, the first part explores the tradition of rūpāntara and saṃvāda. The author appears to clarify how the term saṃvāda, possibly referring to the "interreligious dialogue" mentioned in the title, helps us understand and analyse the PC tradition.[13] The subsections not only summarise the findings from each segment of the tradition, Sanskrit and Hindi, but also expand the foundation of the argument by examining rūpāntaras in other Indian languages and foreign languages. The second part of the chapter delves into Hindi dramas influenced by the KPC, specifically those that Agrawal would categorise as autonomous allegorical dramas (svatantra rūpaka nāṭaka) and dramas partially influenced by Kṛṣṇamiśra's PC (aṃśataḥ prabhāvita nāṭaka).[14] An interesting element introduced in the discussion of the fifth chapter is the potential connection between the PC and Nauṭaṅkī theatre.[15]

Although Singh rightfully emphasises the significance of bhāṣā nāṭakas in the realm of Hindi literature,[16] the title of his research work, where *Prabodha Candrodaya Nāṭak* refers specifically to the KPC, is somewhat misleading. It may create the impression that Kṛṣṇamiśra's drama is the sole source for the rūpāntaras.[17] However, the following section will demonstrate that several aspects of the KPC

---

11 According to Singh, each author within the tradition adapts the plot of the retellings to suit contemporary times and societal contexts. Singh 2020: vi.

12 Singh 2020: vi. Singh may explain the terminology in a part of the text I could not access, perhaps in subsection 5.1.1.

13 I am uncertain since the adjective "antaḥdharmik" does not appear in the main title and subtitles of the chapter. However, in the introduction, the author mentions that each text within the PC tradition presents dialogues and debates among different sampradāyas or dharmas (referred to as "sampradāyoṁ ke saṃvād-vivād/mat-matāntar" and "dhārmik saṃvād-vivād/mat-matāntar"). For this reason, it appears that antaḥdhārmik should be understood in this context, qualifying the genre of the PC texts. See Singh 2020: v.

14 See footnote 5 above.

15 The table of contents does not clearly indicate whether the author refers to the PC tradition as a whole, the KPC specifically, or other texts within the tradition. The section is titled "Lokanāṭya: Nauṭaṅkī (pratīkātmaka)", which suggests that it may pertain to an allegorical typology of Nauṭaṅkī plays. See also subsection 1.2.3 for our earlier discussion on Nauṭaṅkī theatre.

16 Singh 2020: 126–127.

17 Moreover, the term rūpāntara itself can be misleading. Although I have translated it as "adaptation", its literal meaning is "having another form, of different form". See HSS 4200. By the end

warrant reconsideration. Exploring these facets can not only enhance our understanding of the text itself but also shed light on its relationship with other texts within the tradition.[18] Additionally, it will be clarified that, for some retelling authors, their knowledge and reception of the drama were not solely dependent on the drama itself but also influenced by one of its commentaries.

2.1 Kṛṣṇamiśra's *Prabodhacandrodaya* and its Reception

The KPC, composed in the eleventh century and widely recognised as a paramount example of allegorical theatre,[19] requires little introduction. Scholars and critics of Sanskrit literature have extensively studied and translated the text into multiple European languages.[20] As such, there is no need to reiterate what previous researchers have already demonstrated. Instead, this section will focus on lesser-explored aspects of the text, considering the findings of earlier scholarship and addressing questions regarding its reception.

While previous studies leading up to the translations of the KPC primarily discuss the Vedānta and bhakti aspects of the text, Matthew Kapstein rightly emphasises its literary qualities, which have often been overlooked.[21] Kapstein's concern is valid, and the present book, while primarily focusing on the religious and philosophical dimensions of the PcN, also aims to highlight the text's literary elements, such as the author's use of rhetorical devices and intertextuality connecting the PcN with other early modern works. However, despite most studies preceding the translations identifying the KPC's position as Advaita Vedānta with

---

of this book, it will become evident that, at least in the case of the PcN, the texts of the PC tradition encompass much more than a mere change of form; they possess a distinct "life" of their own. It seems that Singh's work emphasises the uniqueness and interconnectedness of each work within the PC tradition. While he may not explicitly define it in the pages I had access to, it appears that he views rūpāntara as implying a change primarily in form.

18 It should be noted that not all retellings explicitly acknowledge the KPC as their source. For instance, King Jasvant Siṃh (r. 1638–1678) of Mārvāḍ, in his *Prabodha Nāṭaka* (c. 1643), does not seem to mention it. See Siṃh 1972: 81–113.

19 A chapter titled "Vallaṇa, Kṣemendra and Kṛṣṇamiśra" can be found in A. K. Warder's history of literature. For Kṛṣṇamiśra's contribution, see Warder 1992: 543–557. See also Dasgupta/De 2017: 479–487; Keith 1954: 251–256.

20 The first English translation of the text is by James Taylor in 1811/1812, republished in 1893. Subsequently, Nambiar 1971 and Kapstein 2009 also provide translations. One of the earliest editions of the text is Brockhaus 1835, predominantly in Latin and Sanskrit, while the first German translation is Goldstücker 1842. It has also been translated into Italian, French, Dutch, and Spanish. For further details on these translations, see Kapstein 2009: lxii–lxiii.

21 Kapstein 2009: xxxv.

a bhakti inclination towards Viṣṇu, certain issues still require examination or reassessment.

Although the KPC does not explicitly mention Advaita Vedānta and the term vedānta appears only twice, Armelle Pédraglio notes that vedānta primarily signifies the Upaniṣads as the culmination of the Vedas and does not necessarily imply a specific philosophical school.[22] While acknowledging the text's strong "illusionistic" tendencies, which indeed exhibit a pronounced Advaita flavour,[23] questions remain to be explored in this regard. Kṛṣṇamiśra's work extensively draws upon the Upaniṣads, as Pédraglio demonstrates in her critical analysis of the French translation in 1974.[24] However, no study systematically traces which Upaniṣads are quoted and which are not or whether certain Upaniṣads exerted greater influence on the text than others. Furthermore, one can inquire about the placement of Upaniṣad quotations and their exact function(s) within the broader context of the text.

Another important area for investigation pertains to the influence of Ādiśeṣa's *Paramārthasāra* (The Essence of the Supreme Truth; sixth century?). Sita Nambiar acknowledges the indebtedness of the KPC to the *Paramārthasāra*, but the possible connections between the two texts require further clarification.[25] Moreover, she suggests that the bhakti supported by the KPC is similar to that of the BhP, as both texts place bhakti in a superior position. Yet she admits that we do not know if Kṛṣṇamiśra was acquainted with the BhP.[26] Without further parallels, the evidence of a possible influence remains weak.[27] Moreover, the BhP presents various forms of bhakti, with the bhakti in Book X being quite different from that in Book II.[28] While the distinction between "intellectual" and "emotional" bhakti may seem rigid in some instances,[29] it is undeniable that the bhakti of Prahlāda differs from the śṛṅgāra (erotic, romantic love) felt by the gopīs and from the

22 Pédraglio 1974: 52.

23 Nambiar 1971: 14–16.

24 Pédraglio 1974: 59–64.

25 Nambiar 1971: 11.

26 Nambiar 1971: 12.

27 Apart from another reference in Nambiar 1971: 28n1.

28 Where Śuka exhorts Parikṣita to engage in yoga informed by bhakti. See Matchett 2008: 115.

29 In the PcN, the character Viṣṇubhakti/Bond-with-Viṣṇu is depicted as focusing on the pratimā (mental image) of brahman, either Kṛṣṇa or Rāma. However, in the words of the Materialist, she also "ties the anklets [to her] feet and dances in front [of the image]". Her bhakti is not only "intellectual" since it involves mental discipline, but it is also "emotional" as it is connected to embodiment (arcana, vandana, kīrtana) and the taste of the "sweet form" (mādhurī rūpa cākhai). See PcN 2.96. In addition, PcN 5.124 refers to the ninefold bhakti practices delineated in the BhP. See section 4.2.

filial love or friendship of the Braj people for Kṛṣṇa.[30] Therefore, it is unclear what Nambiar precisely means by referring to the bhakti of the BhP.

Indeed, the bhakti depicted in the KPC is not akin to that of the gopīs. This raises the question of why researchers have not explored a possible link between the bhakti in the *Viṣṇu Purāṇa*[31] and, more importantly, the BhG. Pédraglio suggests that despite the KPC's inclination towards Viṣṇu and the central role of Viṣṇubhakti, it is doubtful whether Kṛṣṇamiśra's PC can be classified as a bhakti text throughout.[32] In her view, the fact that the BhG is also portrayed as a character but does not play a significant role in the development of the drama suggests that Kṛṣṇamiśra's PC lacks a distinctly sectarian character.[33] While her conclusion may be well-founded, it seems somewhat unlikely that the assessment of the BhG's influence solely depends on its portrayal as a character.

Even in the PcN, composed seven centuries after the KPC, the strong influence of the BhG's discourse on kingship and the nature of power is evident (and, in places, related to that of the yogi). In the BhG, Arjuna's dilemma on the battlefield places him "in the middle" of conflicting choices between his duty as prince-warrior (kṣatriya dharma) and his attachment to his relatives.[34] Similarly, at the beginning of the PcN, King Kīratibrahma is described as desiring to know his nature but is burdened by his responsibilities as a king and remains "in the middle".[35] Arjuna can be an ideal king if he recognises that the only true relationship is with the divine and behaves like the divine by controlling his nature through Bhakti Yoga.[36] These elements are also present in the KPC where King Manas is taught to relinquish his attachment to both his loved ones and his characteristic fickleness to become a better king.[37] It would be interesting to discover what else we can learn by comparing the KPC with theorisations of bhakti in the BhG.

Regarding the association authors sought to create with Kṛṣṇamiśra and his nāṭaka, Kapstein highlights the identity of Kṛṣṇamiśra as "an important historical

---

30 Friedhelm Hardy builds his argument about the existence of different forms of bhakti by drawing upon previous research by Paul Hacker on Prahlāda's character. See Hardy 1983: 36–43.

31 See Schreiner 2013: 579–767 for an overview of the *Viṣṇu Purāṇa*, including its theology.

32 Pédraglio 1974: 55–58.

33 Pédraglio 1974: 58–59.

34 See Malinar 2007: 150.

35 Verse 1.41 states that "the king was caught again in the web of kingly duty". Verse 1.42 reads: "He, who desired to run away [to avoid] being caught in the web, stayed in the middle [of it] and exulted." Both are spoken by the First Actor to his wife.

36 See Malinar 2007: 200–201.

37 Especially in Act V. See Kapstein 2009: 207–237.

problem".[38] As Kapstein points out, it remains uncertain whether the yati attached to his name in the opening of the KPC means that Kṛṣṇamiśra was an ascetic or a lay scholar and counsellor at the court of Kīrtivarman, the Candella king.[39] Additionally, Kapstein mentions an interesting detail found in one of the commentaries on the KPC, Rāmadās Dīkṣit's *Prakāśa* (Elucidation), which presents a short story regarding Kṛṣṇamiśra. According to the story, Kṛṣṇamiśra purportedly composed his drama, *Prabodhacandrodaya*, to instruct a student who had a fondness for kāvya but disliked and avoided studying philosophy in the conventional manner.[40] The verse from Dīkṣit's *Prakāśa* describing this episode is as follows:

> Now, indeed, wishing to instruct – through the pretext of a drama called *Prabodhacandrodaya* – a pupil, who avoided the much-taught doctrine of Vedānta, and the awakening therein, the extremely compassionate, Śrī Kṛṣṇamiśra composed it.[41]

Kapstein proposes a theory suggesting that the tale must have spread at a time when the actual historical circumstances surrounding the composition of the KPC had already fallen into oblivion.[42] While this may hold some truth, it becomes challenging in some instances to determine whether historical memory was retained or if the historical account was included because the main source of a retelling was the KPC. This is the case with Gulābsiṃh's bhāṣā PC, composed in 1789. Gulābsiṃh not only mentions Kṛṣṇamiśra but also refers to the same historical events that appear in the KPC, specifically the Cedi king Karṇa's victory over Kīrti-

---

38 Kapstein 2009: xxxi.

39 We have already encountered yati as translated as "caesura". In its primary meaning, "yati" refers to an ascetic as someone who exercises restraint, just as caesura blocks or regulates the rhythm of a verse. As previously mentioned, it seems likely that Kīrtivarman is the model for the PcN's Kīratibrahma.

40 Kapstein 2009: xxxii; Dīkṣit 1845. Krishnamachariar 1937: 676 appears to be the first to report this story in secondary literature.

41 "Atha kam apy antevāsinaṁ bahuśo 'dhyāyita-vedānta-siddhāntam api tatra 'avabodha-parāṅmukhāṁ nāṭakādi-rasikam parama-kāruṇikāḥ śrī-Kṛṣṇamiśrāḥ Prabodhacandrodayākhya-nāṭaka-vyājena taṁ bubodhayiṣavas, tad racayanti sma." The text appears in the "Scholien" section of the first edition of the text: Brockhaus 1835: 3. It appears immediately after Rāmadās Dīkṣit's nāndī and before he begins the proper commentary on the first śloka of the KPC. Additionally, the desiderative form bubodhayiṣavas could be interpreted more emphatically to imply that the teacher desired to enlighten him.

42 Kapstein 2009: xxxii. The text was composed in the historical context of the definitive victory of Candella king Kīrtivarman of Jejākabhukti (r. 1060–1100) over the Cedi king Lakṣmikarṇa (r. 1041–1073). Kṛṣṇamiśra wrote his PC as a celebration of the former's triumph over the latter, as elaborated in the opening section of the KPC. See Kapstein 2009: 5–13. An inscription at Mahobā (in modern-day Uttar Pradesh) draws inspiration from Kṛṣṇamiśra's praise of Kīrtivarman through the voice of one of the actresses in the KPC. See Kapstein 2009: 9–11; Hultzsch 1892: 217–222.

varman of the Lunar dynasty and his subsequent defeat by Kīrtivarman's ally/minister Gupāla.[43] The two accounts – the historical memory and the concise edifying tale – are not mutually exclusive. In the PcN, for example, Kīrtivarman (referred to as Kīratibrahma) and his ally/minister Gupāla (Gopāla) are both present.[44]

Returning to the disciple-student anecdote mentioned by Dīkṣit, it is absent in the KPC but present in some of the retellings.[45] In the opening section of the GH, Banvālīdās addresses Kṛṣṇamiśra as "Krishnadās Bhat" and mentions that as a wise brahmin teacher, he aimed to guide one of his students towards spiritual knowledge. To achieve this, he attempted to teach the student from books about Sufi doctrine and belief in God. Despite the guru's good intentions, the disciple was not easily convinced as he was strongly inclined towards fantasy tales and dramatic presentations. Consequently, the knowledgeable Kṛṣṇadās decided to persuade him using drama.[46] Banvālīdās's GH served as a primary source for two retellings a century later, the PcN and Nānakdās's PC.[47] The latter includes the story of Kṛṣṇadās and his disciple, following a similar narrative found in the GH, portraying the merciful and skilful teacher's attempt to guide his foolish (mūrkha) student.[48] In the PcN, as mentioned in the previous chapter, the Kṛṣṇadās-disciple narrative forms the intradiegetic layer of Brajvāsī's drama.[49] However, Brajvāsī's version introduces a comparison between the salvific actions of the guru and those of a doctor who, through their expertise, "cures" a student/patient afflicted by fever. This retooling by Brajvāsī establishes a connection not only to Banvālīdās's GH but also to the RcM.[50]

The reception of Kṛṣṇamiśra and his PC was thus not only shaped by the work itself but also influenced by the commentaries on the KPC. Unfortunately, this aspect that has yet to be thoroughly investigated by scholars. In this regard, Rāmadāsa Dīkṣit's commentary is particularly intriguing, despite its brevity. One interesting obser-

---

43 Gulābsiṃh 1905: p. 4, v. 18; p. 6, vv. 25–27.

44 See PcN 1.25. In Kṛṣṇadās Bhaṭṭ's narration: "There was a king called Kīratibrahma of incomparable aspect. [. . .] The king had a minister called Gopāla."

45 This narrative level, which we identified as the intradiegetic level in subsection 1.2.2, does not appear in earlier PCs like Malha kavi's (c. 1544), nor in later ones such as Dhauṁkal Miśra's PC (c. 1799). For the texts, see Malha kavi 1962 and Dhauṁkal Miśra 1962. For an introduction to these authors, see Agrawal 1962: 212–217 (the former) and 236–238 (the latter).

46 See Cappello (forthcoming).

47 Nānakdās composed his PC in 1789. See Agrawal 1962: 232–235. The fact that Nānakdās and Gulābsiṃh both composed their PC in 1789 provides another potential clue indicating the questionable nature of attributing the spread of the story of the foolish disciple solely to the lost historical memory of Kīrtivarman over Karṇa.

48 See Agrawal 1962: 397–398, 400.

49 See "Structure and Narrative Levels" in subsection 1.2.2.

50 See section 4.2.

vation is that while the KPC mentions "vedānta" twice in the sense of the Upaniṣads, the commentary refers to "vedānta-siddhānta", potentially associating the narrative with a specific Vedānta doctrine or established perspective. Furthermore, the inclusion and utilisation of the teacher-disciple tale in the retellings based on Banvālīdās's GH validate the findings of recent studies on the translation processes of Indic texts into Persian. We shall now delve deeper into Banvālīdās's GH and its historical context, specifically exploring the reasons why it is plausible to believe that Brajvāsīdās drew inspiration from it for his PcN.

2.2 Banvālīdās's Persian *Gulzār-i Ḥāl*

In the existing secondary literature on the PC tradition, the GH is not extensively discussed in relation to the seventeenth- and eighteenth-century retellings. Saroj Agrawal briefly mentions it in the fifth chapter of her monograph when addressing the PcN and Nānakdās's version because these texts reference the GH.[51] She also briefly mentions it in the "translations in other languages" section of the same chapter.[52] Similarly, in the introduction to his work, Naveen Singh does not mention the GH among the seventeenth-century adaptations but refers to it in the context of foreign-language translations.[53] It is important to note that these works primarily focus on Hindi and Sanskrit retellings of the PC, and the authors' language expertise plays a role in determining the sources they engage with. However, it is also crucial to recognise the significant role that Persian played in producing and transmitting of Indic knowledge, especially when considering the Persian retelling in recent studies.

Persian served as the language of the Mughal administration, and many Hindus, particularly from the Hindu Kāyastha communities, who worked as scribes and secretaries in the court, had to acquire proficiency in it.[54] Banvālīdās (?–1674), belonging to a kāyastha family, began his career in the service of the Mughal prince Dara Shukoh (1615–1659). Dara Shukoh, favoured by his father Shah Jahan (r. 1627–1658),[55] was one of the primary patrons of Persian translations of Indic

51 PcN 1.18: "Balīrāma made a book of it [in the] Persian language, | [but] that knowledge [as well is] extremely difficult, not easily understandable ||."

52 Agrawal 1962: 263–264.

53 Singh 2020: vi: "anya bhāratīya bhāṣāoṁ tāmil, telugu, baṅglā, malayālam bhāṣā ke sāth sāth fārsī, aṅgrejī, jarman aur freṅc bhāṣāoṁ meṁ is nāṭak ke anuvād yā rūpāntar prāpt hue haiṁ."

54 See Alam/Subrahmanyam 2004: 61–72; Kinra 2015. On the role of Persian in the Mughal empire, see Alam 1998, 2010.

55 See Faruqui 2012: 38–41.

philosophical material, continuing the tradition established by Akbar.[56] He sponsored a fresh translation of the *Yogavāsiṣṭha* in 1656, marking the start of a more profound interest in Hindu philosophy.[57] He also sponsored the *Sirr-i Akbar* (The Greatest Secret), a Persian adaptation of fifty Upaniṣads, written with the assistance of scholars from Benares.[58] His engagement involved not only written sources, as he reportedly engaged in discussions on Hindu doctrine with ascetics and other religious figures such as Bābā Lāl Dās in Lahore.[59] These conversations culminated in his work *Majmaʿal-Baḥrayn* (The Confluence of the Two Seas), where the prince compared Sufi and Hindu ideas, reflecting his encounters with Hindu mysticism.[60]

Dara Shukoh and Banvālīdās shared a significant connection as they both studied under the same Sufi teacher, the Qādirī Mullā Shāh Badakhsī (?–d. 1661). This association with the Mughal court has recently sparked increased scholarly interest in Banvālīdās and his work.[61] At some point, Banvālīdās appears to have distanced himself from serving the heir prince and chose to lead a life away from the court.[62] As a writer, Banvālīdās, also known by his pen name Walī, composed several works, including Persian adaptations that combined non-dualism and Sufi teachings.[63] In the GH, he skilfully employs various literary techniques to infuse the text with Sufi elements, including incorporating Persian Sufi poetry, authored by both himself and earlier Sufi writers, as well as transliterating Indic terminology into Perso-Arabic script and providing explanatory glosses.[64]

However, the precise source or sources of Banvālīdās's GH remain unknown. The manuscript tradition presents two different prologues to the text. In one, the

---

56 Akbar (r. 1556–1605) and his successors all promoted Persian translations of significant works like the *Mahābhārata* and *Rāmāyaṇa*, which were seen as sources of political wisdom and advice. For an overview of these translations, see Mujtabai 1978: 65–66. Regarding a *Mahābhārata* translation commissioned by Akbar, known as the *Razmnāma*, see Truschke 2011: 506–520.

57 See Gandhi 2020a: 194–201 for further details, including information on other translations of philosophical texts commissioned by the prince. For insights into the *Yogavāsiṣṭha* translation commissioned by the future Jahangir, see Nair 2020.

58 See D'Onofrio 2010.

59 The conversation was conducted in the vernacular language but documented in Persian. See Huart/Massignon 1926; Gandhi 2014.

60 For an overview of the text, see Gandhi 2020a: 186–192.

61 See Cappello 2019, 2021; Gandhi 2020b.

62 Gandhi 2020b: 81.

63 In addition to the GH, there are two other noteworthy texts: the *Rāma Gītā*, associated with the *Adhyātma Rāmāyaṇa*, which exhibits Advaita influences, and the *Om-nāma*, an adaptation of the *Yogavāsiṣṭha* that combines elements of Haṭha Yoga within an Advaita and Sufi framework. See Gandhi 2020b: 87–95.

64 See Cappello 2021: 268. For information on how Banvālīdās Islamicised the PC story, see Gandhi 2020b: 84–87.

author states that he has made a tarjuma (translation or redaction) of the GH from Sanskrit into Persian. In the other, Banvālīdās explains that the *Prabodhacandrodaya Nāṭaka*, originally composed by Krisnadās Bhat, was rendered into bhākhā, that is bhāṣā, by "Svāmī Nanddās" (the Vallabhite poet). Banvālīdās goes on to say that he accomplished his Persian translation by studying Nanddās's work and with the assistance of Bhavānīdās, an expert in bhākhā literature, who was likely the chief astrologer at the Mughal court. This collaboration with Bhavānīdās confirms the collective "translation process" within the Mughal atelier.[65] Carl Ernest has clarified that these translations were conducted by a team of Persian scholars, aided by Sanskrit scholars, with communication likely occurring in a shared bhāṣā (regional language).[66] It is thus plausible that bhāṣā versions of the texts were also available to the translators. In the case of the GH, the circumstances are further complicated by the inclusion of the disciple-student tale mentioned earlier, which raises questions about whether and how the Sanskrit commentaries on the KPC were employed to elucidate Kṛṣṇamiśra's drama.

It is possible that Brajvāsīdās had access to a GH manuscript that did not mention Nanddās's PC, as he does not refer to it in his own PcN. In the only dohā where he mentions Banvālīdās, Brajvāsīdās maintains a relatively neutral stance:

Balīrāma [Banvālīdās] made a book of it [in the] Persian language,
[but] that knowledge [as well is] extremely difficult, not easily understandable.[67]

The difficulty of understanding mentioned here should not be interpreted as a specific critique of the Persian language since, in the previous dohā, Brajvāsīdās also says that to reflect in Sanskrit or Prakrit, one needs an uncommon degree of knowledge. For this reason, we cannot conclude from this verse that the author of the PcN intended to "rectify" the Persian Sufi adaptation. Instead, Brajvāsīdās simply wanted to provide his own perspective on the story. Additionally, we can speculate that he may have also belonged to the community of Persian-speaking Hindus, albeit a century later than Banvālīdās, as such a group existed until the early twentieth century.

Ronald S. McGregor's observation that Persian was considered a legitimate medium for transmitting Indic knowledge, particularly philosophical and religious knowledge, is convincing.[68] If we look again at the figure of Banvālīdās, we find that

---

65 See Cappello (forthcoming) for a translation and ampler discussion of the two kinds of prologue.

66 Ernst 2003: 179–180.

67 PcN 1.18: balīrāma tākī karī bhāṣā yamana kitāba | soū vidyā ati kaṭhina samujhi na parai sitāba ||.

68 McGregor 1986: 137. The topic of translations from Persian into bhāṣā remains understudied, although some examples exist in the medical field. See Friedlander 2011: 50–52.

he defied easy categorisation as either Muslim or Hindu.[69] As Supriya Gandhi highlights, the contemporaneous work *Dabistān-i Maẕāhib* (School of Religious Sects, c. 1645–1658) by Mīr Ẕū al-Fiqār Ardistanī, penname Mūbad Shāh, portrays Banvālīdās as someone familiar with both the mosque and the temple. Another contemporaneous Persian account, the *Nuskha-i aḥwāl-i Shāhī* (Account of the Shah's Ecstatic States, c. 1666–1667), dedicated to the life of Mullā Shāh, presents a somewhat ambiguous report by Tawakkul Beg. According to him, when Mullā Shāh requested that Banvālīdās convert to Islam, Banvālīdās declined, stating that he had overcome both Islam and "infidelity", referring to non-Islamic religions. However, Beg's account later confirms that the GH author eventually did embrace Islam. A third account, perhaps the *Safīna-i Khushgū* (Compendium of Khushgu, c. 1724–1735) composed by Brindābandās, reports that Banvālīdās moved to Haridwar, a sacred spot for Hindus, and that his body was committed to the waters of the Ganges after his death.[70] Banvālīdās's fluid identity and the versatility of his work may have made his GH appealing to Brajvāsīdās.

However, regardless of the reasons, it is important to recognise the GH as an integral part of the PC tradition and not dismiss it as a mere coincidence. The Persian text actively engages with both Sanskrit and regional-language compositions of the tradition before and after. As mentioned earlier, Brajvāsīdās shows no awareness of Nanddās's PC, which is also rooted in the Puṣṭimārga tradition, nor does he indicate that the GH was based on Nanddās's PC. Therefore, it is worth briefly highlighting that Nanddās's PC, which is more focused on affirming Vallabhite theology, would likely not have served as a source for Brajvāsīdās's project.[71]

2.3 Nanddās's *Prabodhacandrodaya*

Ronald S. McGregor points out that Nanddās's PC has remained elusive in the histories of Indian literature. Indeed, this celebrated composition by the Vallabhite author has survived solely in manuscript form, with no published editions.[72] McGregor examined the manuscripts he discovered in European libraries and collections and,

69 On this topic, see Talbot 2009. See also the articles in Dalmia/Faruqui 2014.

70 The three accounts mentioned are discussed in Gandhi 2020b: 82. In the PC, Nānakdās acknowledges the influence of the GH and refers to Banvālīdās as "Balīrāma hari santa". See Nānakdās 1962: 401.

71 McGregor attributes this PC retelling to the Vallabhite Nanddās, presenting three preliminary considerations. One of these considerations is Brajvāsīdās's use of the GH as a written source, indicating an awareness of the GH's connection to the Vallabhite milieu. See McGregor 1971: 493.

72 See McGregor 1971: 487. Agrawal 1962 does not refer to it either.

attributing them to the Vallabhite poet, provides some clues indicating that Nanddās's PC must be distinguished from the PcN. McGregor demonstrates that Nanddās's PC espouses a Śuddhādvaita (Pure Non-Dualism) viewpoint, evident through various elements: a strong emphasis on Viṣṇubhakti; references to Kṛṣṇa while omitting any mention of Rāma;[73] a preference for the child Kṛṣṇa and Kṛṣṇa as Śyāma (the Dark or Dusky One); the portrayal of Kṛṣṇa as eternally young (Kṛṣṇa nitya kiśora) engaging in the rāsa or circular dance with the gopīs, alongside the acceptance of Rādhā as Kṛṣṇa's consort, a belief established under Viṭṭhalnāth's leadership; and an emphasis on puruṣa (self) being inherently ānanda, vilāsa, and cit (bliss, delight, and consciousness).[74]

The PcN briefly mentions vilāsa in connection with Mana (Mind), but it remains uncertain whether it carries the same significance as in Nanddās's PC.[75] A more convincing similarity with Nanddās's PC can be found in the references to Kṛṣṇa, renowned for his exploits in the Braj region, along with Rādhā.[76] However, it is important to note that these references should be understood as the author's preferred deity rather than indicating a definite inclination towards Śuddhādvaita throughout the entire PcN, which explicitly promotes a broad form of Vaiṣṇavism, incorporating several other mentions of Rāma and his caritas (deeds).[77] While these elements are insufficient to establish a comprehensive comparison or draw definite conclusions, they provide valuable indications that diverse interpretations of the same story could exist within the same religious tradition. Furthermore, they also prompt us to recognise that Nanddās's PC and the PcN are separated by two centuries, suggesting that the concerns of the respective authors, perhaps reflecting those of the paramparā, evolved over time.

73 Śrīkṛṣṇa bhagati/bhakta in the introduction to the drama. See McGregor 1971: 489–490, 492.

74 See McGregor 1986: 141–142.

75 PcN 1.135: [Good Reason:] "Mind's own nature is described [as] incomparable, extremely effulgent, infinite greatness. His light [is] the brilliance in the universe. King Bewilderment's people are not fit [to be] with him. They cheated him, [but] this matter is kept secret." In this verse, I translate vilāsa as "brilliance" since it deals with the luminous nature of Mana (Mind). For a discussion of this verse in the broader portrayal of the relationship and nature of Paramātama Puruṣa, Jīvātama Puruṣa, and Mana, see subsection 4.1.1.

76 PcN 1.38: [First Actor:] "For the sake of the devotees [he took] a form with qualities, the One living in Braj, fond of worldly enjoyment, a root of joy. | Nanda's son [Kṛṣṇa], delighting in the rāsa dance [in] a matchless couple [with] the venerable, fair Rādhā ||."

77 See subsection 4.1.1.

Chapter Summary

This chapter has introduced the PC tradition, primarily focusing on texts closely related to the PcN, namely the KPC, the GH, and Nanddās's PC. Its main objective was to examine the existing scholarship's understanding of the relationships between these texts. The chapter highlighted that Agrawal's monograph and Singh's recent PhD dissertation tend to overlook the unique aspects of each Sanskrit/Brajbhāṣā/Hindi retelling by placing undue emphasis on the KPC as the sole criterion for their literary value. Furthermore, these scholars' classification systems oversimplify the complex dynamics of the retellings' relationships. In this regard, the first section of the chapter briefly emphasised the importance of considering commentaries on the KPC when studying transmission dynamics. Additionally, it questioned the assumptions about the KPC, urging a more thorough examination of its philosophical and religious stances. Notably, the chapter identified the surprisingly underestimated influence of the BhG on the KPC.

Agrawal's and Singh's studies also neglect to adequately acknowledge the significance of Banvālīdās's GH within the tradition. The second section of the chapter explored the historical and cultural backdrop underlying Banvālīdās's work, drawing on existing secondary literature. It particularly addressed McGregor's inquiry, as a pioneer in the field, regarding the reasons behind Brajvāsīdās's choice of the GH as a source for his PcN and his explicit declaration of such. The proposed hypothesis suggests that due to the "fluid" identity of its author, Banvālīdās's composition, despite its Islamic Sufi inspiration, was perceived as an appropriate foundation for the PcN.

The subsequent section supports this hypothesis by clarifying that while Nanddās's PC exhibits a pronounced Śuddhādvaita inclination, as McGregor has argued, the same cannot be said for the PcN, which embraces a broad Vaiṣṇava viewpoint accommodating Rāma alongside Kṛṣṇa as the chosen deity. These observations challenge McGregor's conjecture that Brajvāsīdās, as a follower of Vallabha's Śuddhādvaita Vedānta, employed the GH as a source due to his awareness of the Persian text's reliance on Nanddās's PC in Brajbhāṣā. The next chapter takes Brajvāsī's drama out of the pathway of the PC tradition and maps out its connections with the RcM.

3 Tulsīdās's *Rāmcaritmānas* and the Prabodhacandrodaya Story

This chapter aims to illuminate the connections between the PcN and the RcM, focusing on the significant role of viveka/Viveka, the central concept/character of the PC kathā, within the RcM. It will demonstrate how the RcM incorporates stories about viveka, utilises metaphors that personify Viveka and other characters from the PC narrative, and underscores the integral role of viveka in Tulsīdās's depiction of bhakti and dharma. To explore this last aspect, particular emphasis will be placed on analysing the character of Bharata as portrayed in the *Ayodhyā Kāṇḍa* book of the RcM.

This chapter serves as the foundation for Chapter 4, which will establish that an analysis of the philosophical and religious vision presented in Brajvāsīdās's PcN cannot be complete without considering the RcM. Consequently, the PcN holds an important place in the history of the reception of Tulsīdās's magnum opus. Therefore, it is crucial to begin this chapter by addressing the influence of Tulsīdās's RcM on subsequent texts. Notably, the investigation will argue that several authors of these texts recognised the association between the RcM and the PC kathā or its characters. While the initial part of the chapter explores the PC story within the context of the Rāmāyaṇa tradition, the final section establishes a dialogue between the RcM and the PcN, considering references to the Rāma story and instances of adaptive reuse of the RcM in Brajvāsī's work. Since the examination of adaptive reuse concerning the philosophical and religious dimensions will be addressed in Chapter 4, the concluding section of this chapter will focus on literary instances unrelated to the Vedānta and bhakti aspects of the PcN.

3.1 The Influence of the *Rāmcaritmānas* on Early Modern Works

The scholarly exploration of texts influenced by Tulsīdās's RcM remains limited. Existing studies primarily focus on the tradition of the Rāma kathā itself, both before and after Tulsīdās, dedicating attention to the various adaptations of Rāma's story in Brajbhāṣa, Avadhī, Khaṛī Bolī, or modern Hindi.[1] Considering the widespread

1 A notable and comprehensive study that explores the story of Rāma across different languages and periods is Bulcke 1999. In my investigation of early modern texts influenced by or based on Tulsīdās's RcM, I primarily rely on Bulcke's study, along with two other works: Pandey 1972 and Stasik 2009.

https://doi.org/10.1515/9783111063836-004

popularity and success of the RcM in North India, one would expect a substantial number of works based on or inspired by Tulsīdās's composition.[2] However, within the extensive corpus of texts listed – and sometimes briefly analysed – in these studies, only a few are recognised as influenced by Tulsīdās's work. This number further diminishes when focusing on texts from the early modern period, which is particularly relevant to my analysis. For example, Camil Bulcke only mentions Naraharidās's *Avatāracarita*, composed in 1676 and apparently influenced by both the RcM and the *Rāmacandracandrikā* by Keśavdās, a poet in the rīti or mannerist tradition.[3]

Ram Lakhan Pandey provides an initial overview of "authorless" texts often attributed to Tulsī, created to "complete" the storyline of the RcM, which imitate its language, prosody, and mood.[4] He then examines the compositions of known authors across genres such as prabandha (long narrative poem), abhinaya (poetry composed for acting purposes), sphuṭa kāvya (miscellaneous poetry), and Puranic. These texts also demonstrate traces of influence from the metrical structure and occasionally the language of Tulsī's text.[5] Noteworthy examples include Madhusūdandās's *Rāmāśvamedha* (c. 1782) and the *Susiddhāntottama Rāmakhaṇḍa* composed by King Rudrapratāp Siṃh in the first half of the nineteenth century.[6] Another instance of the RcM's literary influence can be found in Raghunāthdās's *Viśrāma Sāgara* (c. 1860) from the Rāmsnehī sampradāya. In this case, the influence encompasses not only prosody and language but also content and mood.[7] Another example is the *Rāmcandrodaya Kāvya* by Rāmnāth Jyotiṣī (1874–?), an

2 For example, Bulcke restates the RcM's unparalleled status as *the* Hindi retelling of the Rāma story. Bulcke 1999: 199.

3 Bulcke 1999: 200. The *Rāmcandracandrikā* (The Moonlight of Rāmcandra; 1601) is generally opposed to the RcM by Indian critics as it is considered to be lacking the devotional inspiration and appeal of Tulsīdās's composition. Scholars like Ramchandra Shukla, aligning with this perspective, argue that Keśavdās retold Rāma's story primarily to showcase his literary "mannerist" skills, capitalising on its popularity during that era. See Shukla 1942: 209–212. However, recent scholarship by Cavaliere 2020 challenges the notion that Keśavdās's *Rāmcandracandrikā* lacks a devotional aspect. Pandey 1972: 33 acknowledges the presence of a dialogic structure similarity between the RcM and the *Rāmcandracandrikā*.

4 An example of such texts is the *Lavakuśa Kāṇḍa*, which narrates the story of Rāma's sons and is often included as an appendix after the *Uttara Kāṇḍa* of the RcM in certain editions. See Pandey 1972: 25–27.

5 A comprehensive overview can be found in Pandey 1972: 30–65.

6 The *Rāmāśvamedha* draws primarily from the *Padma Purāṇa* version of the Rāma kathā. Shukla 1942: 374–376 briefly describes it in glowing terms. The complete text, along with an introduction, is published in Pandey/Mishra 1993. The *Susiddhāntottama Rāmakhaṇḍa* is analysed in Pandey 1955.

7 See Pandey 1972: 51–53; Lamb 2002: 44.

astrologer and poet at the court of Ayodhya.[8] According to Pandey, this work incorporates teachings and descriptions related to nīti (conduct/political wisdom) and dharma that align with the RcM.[9]

A later text is the *Kauśalendra Kautuka* by Bihārīlāl Viśvakarmā "Kautuk", composed in 1936. The style of this work resembles Tulsī's *Kavitāvalī*, consisting of a series of quatrains (kabitta) referencing selected events from the Rāma narrative. Bihārīlāl Viśvakarmā explicitly acknowledges his indebtedness to Tulsīdās in the conclusion of his *Uttara Kāṇḍa*:

kachuka prabhūti karatūti hai na meri yaha
kauśalendra kautuka prasāda tulsī kī hai[10]

This is not a creation or effort of mine:
the *Kauśalendra Kautuka* is Tulsī's blessing.

Bihārīlāl Viśvakarmā not only venerates Tulsīdās as the originator of Rāma's story, which he considers authoritative, but also elevates the author of the RcM to a divine level. He uses the term prasāda (blessing), likening Tulsīdās to a deity whose sacred remnants are the source of his own works. Viśvakarmā sees himself as the recipient of Tulsīdās's blessings, which have manifested in the form of his literary creations.[11]

In a similar vein, the slightly earlier *Ubhaya Prabodhaka Rāmāyaṇa* (The *Rāmāyaṇa* Illuminating Both [Nirguṇa and Saguṇa Brahman]; c. 1882), composed by Banādās in Avadhī, dedicates an entire book to Tulsīdās and his RcM.[12] Banādās (1821–1892), revered as mahatma within the Rāmrasik sampradāya, frequently paraphrases and quotes from the RcM in his retelling.[13] It is noteworthy that Banādās

8 While the text's title, "The Rise of Rāma's Moon", recalls the PC story, it is uncertain if there is a direct influence in content or other aspects.

9 Pandey 1972: 54.

10 A brief, unnumbered text excerpt is quoted in Pandey 1972: 54–55.

11 Various Urdu adaptations of the Rāma kathā exist, some of which have been influenced by or based on the RcM. See Yardi 1994: 173–174. One notable example is the *Rāmāyaṇa*, composed by Munshi Jagannath Khushtar in 1864, which directly references Tulsīdās in its explanation for the composition of the text. For a discussion on the significance of Tulsīdās's mention by Munshi Khushtar, see Phillips 2014.

12 The *Ubhaya Prabodhaka Rāmāyaṇa* was first published by Naval Kishore Press in 1892. See Pandey 1972: 82. Citations to this text are from Banādās 1980, accompanied by my English translation. Other studies in Hindi about the author and his works exist: see B. P. Singh 1976; H. S. Singh 1996. A collection of Banādās's works can be found in B. P. Singh 1990. Stasik 2009: 149–160 offers insights into the author's life, an overview of the text, its religious vision, and further references in Hindi.

13 Banādās 1980: 20.

upholds the theology of the divine name presented by Tulsīdās, as the title of his work alludes to a verse of the *Bāla Kāṇḍa* of the RcM:

nāma rūpa gati akatha kahānī
    samujhata sukhada na parati bakhānī
aguna saguna bica nāma susākhī
    ubhaya prabodhaka catura dubhāṣī[14]

The path [of] name and form [is] an indescribable tale,
    it brings joy to those who understand it but is beyond words.
The name [is] a faithful witness between the qualified and the unqualified [brahman],
    a brilliant interpreter that illuminates both.

These verses refer to the name of "Rāma", which encompasses the dual aspects of the Absolute, similar to a bilingual person or skilled interpreter revealing the meaning of two languages. Banādās wishes for his *Ubhaya Prabodhaka Rāmāyaṇa* to be as enlightening as the name of Rāma's itself. Furthermore, he proclaims that his book possesses "the power to obliterate the world and worldly affections, to make peace enter one's heart, and is a beautiful abode of discrimination".[15]

In the introductory section (mūlakhaṇḍa) of his work, specifically in verses 22 to 24, Banādās allegorically summarises the plot and characters thus:

bhai laṅkāgaṛha agama moha dasakandhara vīrā
    kumbhakarna hai krodha sahaja hī dahai sarīrā
meghanāda hai kāma mahodara puni haṅkārā
    lobha jānu atikāya akampana māna bicārā
anī ādi āscarya hai so mātsarya hi māniye
    kaha banādāsa bahu bāsanā tṛṣnā kaṭaka hi jāniye

The fortress of Fear of Laṅka is impassable, and Bewilderment-Rāvaṇa [is its] chief,
    Kumbhakarṇa is Anger, that quickly consumes the body.
Meghanāda is Desire, then Mahodara is the I-maker,
    Atikāya is Greed, and Akampana embodies helpless Pride.
The army at first is strange;[16] it is indeed a manifestation of Intoxication.

14 RcM 1.20.4. The Avadhī text and numbering are from Tulsīdās 2016–2023.

15 sānti ura āvata lagāvata na neha kahūṁ jagata nasāvata viveka suṭhi dhāma hai. Quoted in Banādās 1980: 15.

16 Or "The first part of the army is strange".

Banādās declares: "Craving and Thirst, the brides,[17] are to be seen as the
[remaining forces of] the army."

akṣaya rāga ati prabala dekha makarākṣa hi māno
vidhi prahasta ko kahi niṣedha hi durmukha jāno
viddalpisvā[18] kapaṭa dambha kahiye suraghātī
trisirā prabala pakhaṇḍa hai manuja arātī

Behold the indestructible Passion, immensely powerful, and consider him Makarākṣa,
Calling Prahasta Prescription, recognise Durmukha as Proscription.
Deceit, the fraud, is named Suraghātī,[19]
and Triśira, the terrible Heretic, is an enemy of humanity.

ihāṁ gyāna kapirāja rīcha kahiye vijñānā
dhīraja aṅgada acala virati atisaya hanumānā

Here Sugrīva is to be identified as Knowledge and Jāmbavan as Learning,
Aṅgada is Persistence, unwavering, [while] the eminent Hanumān is Dispassion.[20]

By portraying characters as personifications of abstract concepts, the author transports the reader into a more profound interpretive realm. The reception of the RcM by the Rāmrasik author seems to emphasise a preference for a philosophical interpretation of the text.[21] Moreover, it reinforces the connection between the RcM narrative and the fundamental ideas within the PC story to such an extent that, in Banādās's Rāma story, the war against Rāvaṇa of Laṅka becomes a battle against Bewilderment residing in the fortress of Fear.[22]

17 Bahu interpreted as bahū (Skt. vadhū).
18 The meaning of this word is unclear.
19 Meaning "killer of the gods", which I take to refer to Devāntaka, Rāvaṇa's brother.
20 The quoted verses are in Banādās 1980: 15, while the sequence of the topics for each section can be found on pp. 23–30.
21 It appears that the author deliberately composed the text to be approached from both the laukika (worldly) and ādhyatmika (spiritual) perspectives. See Singh 1976: 215.
22 There is a work called *Jñānadīpikā* (The Lamp of Knowledge) attributed to Tulsīdās (now considered spurious), which claims in the colophon to its second chapter, "Description of Knowledge", that it was composed in agreement with the *Prabodhacandrodaya* (the specific text version is unknown). Katre defends the *Jñānadīpikā* as a pre-RcM composition by Tulsī, going against the opinion of the Triennial Reports on Manuscripts published by the Nagari Pracharini Sabha and also of other notable scholars like Mataprasad Gupta. While the attribution of the work is not directly relevant to our present discussion, it is interesting to note the perceived asso-

In a broader sense, the spiritual interpretation of the Rāma story finds its roots primarily in the *Adhyātma Rāmāyaṇa* (fourteenth–fifteenth century), one of the principal sources of the RcM.[23] Banādās's decision to provide a partial summary of his work, drawing from Tulsī's composition, is likely because the RcM itself contains numerous passages exploring the themes of viveka and other concepts in an allegorical manner. Consequently, the next section aims to illustrate further the intricate interplay between the Rāma story in the RcM and the PC story.

3.2 The Place and Significance of Viveka in the *Rāmcaritmānas*

The significance of Bharata, being Banādās's ideal bhakta,[24] might not be a mere coincidence. In this section, I will explore the distinct connection between Bharata and viveka, aiming to demonstrate his pivotal role in the context of discernment. First, I will examine the references, allusions, and metaphors pertaining to viveka within the RcM. Subsequently, the focus will shift to the interaction between viveka, bhakti, and dharma, specifically on Bharata. By utilising viveka as a case study, we can elucidate the various ways in which the RcM expresses its relationship with the PC story while also underscoring an aspect of Tulsī's work that serves as a potential source of inspiration for Brajvāsīdās's PcN.

3.2.1 References, Allusions, and Metaphors: Viveka and the Prabodhacandrodaya Story

The concept of viveka is prevalent throughout the seven books of the RcM, although its frequence is inconsistent. The term bibeka (Skt. viveka) is predominantly employed in the *Bāla Kāṇḍa*, *Ayodhyā Kāṇḍa*, and *Uttara Kāṇḍa* of the RcM. In addition to the noun form, we find its opposite, abibeka (lack of discrimination), and its derived adjective, bibeki (Skt. vivekin). Bibeka is often paired with another noun, birati (Skt. virati), which means "dispassion" and can be considered synonymous with vairāgya. It is also accompanied by the adjective bimala (Skt. vimala), meaning "pure". The presence of birati and bimala not only facilitates the construc-

ciation between the RcM and the PC kathā. For information on the colophons of the recovered manuscript and further references, see Katre 1963.

23 On the sources of the RcM, see Vaudeville 1959.

24 Stasik 2009: 150.

tion of alliteration with bibeka, a rhetorical device favoured by Tulsīdās,[25] but they also play an integral part in shaping the role and meaning of viveka in the RcM.

While the second question will be addressed later, Tulsī's text contains verses that allude to the existence of stories or teachings related to viveka, among other topics. For example:

karama dharama itihāsa anekā
karai nirūpana birati bibekā
(1.162.3)

[Through] numerous stories [of] ritual actions and duties,
he defines dispassion and discrimination.

lage kahana upadesa anekā
sahita dharama naya birati bibekā
(2.277.2)

They began imparting various teachings
connected with duty, wise conduct, dispassion, and discrimination.

kahata anuja sana kathā anekā
bhagati birati nṛpanīti bibekā
(4.12.4)

He told his younger brother many tales
[of] devotion, dispassion, kingly conduct, and discrimination.

Verse 1.162.3 is part of the story of King Pratāpbhānu, who encounters a disgraced enemy king disguised as an ascetic. The ascetic impresses Pratāpbhānu through various actions, including narrating wondrous tales. Verse 2.277.2 refers to the family priests of King Daśaratha and King Janaka, while verse 4.12.4 pertains to Rāma instructing Lakṣmaṇa.

These verses may allude to a collection of tales, which could include Puranic narratives, as well as collections of animal tales on political wisdom like the *Pañcatantra* (The Five Books, c. 300 CE) by Viṣṇuśarma and Nārāyaṇa's *Hitopadeśa* (Beneficial Advice, c. 800–950?). However, it is also possible that Tulsīdās envi-

25 For example, kāma koha mada moha nasāvana | bimala bibeka birāga baṛhāvana || (1.42.3); bharata bibeka barāhaṁ bisālā | anāyāsa udharī tehi kālā || (2.296.2); birati bibeka binaya bigyānā | bodha jathāratha beda purānā || (3.45.3); āsa trāsa iriṣādi nivāraka | binaya bibeka birati bistāraka || (7.34.3); bhūta dayā dvija gura sevakāī | bidyā binaya bibeka baṛāī || (7.125.3).

sioned the story of the Prabodhacandrodaya as part of such an enlightening narrative set.[26] This idea becomes more evident through references to its events and characters:

brahmacarja brata saṃjama nānā
    dhīraja dharama gyāna bigyānā
sadācāra japa joga birāgā
    sabhaya bibeka kaṭaku saba bhāgā
(1.83.4)

The various self-restraints like Celibacy, Vows,
    Firmness, Duty, Knowledge, Insight,
Good Conduct, Prayer, Yoga, and Dispassion –
    all of Discrimination's frightened army escaped.[27]

bhāgeu bibeka sahāya sahita so subhaṭa saṃjuga mahi mure
    sadagraṃtha parbata kaṃdaranhi mahũ jāi tehi avasara dure
(1.83.5)

Discrimination fled along with his allies; the noble warriors retreated from battle
    and sought refuge in the caves in the distant mountains of right scriptures at that moment.

hoi bibeku moha bhrama bhāgā
    taba raghunātha carana anurāgā
sakhā parama paramārathu ehū
    mana krama bacana rāma pada nehū
(2.92.3)

If Discrimination arises and the delusion [of] Bewilderment dissipates,
    there will be love for the feet of the Lord of the Raghus.
Friend, the supreme Highest Goal is simply this:
    love for Rāma's feet in thoughts, actions, and words.

---

26 In the fourth volume of the *Tulsī Granthāvalī*, Ramchandra Shukla lists several secondary stories mentioned in the RcM by referring to their main characters. However, this series does not include the references I believe are made to the PC kathā. For further details, see Shukla 1973–1977, vol. 4: 173–340.

27 Regarding this second set of extracts relating to viveka, it is evident that they unmistakably refer to the personified tale of viveka as a king involved in a fratricidal struggle. This is why I have translated these notions using capital letters, which differs from Lutgendorf's approach in Tulsīdās 2016–2023 and the Gita Press translation in English (Tulsīdās 1972).

saciva birāgu bibeku naresū
　　bipina suhāvana pāvana desū
bhaṭa jama niyama saila rajadhānī
　　sānti sumati suci sundara rānī
sakala aṅga sampanna su rāū
　　rāma carana āśrita cita cāū
(2.234.3–4)

Dispassion [served as] the minister, and Discrimination [as] the king,
　　the forest [was] the beautiful, pure country.
Moral Principle and Observance [stood as] soldiers, the mountain [as] the capital,
　　Peace and Good Reason [as] the virtuous, charming queens.
The king was possessed of all the qualities [of a prosperous reign],[28]
　　the mind focused on the shelter of Rāma's feet.

jīti moha mahipālu dala sahita bibeka bhuālu
　　karata akaṇṭaka rāju puraṁ sukha sampadā sukālu
(2.235)

King Discrimination emerged victorious over King Bewilderment and his army,
　　bringing peace to the kingdom,[29] filled with joy and wealth at that time.

Verses 1.83.4 and 1.83.5 depict an episode from the story of Śiva and Pārvatī, in which the god of desire, Kāmadeva, is dispatched by the other gods to ignite love in Śiva's heart for Himālaya's daughter, Pārvatī. These verses specifically illustrate the profound impact Kāmadeva's arrows have on individuals, including here, importantly, Discrimination's allies. In verse 2.92.3, Lakṣmaṇa, Rāma's brother, explains to the Niṣāda king that the circumstances of Rāma's exile were not the result of anyone's wrongdoing but rather the outcome of the universal law of karma (karman). Lakṣmaṇa further draws a parallel between the plight of an individual lost

28 There is a sequence of six elements that resemble the seven constitutive "limbs" of an ideal realm mentioned in 6.1.1 of Kauṭilya's *Arthaśāstra*: svāmin, amātya, janapada, durga, kośa, daṇḍa, and mitra (Lord, minister, countryside, fort, treasury, army, and ally are the constituent elements). Translated from Olivelle 2013: 272. On the controversial compositional history and dating of the text, see Olivelle 2013: introduction.

29 Once again, Tulsīdās deliberately employs terminology from political science. The term akaṇṭaka, which literally means "without thorns", is a reference to the elimination of unlawful activities and serves as the title for chapter 4 in Kauṭilya's *Arthaśāstra*: kaṇṭaka śodhana, translated by Olivelle as "eradication of thorns". For a detailed discussion of the chapter's contents, see Olivelle 2013: 223–252.

in the cycle of rebirth and the state of dreaming. Conversely, a liberated person is depicted as one who has awakened from the dream of existence and wholeheartedly dedicates themselves to serving Rāma.[30] Verses 2.234.3–4 and 2.235 are found within a description of Rāma's governance of the āśrama (hermitage, ashram) of Citrakūṭ. These verses draw a parallel between Rāma's reign and the virtuous rule of King Discrimination.

These verses, embedded in different sections of the narrative, indicate that viveka not only possesses a distinct role in the RcM but also serves as a conceptual link between the PC story and the RcM. It is important to note that I am not referring to a specific version of the story with a fixed time frame and a single author but rather to the overarching narrative of the PC. There likely existed a long-standing tradition of narrating, listening to, and discussing the PC story within the context of the satsaṅga, as attested by Brajvāsīdās.[31] While Tulsī himself may not have explicitly referred to a particular written retelling, the story of the PC encompasses all the elements he associates with viveka.[32] As demonstrated by the earlier quotations, viveka relates to dispassion, as well as to duty, good conduct, and bhakti.

In the PC story, Vairāgya serves as a faithful counsellor to King Viveka. He is the long-forgotten son of Mind, who, sent by Bhakti, appears only at the end of the fourth act of the drama and plays a vital role in King Mind's awakening in the final act. Furthermore, the reference to the "mountains of right scriptures" in RcM 1.83.5, where Discrimination and his army seek refuge, may allude to Mount Mandara, the dwelling place where Upaniṣad's daughter, Bhagavadgītā, is said to reside in the PC story.[33] In RcM 2.234.3–4, Vairāgya is confirmed as Discrimination's minister, and his queens are mentioned, albeit with a variation: in the PC, Sumati and Upaniṣad are portrayed as Viveka's wives, while Śānti serves as an aide. Tulsīdās introduces the references to Rāma and devotion to him within the RcM.

---

30 This dialogue is commonly known as *Lakṣmaṇagītā* due to its similarity to the teachings of the BhG.

31 See subsection 1.2.3.

32 Before Tulsī's time, besides the Sanskrit version by Kṛṣṇamiśra and other Sanskrit renditions, we only have knowledge of the retelling by Malha kavi, composed in 1544 in Brajbhāṣā. For further information, see Agrawal 1962: 212–217.

33 See 6.62 in the KPC; Kṛṣṇamiśra 2009: 248–249. Similarly, in PcN 6.65, Faith says: "The mount Mandara, where the worship [of] Hari takes place in many beautiful ways, is distant. There dwelled Upaniṣad, together with [Bhagavad]Gītā, out of fear of Sophistries." In contrast to Discrimination and his army in RcM 1.83.5, the *Uttara Kāṇḍa* 70.1–70.4 and 71a describe Māyā's powerful family (parivāra). The latter is also defined as an army (kaṭaka) composed of Kāma (the general), Dambha, and Pāṣaṇḍa, among others. These personifications are also present in the PC kathā, where the pāṣaṇḍa category includes the Deniers of the Vedas (nāstikas).

Furthermore, Tulsīdās not only employs the allegory of the PC story but also metaphorically utilises viveka on other occasions:

soka kanakalocana mati chonī
    harī bimala guna gana jagajonī
bharata bibeka barāhaṁ bisālā
    anāyāsa udharī tehi kālā
(2.296.2)

Anguish, a Golden-eyed demon, had stolen the earth [of his] Mind,
    womb of the world, repository [of] pure qualities.
[But] Bharata's great Discrimination-boar
    effortlessly saved it at that moment.[34]

sauraja dhīraja tehi ratha cākā
    satya sīla dṛṛha dhvajā patākā
bala bibeka dama parahita ghore
    chamā kṛpā samatā raju jore
(6.79.3)

Courage and fortitude [are] the chariot['s] wheels;
    truth, good conduct, and firmness [are its] staff and flags;
Strength, discrimination, self-restraint, and concern for others [are its] horses,
    harnessed [together] by forbearance, compassion, and equanimity.

The context of the first quotation is Bharata's visit to Citrakūṭ, where he endeavours to persuade Rāma to return and assume the throne in Ayodhyā. Throughout this episode, Tulsīdās repeatedly extols Bharata, associating him with the virtue of viveka and likening him to Viṣṇu's avatāra (lit. the "descent" of a deity on earth) who liberated the earth from the demon Hiraṇyākṣa, as depicted in the quoted verse.[35]

---

34 Following Lutgendorf's interpretation in Tulsīdās 2016–2023, vol. 4 (2018): 249, I translate joni as "womb", but I keep the term jaga, which Lutgendorf's rendition left implicit: "The gold-eyed demon of grief had stolen the earth of intellect, womb of pure virtues, but the great boar-avatar of Bharat's discernment effortlessly liberated her at that moment." Retaining jaga highlights what I believe is an intentional double meaning, suggesting that not only is the earth the foundation of the world, but also that one's mind – with its faculty of thinking – is the ground for the phenomenal world constructed through sensory perceptions.

35 In RcM 7.1a, the beginning of the *Uttara Kāṇḍa*, Bharata's mind is once again described as sinking in the ocean of separation from Rāma, but this time, he is saved by Hanumān, who is likened to a vessel. The difference between these two verses lies in their contexts: the verse in

Verse 6.79.3, written in the caupāī metre, appears in a segment where Rāma describes his weaponry against Rāvaṇa to a fearful Vibhīṣaṇa. Tulsīdās intertwines Rāma's narrative with a Puranic myth and further embellishes the well-known metaphor of the chariot and its components to construct his own vehicle of spiritual triumph. In the *Kaṭha Upaniṣad*, the chariot mentioned in verses 3.3 to 3.9[36] represents the body, with the passenger being the individual self (ātman). Its spiritual victory lies in the awakening and liberation of the individual self from the cycle of rebirths. In the RcM, Rāma delineates the qualities and attitudes essential for traversing and conquering the ocean of rebirth as the constituents of a virtuous vehicle (dharmamaya) (6.79.6).[37] Driving this chariot of virtue is the responsibility of the individual self. In Tulsī's work, viveka emerges as a total intellectual capacity to be cultivated incessantly.

3.2.2 The Significance of Viveka

The verses cited in the preceding subsection are derived from various books and contexts of the early modern retelling of Rāma's tale. Through these passages and several others, I aim to explore the role of viveka in the RcM and shed light on its more profound significance as a concept that unites the realms of bhakti and dharma within the narrative.

In the opening of the *Uttara Kāṇḍa*, specifically in 7.14.4, Bhuśuṇḍi the crow engages in a conversation with Garuḍa, praising the virtues of the Rāma story:

---

the *Ayodhyā Kāṇḍa* is part of a context where Bharata's character is carefully constructed to exemplify his qualities as, first, a devotee, and then, a prince and a brother. On the other hand, the *Uttara Kāṇḍa*'s focus seems to be more on the progression of events. This verse appears at the beginning of the final book, emphasising Rāma's return and blessed rule. Therefore, it is understandable that the author has reworked the metaphor to introduce Hanumān, who will announce Rāma's return to Bharata and set the seventh book of the epic in motion. From a religious perspective, we can interpret the verse as depicting the comforting and salvific power that a model devotee (Hanumān) exerts on another. Furthermore, in support of my argument, the *Uttara Kāṇḍa* dedicates relatively less space than *Ayodhyā Kāṇḍa* to Bharata.

36 See Olivelle 1998: 388–389. Several passages in the *Mahābhārata* employ the symbology of the chariot, e.g. 12.228.8–12, 12.238.1–2, and 14.50.1–6. See also BhP 4.29.18–19. The text I refer to throughout is Shastri 1983.

37 The description of the chariot, occupying vv. 79.3–79.6, is worth quoting in its entirety: sauraja dhīraja tehi ratha cākā | satya sīla dṛṛha dhvajā patākā || bala bibeka dama parahita ghore | chamā kṛpā samatā raju jore || īsa bhajanu sārathī sujānā | birati carma santoṣa kṛpānā || dāna parasu budhi sakti pracaṇḍā | bara bigyāna kaṭhina kodaṇḍā || amala acala mana trona samānā | sama jama niyama silīmukha nānā || kavaca abheda bipra gura pūjā | ehi sama bijaya upāya na dūjā || sakhā dharmamaya asa ratha jākeṁ | jītana kahā̃ na katahũ ripu tākeṁ ||.

khagapati rāma kathā maiṁ baranī
    svamati bilāsa trāsa dukha haranī
birati bibeka bhagati dṛṛha karanī
    moha nadī kahaṁ sundara taranī

O king of the birds, I narrated Rāma's story,
    a delight for my mind that dispels fear and suffering.
It strengthens dispassion, discrimination, and devotion –
    a beautiful vessel to [navigate] the river of bewilderment.

In this verse, Bhuśuṇḍi emphasises the transformative power of the Rāma kathā, highlighting its ability to bring solace and alleviate suffering. Moreover, it serves as a catalyst for cultivating dispassion, discrimination, and devotion within individuals. These qualities, represented by viveka, play a crucial role in navigating the treacherous waters of confusion and attachment. In the hands of Tulsīdās, the narrative of Rāma's story emphasises the importance of discrimination both as an inherent virtue and through the portrayal of characters who embody this quality. Discrimination is presented as a distinctive attribute of a guru (1.44.2, 1.181.1), and it is frequently associated with the god Śiva, who acts as a teacher to his first wife Satī and later to Pārvatī. Śiva demonstrates discernment (1.46, 1.110.2) and encourages his consort to cultivate the same (1.51.2).

Discrimination is depicted as a determining factor in shaping one's thoughts, as evident in the recurring phrase "bibeka bicari" (thinking [with] discernment). It also extends to one's speech, as seen in the expression "bibekamaya bacana" (discerning words). Tulsīdās underscores the redemptive potential of viveka even in seemingly ordinary situations. For instance, when Daśaratha implores his wife Kaikeyī to reconsider her request for Rāma's banishment, he appeals to her to set aside any mocking or resentful attitude and instead contemplate her decision with discrimination:

priyā hāsa risa pariharahi māgu bicāri bibeku
    jehiṁ dekhauṁ aba nayana bhari bharata rāja abhiṣeku
(2.32)

Dear, let go of your mocking laughter and resentment and deliberate on your request with discrimination,
    so that I may look upon Bharata's consecration as crown prince favourably.[38]

38 I translated nayana bhari, literally "eyes filled [with tears]", as "favourably" to emphasise that King Daśaratha's emotional involvement in Bharata's consecration signifies a change of heart in "favour" of Kaikeyī's son.

While Daśaratha urges Kaikeyī to exercise discernment, her words, unlike his kindly utterances,[39] are described as kaṭu, meaning "bitter" or "displeasing" (2.29.4, 2.34.2), and her understanding/intellect as kumati "wicked" (2.29.1, 2.32.2, 2.161.1) or kubuddhi "injudicious" (2.30.1).[40] In practical circumstances, viveka entails thinking, speaking, and acting in accordance with one's social duty or dharma. The characters who possess discrimination and express it are virtuous individuals and devout followers of Rāma.

Simultaneously, viveka is closely intertwined with bhakti and the overall religious perspective of the RcM. I will demonstrate this by examining the portrayal of Bharata, the primary character associated with discrimination. In the Sanskrit *Rāmāyaṇa* (The Journey of Rāma; perhaps between 500 BCE and 300 CE) Vālmīki describes Bharata as "satyadhṛtiḥ" (steadfast in truth), "dharmātman" (righteous), "bhrātṛvatsala" (affectionate towards his brother), and "dharmaprekṣa" (observant of righteousness).[41] The young prince refuses to accept what he believes rightfully belongs to his elder brother, adhering to his royal and familial duties.[42] In the RcM, the emphasis is placed not so much on Bharata's adherence to kingly conduct but rather he is reinterpreted as a devotee who practises his duty, which is nourished by viveka. In this regard, my reading of Bharata aligns with that of Charlotte Vaudeville, highlighting the priority of bhakti over rāja dharma (royal duty) in shaping his character.[43]

To my mind, it is crucial to acknowledge the central role of bhakti in comprehending Bharata's character.[44] Looking again at the metaphor in 2.296.2, the Puranic boar represents Viṣṇu's descent to earth as Varāha. Thus, we can further interpret the verse: from Discrimination-boar to Discrimination-Viṣṇu. If discrim-

---

39 See mṛdu bacana (2.32.2); bani sabinaya (2.30.2).

40 On the other hand, in the PC kathā, Mati or Sumati represents Viveka's wife, as discrimination is initially paired with good and right understanding. In RcN 2.30.1, Kaikeyī is compared to an unsheathed sword of rage, and likewise, Brajvāsī likens a woman to Kāma's sword in PcN 1.62.

41 Respectively at 2.76.27, 2.76.19, and 2.79.16. See Vālmīki 1986.

42 Tulsī's characterisation of Bharata also differs from that in the *Adhyātma Rāmāyaṇa*. For more information, see Vaudeville 1959, vol. 1: 171. On the multidimensional meanings of dharma in the Sanskrit *Rāmāyaṇa*, see Brockington 2004. For an interpretation of the Sanskrit *Rāmāyaṇa*'s characters from social and political perspectives, see Pollock's introduction to Vālmīki 1986: 3–76.

43 Vaudeville 1959, vol. 1: 174.

44 This differs from Danuta Stasik's approach, which explores the social significance of the RcM but does not seem to consider the bhakti aspect of the text. See Stasik 2013: 237–248. While I do not deny that Tulsī's work contains elements related to political theory and social norms (see e.g. footnotes 28 and 29 in this chapter), I believe that an interpretation of Bharata, at least, should integrate the religious perspective.

ination saves the mind, Hari (Viṣṇu) saves the earth – a transition from psychological value to cosmic significance. The importance of viveka is not downplayed, as the comparison brings a sense of identification: if discrimination is the boar, then discrimination is Viṣṇu. Consequently, contemplating, speaking, and acting with viveka implies engaging with God. Bharata's viveka is often coupled with references to spiritual advancement or Rāma:

bimala bibeka dharama naya sālī
bharata bhāratī mañju marālī
(2.296.4)

Endowed with pure discrimination, propriety, and humility,
Bharata's expression was beautiful, like that of a gander.

nirakhi bibeka bilocananhi sithila saneham̐ samāju
kari pranāmu bole bharatu sumiri sīya raghurāju
(2.297)

[With] discerning eyes, observing the assembly with tender affection,
he paid obeisance and, remembering Sītā and the Lord of the Raghus, spoke.

rāma pema bhājana bharatu baḍe na ehi karatūti
cātaka haṃsa sarāhiata ṭem̐ka bibeka bibhūti
(2.324)

Yet for Bharata, the receptacle of Rāma's love, his endeavours were not significant:
the cuckoo is celebrated for its resolve, the gander for the potency of its discrimination.

Philip Lutgendorf points out that Bharata's episode occupies the entire second half of the *Ayodhyā Kāṇḍa*. To this day, Kaikeyi's son is considered a model devotee primarily because he is "a model brother".[45] The bond of fraternal love indeed provides a solid foundation for expressing bhakti. However, in my view, as I have attempted to demonstrate through the interpretation of the verses, Bharata is a

45 See Lutgendorf 1991: 348–349. Bharata plays a central role in two of the four passages that Ramchandra Shukla identifies as the most emotionally impactful in the entire RcM. For more details, see Shukla 1977: 196. The four episodes include Rāma's banishment from Ayodhyā and his departure, the meeting of Rāma and Bharata, the hospitality of Śabarī, and Bharata's anticipation of his brother's return. The enduring popularity of these episodes is further evidenced by the movie *Bharat Milap* (1942), which is based on both Vālmīki's *Rāmāyaṇa* and the RcM. See Dwyer 2006: 35.

model brother because his actions are guided by his discrimination, which is closely connected to his bhakti. Bharata's viveka is specifically highlighted in the verses preceding his interactions with his elders and subjects (2.296.4 and 2.297), that is, before he is required to fulfil his duties in the political sphere. Therefore, viveka becomes the point where bhakti intersects with dharma.[46]

Recalling how Tulsīdās meticulously crafts Rama's younger brother as an ideal devotee throughout the *Bharat Carit* is worth remembering in this context.[47] The supremacy of bhakti over one's duty as the ultimate good sets the tone for Bharata's arguments in a discussion at the court of Ayodhyā. Upon returning to the royal palace and learning of his father's demise, his brother's banishment, and his expected consecration as heir to the throne, Bharata is initially confronted by the sage Vasiṣṭha, Queen Kausalyā, and the ministers who advocate for conventional kingly dharma. They request that Bharata keep his father's word and assume kingship (2.173.3), presenting it as to his own benefit (hita, 2.175.1). In response, the young prince challenges his own dharma, stating that his spiritual health takes precedence and asserts that his true well-being lies in serving Rāma and Sītā (hita hamāra, 2.177.1; mora hita, 2.177.4; bhala mora, 2.260.4).[48]

The primacy of bhakti is illustrated by the journey to Citrakūṭ, which also occurs in Vālmīki's *Rāmāyaṇa*. Still, in the RcM, it takes on the dimension of Bharata as a devotee who suffers from separation from his Lord.[49] The devotional essence of the journey to Citrakūṭ revolves around viraha, the intense feeling of separation from Rāma, which had already led to Daśaratha's demise (2.173.2). The symptoms of this "poison of separation" (biyoga biṣama, 2.183.1) are intertwined with referen-

---

46 Francesca Orsini explores the factors that led Hindi intellectuals to consider the RcM a classic of Hindi literature in the twentieth century. She observes that the significance of the text for literary critics also lies in the fact that Tulsī "had showed a way of reconciling reform and social order, *bhakti* and authority". While I do not specifically address the aspect of bhakti as reform, my ongoing analysis of Bharata's character highlights a particular articulation of the discourses on bhakti and dharma in the RcM, which could have formed the foundation for the interpretations of Hindi scholars. See Orsini 1998: 126. For contrasting and competing interpretations of Tulsīdās by Hindi literary critics, see Mangraviti 2019: 138–171.

47 In the conclusion of the *Ayodhyā Kāṇḍa* 2.326, Tulsī refers to Bharata's actions in this manner. This expression has since become a classic phrase critics use to refer to Bharata's narrative. For more details, see Chaturvedi 1953: 120.

48 See also 2.290.4, where Vasiṣṭha asks Rāma to decide who should rule in Ayodhyā, whether Rāma should return or let Bharata rule during his exile. In this case, like Bharata's situation, dharma seems to be connected with, yet somewhat subordinate to, hita (well-being). The verse reads: mahārāja aba kījai soī | saba kara dharama sahita hita hoī || (Now, O great king, please act so that well-being may be [achieved], along with everyone's duty).

49 Vaudeville 1959, vol. 1: 176 describes Bharata's journey in the forest as a tīrthayātrā (pilgrimage).

ces to devotional practices and attitudes. In the context of experiencing viraha from God, Tulsī highlights the importance of remembrance, hearing, and focusing on the Lord: before embarking on the journey, Bharata and Śatrughna concentrate their thoughts on Rāma and Sītā's feet (sumiri, 2.187; sumirata, 2.201.1; sumire, 2.202.2); during the journey, Bharata listens to accounts of Rāma's noble deeds narrated by the forest dwellers (2.223.4). Additionally, there are acts of worship (arcana) and prayer (vandana). Upon reaching the Ganges, Bharata reveres the spot where Rāma used to bathe and becomes so absorbed in thoughts of his God-brother that he feels as if he is in Rāma's presence (2.196.2). He longs to behold the sleeping place of Rāma and Sītā, as merely glimpsing it soothes the burning agony of separation (2.197.4).

Furthermore, numerous physical manifestations illustrate the impact of Bharata's love for Rāma and the anguish of their separation. These signs include falling (2.163.1), lamenting (2.166), teary eyes or shedding tears (2.175.5),[50] trembling limbs and a quivering body (2.196.1),[51] rushing upon hearing of the arrival of Rāma's friend (2.192.4), a voice choked with emotion (2.210), and even losing consciousness upon meeting Rāma (2.240–2.240.1, 2.241.4).[52] The narrative also presents the repetition of God's name, which holds great reverence in Tulsī's work. When Bharata enters Prayāg, a significant moment before his encounter with the sage Bharadvāja, he repeats "Sītā-Rāma" (2.203; also mentioned in 2.219.3). Another poignant moment of the episode is Bharata's affectionate welcome of the tribal king, Guha, due to his close relationship with Rāma as being "dear to Rāma" (rāmapriya, 2.192.3) and "Rāma's friend" (rāma sakhā, 2.192.4).

The voyage is emotionally charged not only due to the depiction of Bharata's intense emotions but also because of the intertwining perspectives within the narrative. On the one hand, the events are portrayed from Bharata's point of view, while on the other hand, there is the perspective of his entourage (and at times even the gods) and the people he encounters along the way.[53] We have already explored Bharata's viewpoint to some extent, so now let us shift our focus to the second perspective.

First and foremost, the author emphasises the distress the court and the general population feel in Rāma's absence,[54] highlighting that they are influenced by

---

50 See also RcM 2.210, 2.219.4, and 2.237.1.

51 See also RcM 2.197.3, 2.210, 2.233.2, and 2.259.2.

52 Consider the depiction of someone deeply absorbed in devotion to the Lord, as described in BhP 11.14.23–24, 11.2.40, 11.3.31–32, and 1.6.16–17.

53 To these we could add the perspective of the author/narrator, as Tulsī's narration occasionally comments on the statements made by characters or the unfolding of events.

54 See RcM 2.175.5, 2.183.1, and 2.187.1; "rāma biraha byākula", 2.213 and 2.224.

Bharata's words and actions. They imitate him by sharing his emotions and replicating his actions.[55] An example of this can be seen when people worship the places where Rāma, Sītā, and Lakṣmaṇa slept, bathed, or rested immediately after Bharata does the same.[56] Second, there are numerous explicit declarations stating that Bharata is "dearest to Rāma" (2.204.4), the "receptacle of Rāma's love" (pemu pātru, 2.207.2), and the "embodiment of Rāma's love" (dhare deha janu rāma sanehu, 2.207.4). Such statements, voiced by both ordinary people and sages, contribute to the portrayal of Bharata's devotion and intensify the emotional impact of the events. Additionally, they invite the audience or reader of the story to identify, if not with Bharata directly, then at least with those around him. The entourage consists of participants who witness the Lord's actions on earth and the acts of His beloved devotees. The entire episode serves to exalt, albeit implicitly, the value of being among God's closest associates.

Undoubtedly, Bharata suffers, and as we have just witnessed, those around him suffer with him. However, he possesses something that others, even sages, do not possess, setting him apart from the rest: his redemptive power. His words act as a soothing balm to the distressed courtiers (2.183.1), he serves as a support against anguish (2.184), he embodies good fortune for Bharadvāja (2.205.2), his mere sight liberates people from the cycle of rebirth (2.216.1), and narrating or hearing about him dispels sin and suffering (2.222.1). While the text does not explicitly prescribe such qualities, Bharata's actions assume an exemplary status as the actions of the crown jewel among devotees (bhagata siromani, 2.219). His qualities resemble the Lord's qualities, subtly suggested by Tulsī at 2.221.1, where ordinary women, conversing among themselves, are struck by the physical resemblance between Bharata-Śatrughna and Rāma-Lakṣmaṇa. This resonates with another passage that compares Bharata's moon to Rāma's sun (2.208.1–2), drawing attention to their complementarity and emphasising Bharata's role as an intermediary, bringing Rāma's nectar-like love to the devotees on earth (2.208.3). Particularly intriguing are verses 2.221.1 and 2.221.2:

kahahiṁ sapema eka eka pāhīṁ
    rāmu lakhanu sakhi hohiṁ ki nāhīṁ
baya bapu barana rūpa soi ālī
    sīlu sanehu sarisa sama cālī

beṣu na so sakhi sīya na saṅgā
    āgeṃ anī calī caturaṅgā

55 See RcM 2.175.5, 2.224.1–2, 2.237.3, and 2.240.

56 See RcM 2.196.3, 2.201.1, and 2.220.4.

One lovingly said to another:
"Friend, are they Rāma and Lakṣmaṇa?
Their age, body, colouring, and shape are the same, friend,
so their affectionate conduct and gait."

[The other replied:] "The guise is not the same, friend; there is no Sītā
and an army marches ahead of them."

Although the author does not extensively delve into metaphysical reflections on the relationship between the Absolute, in this case, Rāma, and the individual self from a perspective other than bhakti, occasional allusions can be found. In this context, the first verse above emphasises the similarity of various features, particularly the aspect of rūpa, which signifies "outer form", "shape", or even "beauty".[57] The second verse establishes the difference between the brothers in terms of "guise", as beṣa conveys the idea of an "outward appearance" or "garb". Intriguingly, its verbal form implies "taking a disguise", suggesting the act of assuming an external aspect, like wearing a costume for a theatrical role.[58] Both terms are employed to indicate the visible appearance that God assumes when manifesting on earth.[59] Rāma, the Lord descended to earth to ease the burden of evil, possesses a rūpa, a worldly appearance connected to an otherworldly reality (svarūpa).[60] The usage of rūpa and beṣa, far from being casual, alludes to a higher level of reality, serving as a clue left by the author to prompt listeners or readers to reflect on the paradoxical relationship between the Supreme Self and the individual self – similar to and yet different from each other simultaneously. The previous instances highlighting Bharata's redemptive power, akin to God's, reinforce the similarity between God and His devotee.[61]

Viewed in this light, the fact that Bharata is not consecrated as the prince regent holds little significance when compared to his consecration as the ideal devotee attained in the *Ayodhyā Kāṇḍa*. The journey to Citrakūṭ appears to be motivated not solely by the desire to resolve a dilemma of political duty but rather by the profound longing of a humble bhakta for union with his Lord and seeking refuge in

57 MW 855.

58 MW 1019. beṣa dhar- is the nominal verb in Brajbhāṣā. See also RcM 1.54.2.

59 For the concept of rūpa, see e.g. *Viṣṇu Purāṇa* 1.4.8. For a study of how the Lord's descent on earth is indicated, see Hacker 1960. For a discussion highlighting the connections between Viṣṇu's descent and dramatic theory, see Couture 2001.

60 I refer to the difference between rūpa (as worldly "form") and svarūpa (as "inner core" or essence) proposed by Goudriaan 1994: 440–441.

61 According to Vaudeville 1959, vol. 1: 175, Tulsī intended to demonstrate the identity (tādātmya) between Rāma and Bharata but chose a more concise approach.

Him (sarana, 2.182.2). While Tulsī upholds the primacy of bhakti, this does not entail completely forsaking one's worldly dharma. Instead, as I previously mentioned, it emphasises the dependence of dharma on bhakti. Several instances portray a Bharata who never forgets his dharma: before he departs from the royal palace, he entrusts the city to capable and reliable servants, recognising the duty of honour to leave it under proper care (2.185.2–3).[62] Furthermore, during the journey, he is depicted caring for his entourage (2.186)[63] and obeying his guru (2.212.2).[64]

Bharata retraces Rāma's steps and leads his people to embark on the boat of viveka, a path to overcome separation from the Lord (2.220). His viveka is a clear indication of his spiritual advancement. Tulsī aptly compares him to a gander (haṃsa) to convey this aspect of his personality:

bharatu haṃsa rabibaṃsa taḍāgā
    janami kīnha guna doṣa bibhāgā
gahi guna paya taji avaguna bārī
    nija jasa jagata kīnhi ujiārī
(2.231.3–4)

Bharata, gander on the Solar dynasty's lake,
    since birth, he has distinguished merit [from] evil:
retaining pure milk and rejecting impure water,
    he illuminated the world [with] his glory.

This metaphor is repeated in 2.313.4, underscoring Bharata as an exemplary embodiment of discernment, as the haṃsa traditionally represents. Moreover, he also embodies the qualities of a yogi.[65] Viveka is a characteristic of yogis who dispel ignorance by discerning/discriminating between puruṣa (self) and prakṛti (material nature), as outlined in the *Yogasūtras* (YS; 350–400 CE) of Patañjali.[66] However, Bharata's goal as a yogi is not aloneness (kaivalya), as the YS states as the objective of yoga, but rather union with his Lord, as his suffering ends only in close proximity to Rāma:

62 In this context, Vaudeville 1959, vol. 1: 181–182 compares between Bharata and King Janaka.
63 See also RcM 2.197 and 2.232.3.
64 It is only natural for a devotee who is referred to as "nīti prīti pālaka" (protector of right conduct and love) in 2.303.3. In this regard, verse 2.232.1 is telling: jauṃ na hota jaga janama bharata ko | sakala dharama dhura dharani dharata ko || (If Bharata weren't born in this world, who would sustain the earth, committed to all duties?).
65 Vaudeville 1959, vol. 1: 183 renames Bharata's carita as sādhu carita (the actions of a wise, spiritually accomplished individual).
66 YS 2.26, 28; 3.52, 54; 4.26, 29.

rāma saila sobhā nirakhi bharata hṛdaya ati pemu
tāpasa tapa phalu pāi jimi sukhī sirāneṁ nemu
(2.236)

Looking at the splendour of Rāma's mountain, Bharata felt profound love
Like an ascetic who, having gained the fruit of his self-discipline, [is] content
[because] his restraints are concluded.

karata prabesa miṭe dukha dāvā
janu jogīṃ paramārathu pāvā
(2.238.2)

As he entered [Rāma's refuge], the fire of [Bharata's] suffering was extinguished,
like a yogi attaining the Highest Goal.

Bharata's path as a devotee is characterised by a unique combination of yoga, asceticism, and bhakti, which distinguishes his particular way of being a bhakta. Tulsī juxtaposes his nema (observance, restraint) and prema (love) (2.311.1) throughout the *Ayodhyā Kāṇḍa*, culminating in Bharata's exile that mirrors Rāma's journey. This exile is marked by qualities such as tranquillity (sama), restraint (dama), self-control (sañjama[67]), observance (niyama), and meditation (upāsā) (2.324.2). It appears that the author intends to emphasise, through Bharata's character, that the cultivation of yoga, austerity, and knowledge is truly valuable only when accompanied by love (2.291.1).

Tulsī's depiction of the bhakta aligns with the perspective advocated by Patton Burchett, which advocates for a broader understanding of bhakti that includes a reconsideration of its relationship with yoga and asceticism. This challenges conventional categorisations and encourages us to expand our viewpoint regarding bhakti and its practitioners.[68] I have discussed elsewhere how such categorisations are challenged when reading Brajvāsīdās's PcN.[69] In Chapter 4, I will delve deeper into these connections, exploring the philosophical and religious dimensions of the PcN and their resonance with the insights gained from studying the RcM.

---

67 The term sañjama (Skt. saṃyama) could also be understood in the Pātañjala sense, referring to the combination of dhāraṇā (concentration), dhyāna (meditation), and samādhi (contemplation). See YS 3.4.

68 While tantra is mentioned in Burchett's discussion, it plays a more peripheral role in my study. For a more comprehensive exploration of tantra, see Burchett 2019.

69 In a paper I presented at the thirteenth ICEMLNI Conference in Warsaw in July 2018, to be published as Pastore (forthcoming a).

3.3 The Role of the *Rāmcaritmānas* in Brajvāsīdās's *Prabodhacandrodaya Nāṭaka*

The preceding sections of this chapter have demonstrated that the Prabodhacandrodaya story is part of the reception of Tulsīdās's *Rāmcaritmānas* by later authors. We have seen how nineteenth-century authors perceived the RcM and the PC kathā as interconnected stories and, to some extent, texts. Furthermore, I have argued that the RcM and the PC kathā *are* indeed related, as evidenced by the RcM's reference to the characters and plot of the PC kathā, as well as the significant role of viveka, which is personified in the PC story as the main protagonist, within its bhakti perspective. These elements are crucial to my central argument, which posits that the RcM is a source of Vedānta and bhakti in the PcN through adaptive reuse.

As mentioned at the outset of this chapter, identifying the influence of the RcM on other early modern texts may not be straightforward. One reason for this is that the dynamics of adaptive reuse or simple re-use may not be readily apparent if the RcM is not explicitly mentioned by the authors drawing from it. This is the case with the PcN, but other authors employ various, more or less implicit ways of referring to it. For instance, Sabal Siṃh Cauhān's *Mahābhārata* (seventeenth–eighteenth century) reuses and evokes Tulsī and his work by invoking its *Sabhāparva*, claiming to have been composed on the same date as Tulsī's RcM, alluding to Tulsī through wordplay involving his name, adopting the caupāī-dohā metrical scheme typical of the RcM, and extolling the power of repeating Rāma's name.[70] While these factors contribute to more immediate identification of the RcM as a source of inspiration, the true nature and extent of the association require careful textual analysis. Another example is the poet Ṭhākur (1756–1823), whose quatrains, Imre Bangha observes, exhibit idioms found in the RcM that develop and adapt themes present in the RcM and other works by Tulsī. Notably, Bangha observes that Ṭhākur held "the poetry of Tulsīdās in high regard".[71]

In his history of Hindi literature, McGregor attributes the enduring popularity of the RcM to the apparent absence of Rāma kathā retellings in seventeenth- and eighteenth-century North India following Tulsīdās's work.[72] While complete retellings during this period may be elusive, the aforementioned examples, including the PcN, provide evidence that authors in these centuries drew inspiration from the RcM for their own original compositions. Before delving into the

70 See Pillai 2021: 266–272. The caupāī-dohā is among the typical prosodical schemes employed in long narrative poems. The sequence of eight to ten caupāī lines develops the narration, while the dohā following them has a summarising function.

71 Bangha 2014: 47.

72 The exception is the sectarian literature produced by the Rāmrasikas. See McGregor 1984: 167.

discussion of how the PcN adaptively reuses the RcM in philosophical and religious contexts, it is important to highlight that the PcN not only references the RcM story but also incorporates elements from it into other narrative contexts. This section establishes the foundation for the subsequent chapter by illustrating that the RcM story was well known to Brajvāsīdās and served as a literary model for his PcN.

3.3.1 Literary Adaptive Reuse and Simple Re-use of the RcM in the PcN

While the next chapter will delve into the references to Rāma himself in the PcN, this subsection focuses on the mentions of other characters and episodes from the prince's tale in Ayodhyā. In Act V of the PcN, Faith narrates the battle between the armies of Discrimination and Bewilderment to Bond-with-Viṣṇu. In the KPC, this part of the text is brief, with Faith simply listing the victories of Discrimination's warriors over Bewilderment and his army.[73] However, in the PcN, the conflict between these opposing concepts/characters is further developed. In verse 5.62, Investigation-into-Reality (bastubicāra) sets out to defeat Desire (kāma):

binā hathiyāra bina saṅgī
    mano hanumāna bajaraṅgī
dapaṭikai sāmuheṁ āyo
    jhapaṭikai madana para dhāyo

Without any weapon, without companions, just like Hanumān [with his] diamond-like body, he came before [Desire], threatening [him]. Springing, he leaped on the intoxicating Desire.

This verse can be seen as an example of adaptive reuse, as it evokes Hanumān's name and attributes to depict the character of Investigation-into-Reality. The comparison to Hanumān likely alludes to his solitary feat in Laṅka, where the son of the wind single-handedly threw Rāvaṇa's palace into burning chaos. Brajvāsīdās not only names Hanumān but also extends the parallel by describing the swift and sudden actions of Investigation-into-Reality, akin to the agile movements of a monkey. The absolutive verb "jhapaṭikai" (springing) and the perfective verb "dhāyo" (he leaped) convey a rapid advancement, akin to a gust of wind.[74]

---

73 See v. 5.46 in Kṛṣṇamiśra 2009: 201: "Then Analyst killed off Lust; Patience brought down Anger, Coarseness, Harm and their kind; Contentment defeated Greed, Craving, Self-Pity, Untruth, Slander, Theft, Fraud and more; Generosity conquered Envy; Altruism demolished Self-Intoxication; while Admiration-for-others destroyed Pride."

74 HSS 1828 defines jhapaṭanā as "kisī (vastu yā vyakti) kī ora jhoṁka ke sātha baṛhnā". Jhoṁka is a "gust or blast of wind" (OHD 397). See HSS 2452 for dhānā.

A second instance of adaptive reuse from the RcM, specifically from its *Yuddha Kāṇḍa*, can be found in the context of warfare in the PcN. In PcN 5.67, Greed's (lobha) presence on the battlefield is described by comparing him to Kumbhakarṇa:

> Then Greed assumed a powerful form, [together] with the formidable Thirst. He stretched his body [out] like a mountain cave, his stomach like a tank full of water. [His long] arms [were] like huge snakes; dreadful, [he was] like Kumbhakarṇa, Rāvaṇa['s] brother. He came in front [of the army] spreading his arms, enfolding everyone at once. It was clear at that time that Greed had become death for all.[75]

Similar to the case of Investigation-into-Reality, certain distinctive characteristics of Rāvaṇa's brother are reused here to depict Greed. In the RcM, Kumbhakarṇa's body is described as having the form (ākāra) of a mountain (bhūdhara), and he is equated to a mountain cave (giri guhāṁ).[76] Likewise, Greed's body expands, resembling a mountain cave (giri kandara sama). Furthermore, the PcN underscores the terrifying nature of Greed, portraying him as deadly as death itself, echoing the depiction of Kumbhakarṇa in the RcM.[77]

An additional example can be found in PcN Act III, where Peace is distressed because she cannot find her mother, Faith. Her friend Compassion serves as her interlocutor, and Peace expresses her feelings as follows:

jo bidhi basa pākhaṇḍa bīca
    jiyata hoya pari bivasa nīca
hota samaya samarathahu dīna
    jimi siya niścara basī malīna
tāte tina madhi lijai śodha
    binu saradhā muhi nāhina bodha
(3.11)

75 Bahuri lobha dhari rūpa viśālā | tṛṣṇā saṅga māha vikarālā || giri kandara sama badana pasārī | udara taḍāga mano jalabārī || ahi viśāla sama bhuja bhaya dātā | kumbhakarana mano rāvaṇa bhrātā || sanmukha āya bāhu vistārī | liye lapeṭa sabai ika bārī || samujhi parī aisī tehi kālā | bhayo lobha aba saba ko kālā ||.

76 RcM 6.64.1: nātha bhūdharākāra sarīrā | kumbhakarana āvata ranadhīrā ||; 6.66.1: koṭi koṭi kapi dhari dhari khāī | janu ṭīḍī giri guhāṁ samāī ||. Translated by Lutgendorf in Tulsīdās 2016–2023, vol. 6 (2022): 137 and 141: "'Master, he whose body is like a mountain and who is resolute in war—Kumbhakaran—approaches'"; "Seizing millions of monkeys, he swallowed them, like locust swarms vanishing into a mountain cave."

77 [. . .] | bhayo lobha aba saba ko kālā || (Greed had become death for all) in PcN 5.67. The RcM's line (6.66.1) reads: kumbhakarana rana raṅga biruddha | sanmukha calā kāla janu kruddhā ||, translated by Lutgendorf in Tulsīdās 2016–2023, vol. 6 (2022): 141 as "Raging in his battle frenzy, Kumbhakaran came on like an embodiment of Angry death."

> If, somehow, she [Faith] is among the heretics, then the hope that she is alive falls. In such circumstances, even a [morally] powerful [person becomes] wretched, as Sītā [became] impure, living [with] the nightly demon. Therefore, search among them, [for] without Faith there is no solace for me.

This verse intertwines the narrative of the RcM with that of the PcN, establishing a parallel between the heretics (pākhaṇḍa) and the demons (rākṣasas) while also alluding to the context of the RcM. The lament of young Peace, deprived of her mother, may evoke the reader or listener's memory of Rāma's anguish when Sītā was kidnapped by the king of the rākṣasas.[78]

An interesting observation about the instances discussed above is that Brajvāsī's choices are not arbitrary. For example, he repurposes positive characters from the RcM (Hanumān and Sītā) to depict positive characters in the PcN (Investigation-into-Reality and Faith), and an antagonist from the RcM (Kumbhakarṇa) as an antagonist in the PcN (Greed). These deliberate choices highlight the thematic connections and intertextuality between the two texts.

There are additional instances of adaptive reuse in which the respective contexts may be more challenging to associate. One such instance is an aphorism of political wisdom that appears relatively ancient and widespread. It is also found in Viṣṇuśarma's *Pañcatantra,* albeit in a different form.[79] Nevertheless, I believe that the direct source and cultural reference for the PcN is the RcM:

tadyapi nātha nīti asa suniye
ripu kahaṁ ghorai kabahuṁ na guniye
(PcN 1.71)

> [Peace:] O lord [of] my life, I know about the true might you describe. Although all are in your power and no one can defeat you in the world, nevertheless, please listen to such political science: never evaluate the enemy while boasting.

[. . .] nāsahi begi nīti asa sunī
(RcM 3.20.6)

[. . .] all soon come to ruin—such counsel I have heard[80]

rāri roga ripu agni ṛṇa nṛpati tapodhana byāla
itane ganiya na choṭa kari sajaga rahiya saba kāla
(PcN 1.72)

78 See e.g. RcM 3.29.1–9.

79 See footnote 33 of the annotated translation (Act I).

80 Translated by Lutgendorf in Tulsīdās 2016–2023, vol. 5 (2020): 65.

A war, a disease, an enemy, a fire, a debt, a king, a great ascetic, a serpent –
consider them rightly, do not diminish [them], stay alert at all times.

ripu ruja pāvaka pāpa prabhu ahi gania na choṭa kari
asa kahi bibidha bilāpa kari lāgī rodana karana
(RcM 3.21)

"An enemy, an illness, fire, sin, a master, or a snake should never be underestimated!"[81]

The maxim is introduced in PcN 1.71 and RcM 3.20.6, where the former is a caupāī (a line with each pāda consisting of sixteen syllabic instants; see Tables 2 and 3) and the latter is a caupaī (fifteen syllabic instants). In the first pāda of PcN 1.71, "nīti asa suniye" is a reuse of "nīti asa sunī" in RcM 3.20.6. This is accomplished by adding the syllable -ye to account for the difference of one syllabic instant between the two variants.[82] The reuse of "nīti asa sunī" as "nīti asa suniye" ("listen to such political science!" as "please listen to such political science") is easily identifiable, thanks to the fact that both precede the actual aphorism. Nevertheless, I still refer to this case as one of adaptive reuse, not simple re-use, because it is closely connected to the maxim.

The gnomic sentence is expressed through one line, consisting of twenty-four syllabic instants in each of the two pādas, arranged in opposite orders. PcN 1.72 is a dohā with a syllabic instant distribution of 13 + 11, while RcM 3.21a is a soraṭhā with an 11 + 13 arrangement. The second pāda of PcN 1.72 reuses the first pāda of RcM 3.21a. Specifically, the phrase "prabhu ahi gania na choṭa kari" from the RcM is reused as "itane ganiya na choṭa kari" in the PcN's verse, as both contain thirteen syllabic instants.[83] Since "prabhu ahi" (master and snake) from the RcM is already present in PcN 1.72 as "nṛpati and byāla" (king and serpent) in the first pāda, the four missing syllabic instants to reach thirteen are filled by "itane" in the PcN. In this case, although the broader contexts of these verses are dissimilar, there is still a partial connection between them. The speakers of the two verses are women: in the PcN, Pleasure (rati) admonishes her husband Desire, while in the RcM, Śūrpaṇakhā complains to Rāvaṇa about Rāma and Lakṣmaṇa disfiguring her.[84]

81 Translated by Lutgendorf in Tulsīdās 2016–2023, vol. 5 (2020): 65.

82 It should be noted that the verbal forms differ between RcM's "sunī" (which is perfective) and PcN's "suniye" (which is a polite imperative).

83 RcM: "[. . .] a master, or a snake should never be underestimated!"; PcN: "[. . .] consider them rightly, do not diminish [them]".

84 See Tulsīdās 2016–2023, vol. 5 (2020): 64–65.

In two instances that can be seen as examples of simple re-use, the backgrounds are completely different. One instance occurs in the context of the meeting between Friendship (maitrī) and Faith (saradhā) in the PcN, while the other instance in the RcM depicts Maina's reaction upon learning that her daughter Pārvatī will marry Śiva:

āpa gayau tatakālahī svāṁga saradhā nāri
    dekhi maitrī mudita uṭhi milī saneha sambhāri
(PcN 4.5)

The actor playing Faith arrived on the scene. Friendship, happy, met her with love.

bhaī bikala abalā sakala dukhita dekhi girināri
    kari bilāpu rodati badati sutā sanehu sambhāri
(RcM 1.96)

Seeing the grief of the Mountain's wife, all the women grew distraught, while she sobbed and lamented, feeling deep love for her daughter.[85]

Here, Brajvāsī engages in re-use by omitting the word "giri" from the rhyme "nāri . . . sambhāri" since it does not fit within the context of the PcN as the girināri (Maina) is not among its characters. Additionally, he re-uses "sanehu" as "saneha" and, in the second pāda of the dohā, loosely follows the morphological sequence of the RcM's second pāda: one verb in the absolutive form ("dekhi" to "kari"), one noun ("maitrī" to "bilāpu"), two other verbs either in the imperfective participle or absolutive form ("mudita uṭhi" to "rodati badati"), and a term ending with a long vowel before the common ending part ("milī" to "sutā").

In the second instance, Brajvāsī re-uses an entire verse from the RcM almost verbatim. Towards the end of Act III, Peace discovers that her mother is alive and is relieved. In the RcM, the queens in Ayodhā are delighted to hear about Rāma breaking the bow and winning Sītā as his bride:

suni yaha khabara śānti haraṣānī
    manahu sikhini suni bārida bānī
(PcN 3.107)

Hearing the news, Peace rejoiced like a peahen hearing the sound [of] the clouds.

85 Lutgendorf in Tulsīdās 2016–2023, vol. 1 (2016): 199.

suni sandesu sakala haraṣānīṁ [. . .]
(RcM 1.294.1)

They rejoiced to hear the message [. . .][86]

prema praphullita rājahiṁ rānī.,
manahuṃ sikhini suni bārida banī
(RcM 1.294.2)

The queens were as excited with love as peahens hearing the rumble of rainclouds.[87]

In PcN 3.107, Brajvāsī connects the first pāda of RcM 1.294.1 with the second pāda of RcM 1.294.2. The element that does not correspond literally to the RcM is the sequence "yaha khabara śānti". However, in terms of syllabic instants, "śānti" corresponds to "sakala" (both have three) and "yaha khabara" corresponds to "saṃdesu" (five). Additionally, "khabara" and "sandesu" both mean "message" or "information".[88] It is thus evident that in these last two cases, re-use may be harder to spot. Nevertheless, it is important to acknowledge that the author of the PcN consciously decided to employ verses from the RcM to convey his message, even in instances where there does not appear to be a conceptual connection between the two.

Chapter Summary

By focusing on the main argument of this book, which asserts that early modern Vedānta sources incorporated vernacular texts like the RcM, this chapter has set the foundation for the next chapter, which will delve into an analysis of the philosophical and religious perspective of the PcN by extensively considering Tulsīdās's work. The purpose was to consolidate the argument that Brajvāsī composed his work with the RcM in mind and to introduce the concepts of adaptive reuse and simple re-use according to Brajvāsī's philosophical and religious understanding.

The chapter first situated the PcN within the history of the reception of the RcM and then delved into the RcM itself. These were necessary steps to identify the literary and conceptual elements of the RcM that could have appealed to Brajvāsī and enrich our understanding of adaptive reuse between the two texts and narra-

86 Lutgendorf in Tulsīdās 2016–2023, vol. 2 (2016): 219.
87 Lutgendorf in Tulsīdās 2016–2023, vol. 2 (2016): 219.
88 See HSS 1130 and 4903.

tives. The first section examined the existing knowledge regarding the influence of the RcM on later early modern works. Due to the limited available secondary literature on the subject, it is thus difficult to assess how and how far early modern Brajbhāṣā, Avadhī, and Khaṛī Bolī authors retold Tulsī's epic poem. However, the analysis revealed that the RcM held high status as a literary work and that some retellings, such as Banādās's *Ubhaya Prabhodaka Rāmāyaṇa*, associated it with the PC story. In the introduction to his work, the Rāmrasik poet drew parallels between the characters of the RcM and those of the PC, for example identifying Rāvaṇa with Bewilderment. The section then proposed that such overlapping might have originated from the RcM itself, where several personified concepts and events of the PC story are present.

The second section focused on one of these concepts – viveka – and presented arguments for the centrality of discrimination in Tulsī's text. The first subsection emphasised the role of viveka throughout the epic, particularly in its first, second, and seventh books or kāṇḍas, which mark the opening and closing of the story. It highlighted how viveka is an integral part of the moral and ethical universe depicted in the RcM, evident in its portrayal as one of Rāma's weapons against Rāvaṇa. Furthermore, Tulsī compares Rāma's reign to that of Discrimination. The second subsection delved into the significance of viveka as a defining quality of Bharata, Rāma's brother, within the epic. This subsection demonstrated that discrimination is crucial for Bharata as it enables him to behave appropriately both as a devoted follower of his brother-god and as a prince fulfilling his duties. Here, viveka becomes the point where bhakti (devotion) and dharma (duty) intersect. Bharata's way of being a bhakta, intimately bound to God, encompasses other elements beyond discrimination, as he is often likened to a yogi whose practice includes concentration and asceticism. Tulsī describes Bharata's path as one of nema (observance) and prema (love). These two terms are also employed by Brajvāsīdās when describing the figures of bhakta yogis in his PcN. Thus, Tulsī's treatment of viveka in the RcM provides a plausible explanation for why Brajvāsīdās drew inspiration from it for his own composition.

The final section of the chapter redirected the focus to the PcN, examining various instances of reuse and re-use of the RcM within the PcN, with a particular emphasis on their literary significance rather than philosophical or religious themes. These instances of RcM reuse contribute to the PcN's descriptions by incorporating references to Rāma's story. For example, the portrayal of Investigation-into-Reality (bastubicāra) is built upon the character of Hanumān, to whom he is explicitly compared. The cases of re-use are intriguing because the contexts in which these verses appear in their respective texts are entirely different. Brajvāsī's decision to re-use the RcM's text in these instances may have been primarily driven by convenience, but it also reflects the prestige associated with the RcM. People not only enjoyed reading Tulsī's work but cherished the experience of listening to it being recited or sung aloud.

The next chapter, starting from Brajvāsī's text, will establish that the connections between the two texts extend beyond literary similarities and delve into their shared philosophical and intellectual foundations. It will argue that the PcN was indeed influenced by the "expected" texts for an eighteenth-century Vaiṣṇava Vedāntin, while also emphasising that its most immediate source of inspiration was Tulsī's magnum opus.

4 The *Prabodhacandrodaya Nāṭaka*'s Vedānta and Bhakti

In the previous chapter, we explored the concept of viveka, particularly as embodied by Bharata, and how it enables us to uncover connections between the narratives of Tulsīdās's RcM and the PC kathā. The objective of this chapter is twofold: first, to demonstrate that the bhakta yogi represented by Rāma's brother serves as a viable model for Brajvāsīdās, and second, to illustrate the extent of adaptive reuse and simple re-use between the two texts, extending even to the Vedantic perspective of the PcN.

To accomplish this, section 4.1 will delve into the passages of the PcN that explore metaphysics and ontology. We will examine how the personified concepts, images, and metaphors typical of Vedantic discourse display the influence of various sources, including classical texts on Vedānta, the RcM, other regional-language works, the discourses of Vaiṣṇava Vedānta thinkers, and the broader intellectual trends prevalent in seventeenth- and eighteenth-century North Indian Vedānta. Section 4.2 focuses more closely on the path to freedom. The main argument is that the PcN presents a bhakti infused with yoga, as evidenced by the portrayal of bhakta yogis, the character of Bond-with-Viṣṇu, and Discrimination's ministers. Furthermore, it contends that the PcN strategically merges this yoga-infused bhakti with knowledge derived from the study of the Upaniṣads.

4.1 Ontology and Metaphysics

This section explores the Vedantic framework presented by Brajvāsīdās by analysing the characters and concepts of the Supreme Self (paramātama puruṣa), Material Nature/Illusion (prakṛti/māyā), the phenomenal world, and Individual Self (jīvātama puruṣa) and Mind (mana). We need to establish this aspect of the text first because, as Angelika Malinar aptly describes, "cosmology is intimately connected to religious practice as it provides maps and orientations in any religious quest, be it for liberation, empowerment or well-being in this life".[1] The passages in the PcN concerning these concepts and characters are primarily found in the first, fifth, and sixth acts, with occasional references in other acts throughout the play. Thus, our discussion will primarily focus on these acts, supplemented by insights and comparisons drawn from both Sanskrit sources and Tulsīdās's RcM.

1 Malinar 2011: 62.

https://doi.org/10.1515/9783111063836-005

Philosophical discourse in the PcN is expressed in narrative form, employing descriptions, similes, and metaphors embedded within dialogues among the characters.[2] These conversations occur between allies, such as Desire (kāma) and his wife Passion (rati) or Discrimination (viveka) and his first wife Good Reason (sumati) in Act I, for instance. While the interlocutor sometimes adopts a role akin to a pūrvapakṣin (opposing interlocutor),[3] I refer to these exchanges as dialogues (vāda) rather than disputes (vivāda) since the polemical aspect is not predominant, except for Discrimination's rejection of rival systems of thought in the final act. In Act VI, Upaniṣad recounts her ordeal to Individual Self and Discrimination. Discrimination's involvement in the discussion serves to interpret Upaniṣad's words, further consolidate them, and counter the arguments and doctrines of rival systems. His role in the last section of the play is clearly exegetical. It can be seen as part of the threefold method of śravaṇa (listening, hearing), manana (thinking, reflection), and nididhyāsana (deep concentration).[4] The other verses analysed in this section do not have a critical purpose per se but display an eulogistic and explanatory nature.[5]

4.1.1 Paramātama Puruṣa

Brajvāsīdās extensively explores the Highest Principle in his vision, the Supreme Self, the only character or concept solely discussed but never physically appearing on stage. However, at the end of the play, one could argue that the reverse occurs: the actors (along with the audience) exit the stage (and the courtly hall), symbolising their attainment (prāpata) of brahman.[6] This subsection aims to demonstrate that addressing the Ultimate Principle in the PcN is not limited to "paramātama puruṣa" alone. Moreover, this principle is often described both positively and through the *via negativa* method. Subsequently, we will examine its main fea-

2 Narrative philosophy operates through examples and illustrations, often without a strict one-to-one relationship with definitions. The discourse unfolds by presenting one example after another, as evident in PcN 5.94–5.95 and 5.102–5.105.

3 For example, in the early stages of the first act, Actress objects to the idea that a king can attain śāntarasa, arguing that vīra and śṛṅgāra and sneha (erotic/romantic and familial love) would be more appropriate. See PcN 1.47.

4 For a more detailed discussion, see section 4.2.

5 In philosophical literature, dialogues serve various functions, and in the sixth act of the PcN, they exhibit a clear asymmetry. These dialogues do not maintain a neutral point of view but instead favour Vedantic thinking. See Black 2015: 250–251.

6 PcN 6.161: bahurau svāṁga saṅga saba laye | nikasi sabhā te bāhira gaye || yāko bhāva ahai yaha abai | bhaye brahma ko prāpata sabai || (All [the people] exited the hall with the actors; they felt now that all had attained brahman.)

tures, including its light, pervasiveness, the question of its being the antarayāmī (inner controller) and sākṣī (witness), the notion of it possessing eka rasa (a single essence), its simultaneous nature as both nirguṇa (unqualified) and saguṇa (qualified), and briefly touch upon the avatāra (descent/embodiment) issue.

Paramātama's Appellatives

The term "puruṣa", accompanied by the tadbhava "paramātama" (Skt. paramātman), primarily refers to this reality, indicating the Supreme Self. The term paramātama, which also occurs alone, serves to distinguish it from the jīvātama puruṣa, the embodied individual self. Furthermore, the terms "brahman", "hari", "bhagavān", and "īśvara" are employed, with the latter two denoting a personal aspect of the Absolute.[7] Although preferred usage can be inferred from the context, no strict rules in the PcN definitively establish the subordination of one concept to another.[8] As for the characteristics of paramātama puruṣa, Brajvāsīdās presents them as follows:

jinake yaha haṃsā beda praśaṃsā puruṣa prasiddha prakāśamaī
abigata abināśī jagata prakāśī roma roma brahmāṇḍa kaī
sarvā abhirāmaṃ saba sukhadhāmaṃ vyāpaka paramānandā
jo amita anantā aja bhagavantā akala saccidānandā
(1.35)

Whose gander, praised [by] the Vedas, the celebrated Supreme Self, consists of light, inscrutable, indestructible, who manifests the world, [in his] every pore [there are] multiple universes.
Bestowing all happiness, abode [of] every joy, pervading, ultimate bliss, limitless, infinite, unborn Venerable One, without parts, existence, consciousness, and bliss.

jāko śruti gāvai kahata na āvai neti neti kahi hāra rahe [. . .]
(1.36)

7 Other terms used in the text include "daī" (2.137, 4.78, 5.8, 5.90), which is employed in a generic sense referring to deity or fate, and, in one instance, "jagadīśa", meaning "Lord of the World" (6.86). The term "īśa" appears in statements uttered by Skullman, the devotee of Bhairava (3.66, 3.73), suggesting it should be understood as one of Śiva's names. For more on "īśvara", see subsection 4.1.4.

8 The BhP also employs various appellatives. See Sheridan 1986: 23. This pattern can also be observed in Śaṅkara's usage of terms such as brahman, parabrahman, aparabrahman, īśvara, parameśvara, and paramātman. According to Paul Hacker, later Advaitins took care to define īśvara as brahman limited by māyā. See Hacker 1995: 85–96.

The revealed texts praise him, [yet], not being able to define [him], they say, defeated: "[He is] neither this, nor that."

The description here is a combination of adjectives in their negative forms (abigata, abināśī, amita, anantā, aja, akala), following the apophatic style of Vedantic canonical discourse.[9] Alongside these, there are also positive attributes (prasiddha, abhirāmaṃ, sukhadhāmaṃ, etc.). Puruṣa represents both the inexpressible reality of the Vedas and the repository of all happiness, the Venerable or Benevolent One (bhagavantā). Furthermore, compared with the other terms, puruṣa/brahman figures prominently in passages with cosmological significance, such as those above.[10]

The term Hari is ambivalent. It is sometimes used to denote an impersonal Absolute, explicitly identified with brahman.[11] Other times, it is employed in contexts that hint at a personal relationship with a devotee. In the latter sense, it is associated with bhagavān and, particularly, Viṣṇu.[12] The significance of Hari is evident in the full name given to the personification of bhakti, who is referred to as either viṣṇubhakti or haribhakti, Bond-with-Viṣṇu or Bond-with-Hari in my rendering. The cosmological dimension of puruṣa/brahman will be discussed first, followed by a detailed exploration of the personal aspect of the Absolute.

Brajvāsīdās's utilisation of negative descriptions for the Ultimate or Highest Reality indicates his incorporation of verbatim quotations from the RcM, thus showcasing his reuse of key verses and concepts. For example:

brahma jo vyāpaka biraja aja akala anīha anādi
nitya saccidānandaghana māyā pati sarbādi
(PcN 1.94)

That brahman, pervading, pure, unborn, without parts, desireless, beginningless, eternal receptacle of being, consciousness, and bliss, Illusion['s] husband, beginning of all.

brahma jo vyāpaka biraja aja akala anīha abheda
so ki deha dhari hoi nara jāhi na jānata veda
(RcM 1.50)

9 With reference to BU 2.3.6.

10 And, for example, in PcN 1.94 and 1.127.

11 For example, PcN 5.42: [Faith:] "Hari, receptacle [of] bliss, the ocean [that is] brahman, unmanifest, imperishable. [. . .] [When] they venerate the single brahman, being affectionate, they obtain Hari's feet, very fortunate."

12 See e.g. PcN 6.130 and 6.131.

That brahman, pervading, pure, unborn, without parts, desireless, undifferentiated –
Can it take on a body and become a man, whom the Vedas do not know?

The distinction between the two dohās lies in the second pāda and the rhyme scheme. However, both texts employ a similar strategy where the first pāda of each dohā extols an attributeless brahman, while the second pāda points to a brahman with positive attributes, depicting a personal aspect. The effect of the second pāda is achieved differently: Brajvāsīdās juxtaposes adjectives or adjectival forms, whereas the RcM presents an interrogative structure. In the PcN, the substitution of "anādi" for "abheda" could be attributed to the fact that in the RcM, "abheda" rhymes with "veda", and it seems essential to Tulsīdās to emphasise the absence of any mention of an embodied god in the Vedas. In contrast, this distinction did not hold the same importance for Brajvāsīdās at that particular point in the text.[13]

Paramātama's Light

A key attribute in Brajvāsīdās's philosophy is that puruṣa is composed of light (prakāśamaī, Skt. prakāśamaya) and provides light (prakāśī, Skt. prakāśin), as highlighted in verse 1.35.[14] The RcM characterises it similarly, emphasising that it is describable only through the *via negativa* (neti neti, 1.12) and that it consists of light:

[. . .] puruṣa prasiddha prakāsa nidhi pragaṭa parāvara nātha
(RcM 1.116)

The celebrated Self, receptacle [of] light, manifest, lord of the universe.

This feature appears in several other instances in the PcN.[15] Moreover, in the final act, brahman is portrayed as self-luminous (svaprakāśa, 6.144), another significant aspect in Vedantic discourse about the Absolute.[16] In verse 1.150, a revealing simile compares paramātama puruṣa to the sun:

13 It can be speculated that this was an opportunity for Tulsī to assert the possibility of discussing God in a language other than Sanskrit. However, it should not be disregarded that this might simply be a conventional expression of the indescribable nature of God or a reference to the lack of authority of the Vedas in this case, as they cannot support his statement about the embodiment of brahman.

14 The term "puruṣa" is frequently associated with light in the Upaniṣads. See e.g. *Chāndogya Upaniṣad* (CU) 3.13.7 and ŚvU 6.14.

15 PcN 1.127–1.128, 1.135, 1.138, 1.141–1.142, 4.31, 5.118, 6.101, and 6.126.

16 The concept of the "svaprakāśatā" of brahman and its meaning were extensively discussed among Advaita Vedāntins. There were variations in interpretation among scholars and disagreements with other philosophical schools. The Advaita Vedāntin Citsukha, which flourished in the thirteenth century, proposed eleven possible definitions for the term. See Timalsina 2009: 25–29.

ravi ko prakāśa jaise dekhiyata mukura madhya mukura ko prakāśa jaise jala ko
 ābhāsa hai
jala ke prakāśahū te hota jo prakāśa tau dekhyo parai mandira ke bhītara ujāsa hai
taise paramātamā teṁ ātamā bicāri lījai ātamā te mana mana teṁ jagata vilāsa hai
sākṣī paramātamā akhaṇḍita sabahī ke māhiṁ sabahī teṁ nyāro sadā ānanda kī
 rāśa hai

In the way the light of the sun is reflected [lit. "seen"] in a mirror, in the way the light of the mirror is the reflection in the water. Then one must consider the light coming from the light of the water, indeed, [as] the light in the temple. Think that, in that way, from the Supreme Self [comes] the Individual Self, from the Individual Self [comes] Mind, from Mind [comes] the brilliance [in] the universe. The observer, the Supreme Self, undivided, [is present] in all, [yet] separated from all, always a compact mass of bliss.

This quatrain is spoken by Discrimination as he recounts his lineage to his wife, Good Reason. The verse highlights the presence of the sun's light reflected in various surfaces and locations: a mirror, a body of water, and a temple. While the source of the brilliance is the sun itself, the key difference lies in the limitlessness of the sun's light compared to the confined nature of the reflected light upon different surfaces. This imagery of light emphasises a shared quality, albeit varying in degree, among Supreme Self, Individual Self, Mind, and the world.

The term "vilāsa" is not easy to understand or translate. It may evoke a sense of "sport", often associated with a deity's earthly activities. As we have observed, Brajvāsīdās's second work, the *Brajavilāsa*, deals precisely with Kṛṣṇa's life and deeds in the Braj region. However, in the current context, this does not apply. A previous verse that describes the nature of Mind helps elucidate the syntax between "jagata" and "vilāsa":

mana ko svarūpa
 baranyo anūpa
ati tejavanta
 mahimā ananta
jinhako prakāśa
 jaga meṁ vilāsa
 (1.135)

Mind's own nature is described [as] incomparable, extraordinarily radiant, [with] boundless greatness. His light [is] the brilliance within the universe.

Since the passages revolve around ontology, and light holds a central role in both, it is possible to interpret vilāsa as synonymous with light. With this interpretation in mind, the third pāda of verse 1.150 can be reinterpreted by including the implied vilāsa within it:

taise paramātamā teṁ ātamā [vilāsa] bicāri lījai ātamā te mana [vilāsa] mana teṁ jagata vilāsa hai

Think that, in that way, from the Supreme Self [comes the brilliance in] the Individual Self, from the Individual Self [comes the brilliance in] Mind, from Mind the brilliance [in] the universe.[17]

This particular imagery, among many others, has become a common depiction used to explain the nature of brahman as pure consciousness and its relationship with the individual self and the world.[18] Śaṅkara, the eighth-century Advaita Vedāntin, in his commentary on the BS (2.3.50), states that the individual self is an illusion, similar to the reflection (ābhāsa) of the sun in the water.[19] Partly in agreement with Śaṅkara, Rāmānuja (1017–1137), the Viśiṣṭādvaita Vedāntin, interprets this simile with a focus on difference. From Rāmānuja's perspective, the pleasure and pain experienced by the individual self do not affect the Supreme Self: the former is a quality of the latter, which means that they are not identical, just as the light is not the same as the sun.[20] On the other hand, Brajvāsīdās appears to emphasise *both* identity *and* difference by distinguishing the sun, the light of the sun, and the reflection, aligning with the viewpoint of the Gauḍīya thinker Jīva Gosvāmī (c. 1517–1608).[21]

Paramātama's Pervasiveness

In the PcN, the concept of paramātama puruṣa is initially portrayed as a cosmic body in a single line within a longer chanda, mentioning that "[in his] every pore

17 These verses can also be interpreted as hinting at the actual manifestation (vilāsa) of light/consciousness (prakāśa) if we go back to the etymology of vilāsa (MW 985).

18 An entire chapter of the *Pañcadaśī* by the Advaita Vedāntin Bhāratītīrtha/Vidyāraṇya is devoted to this topic. See Fort 2000: 502.

19 Śaṅkara 2009: 515. The sutra of 2.3.50 recites "ābhāsa eva ca" (And it is only a reflection). This viewpoint is also shared in the KPC (6.139). See Kapstein 2009: 269.

20 See 3.2.18–19 and 3.2.27 in Vireswarananda/Adidevananda 2003: 341–342, 346–347.

21 Jīva Gosvāmī introduced a fourth level of differentiation: the sun's rays. For an exploration of his usage and interpretation, see Gupta 2016a: 49–50. See also Gupta 2007: 42–43.

[there are] multiple universes" (1.35).[22] Another verse that extols this characteristic of the Highest Principle reuses the RcM:

jo hari abināśī saba ghaṭa bāsī svayaṁ bilāsī ādi ahai [. . .] (PcN 1.93)

That Hari, indestructible, dwelling in every body, enjoying himself, is the beginning.

jaya jaya abināsī saba ghaṭa bāsī byāpaka paramānaṃdā [. . .] (RcM 1.185.2)

Hail to the indestructible, dwelling in every body, all-pervading, supreme bliss.

The verse in the PcN follows a tribhaṅgī structure, with the second caesura occurring after "bāsī".[23] This aligns the first part of the verse with the RcM, where "jo hari" corresponds to "jaya jaya" in terms of syllabic instants (four). The immanence of the Supreme Principle is expressed through "saba ghaṭa bāsī" in both texts:[24] the brahman is compared to pervasive open space and human bodies to pots. The space inside a pot is "limited" on account of it being in a sense embodied; that is, enclosed and contained by the physical contours of the pot. In the same way, the Supreme Self is embodied within the bodies of individual selves. When the pot is broken, the "limited space" it once contained reunites with the indestructible, boundless, uncontained space. Similarly, when the body of the individual self ceases to be, the space it contained becomes one with the Supreme Self. When the pot is broken, the "limited space" once contained in a pot reunites with the indestructible, boundless, uncontained space.[25] "Ghaṭa" can also be interpreted as a water-jar, and Tulsī employs this image in another verse:[26]

---

22 roma roma brahmāṇḍa kaī ||. See also PcN 6.81. The concept of the "sacrificial body" of puruṣa is well known from hymn 10.90 of the *Ṛgveda*. It evolves in the Upaniṣads, where the idea of a Single Principle that is subsequently individuated is introduced. See Malinar 2018.

23 The tribhaṅgī is a metre consisting of thirty-two syllabic instants, with rhythmic regulation through three caesuras (yati), as the denomination of the metre (tri) indicates. The caesuras occur after the tenth, eighteenth, and twenty-sixth mora. In this case, the first caesura falls after "abināśī", the second after "bāsī", and the third after "bilāsī". For the prosody of the PcN, see subsection 1.2.4.

24 The primary meaning of the noun "ghaṭa" is indeed "jar, pitcher, large earthen water-jar, watering-pot". See MW 375.

25 The concept of space (ākāśa) is often associated with brahman. This association can be found in Gauḍapāda's *Kārikās*, his verse commentary to the *Māṇḍūkya Upaniṣad*. See Karmarkar 1953: 20–21 and e.g. in the introduction of the *Paramārthasāra* (1.24, 1.51). See Śaṅkara's commentary on BS 1.1.22, 3.14, and 3.41 in Śaṅkara 2009: 446–447. Also Jhaveri 1956: 300, 302, 304–305; Chakrabarti 1996; Divanji 1948.

26 See RcM 1.116.3–4. See also footnote 25 above.

sānuja mili pala mahu saba kāhū
kīnha dūri dukhu dāruna dāhū
yaha baṛi bāta rāma kai nāhīṁ
jimi ghaṭa koṭi eka rabi chāhīṁ
(2.243.2)

Along with his younger brother, he met everyone and quenched the intense burning [of their] pain.
This was not difficult for Rāma, [for] he was like the single sun reflected [in] a million water-jars.

I consider this an additional example of shared imagery between the PcN and the RcM. Although the specific context of this simile in the RcM differs from Brajvāsīdās's elaboration on the light of the sun reflected on various surfaces discussed earlier, it may still evoke the notion that Rāma is the undivided Self underlying the multitude of embodied selves and thus can influence their emotions effortlessly.

As we have observed, in PcN 1.93–1.94, brahman is described as pervading (vyāpaka). Another way to convey this characteristic is by defining him as antarayāmī (Skt. antaryāmin). This attribute, which can be found in the Upaniṣads,[27] has been widely adopted and adapted in subsequent texts and extensively discussed by proponents of Vedantic schools of thought.[28] "Antarayāmī" is employed only once, at the end of an extensive eulogy of the impersonal puruṣa manifesting as the personal deity Kṛṣṇa of Vṛndavāna (1.38). The term is employed to signify the Absolute rather than the embodied individual self.[29] Antarayāmī refers to the

27 BĀU 3.7.1–23. See Olivelle 1998: 85–89.

28 Among Vedantic philosophers, Rāmānuja held the aspect of the Absolute as antaryāmin in high regard. He explained the relationship between brahman as the ruler and the body as the ruled or controlled. According to Rāmānuja, the antaryāmin, present in the heart of the devotee, could even inspire devotion. See Carman 1974: 135–140, 186; Lipner 2018. In the Vallabhite interpretation, the antaryāmin, considered a manifestation of Kṛṣṇa in the jīva, is also referred to as adhideva, the inner god controlling the deities of the sense organs. See Smith 2016: 130, 134; Barz 1976: 76–78. According to Narain, Vallabha proposed a plurality of antaryāmins, one for each jīva. See Narain 2004: 100–101. The Gauḍīyas incorporated the concept of antaryāmin into their complex avatāra theory. Specifically, the three puruṣa-avatāras, with cosmogonic functions, each served as the antaryāmin for three different levels of existence: one for the totality of all universes, one for a single universe, and one within the bodies of embodied selves. See Holdrege 2015: 56–60.

29 See Malinar 2007: 114–116 for the BhG's use of antaryāmin. BĀU 3.7.1–23 discusses the ātman as the "immortal inner controller". In this context, a noticeable shift of emphasis in the PcN towards the Lord as antaryāmin, similar to Rāmānuja and the RcM, is clearly discernible.

"inner controller" or "inner guide", and this significance is expressed through the poignant simile of a puppeteer and a puppet show:

nācata pūtarī pekhane loṁ jaga ḍora nacāvanahāra ke hātha hai [. . .] (5.99)

[You] dance, as in a puppet show, [at] the hands of the puppeteer, [through] the strings [of] the world.[30]

jaise dāru pūtarī karakai
  nacai nacāvai bājī garakai
vyāpaka puruṣa akartā joī
  sūtradhāra sama jānahuṁ soī
kartā karma puruṣa jo ahaī
  putarī sama take basa rahaī
karma karata putarī saba dekhai
  sūtradhāra ko kou na pekhai
(6.87)

As with a wooden puppet: it dances, [but] it is the player who, after making it, makes it dance. The inactive Supreme Self, all-pervading; know him as the puppeteer. The active Individual Self stays in his power, like a puppet. Everyone can see the active puppet; no one can glimpse the puppeteer.

In verse 5.99, the goddess Sarasvatī addresses Mind, while in verse 6.87, Discrimination speaks to Individual Self. In both cases, the speaker explains to the interlocutor the latter's fundamental nature. The analogy of the puppet and the puppeteer vividly portrays the relationship between the active individual self and the inactive, all-pervading Supreme Self. Just as a wooden puppet dances under the control of its creator, the active self engages in actions while being guided by the underlying power of the Supreme Self. The active self is observable to everyone, while the puppeteer, representing the Supreme Self, remains unseen and elusive. Although these verses do not specifically mention a personal god (as in 1.38 with Kṛṣṇa), they emphasise the power of the Highest Principle as the one "who holds the strings", the puppeteer (sūtradhāra), or the one "who plays" (bājī), highlighting the control and

30 The word "ḍora" in 5.99 is synonymous with "sūtra". See HSS 1961. Interestingly, the fourth meaning provided by the HSS for the word is "avalamba/sahāra", which translates to "dependence, support, a prop" as well (MW 103). Interpreting "jaga" as "avalamba", a physical support, the sentence could also be rendered as: "[You; i.e. Mind] dance, as in a puppet show, [at] the hands of the puppeteer, [on] the stage [of] the world."

subordinate status of the individual self.[31] While the term antarayāmī is not present in these quoted verses, its presence and influence can be found in the RcM, which employs the same imagery of Rāma as the sūtradhāra who makes wooden puppets dance.[32]

Another term used to explain the quality of pervasiveness is sākṣī (Skt. sākṣin). This term holds significant meaning in Vedantic philosophies, and the PcN incorporates it in its treatment of the Supreme Self.[33] It appears three times: first in verse 1.150, which has been partially analysed above in relation to the luminosity and self-luminosity of puruṣa. The extended simile, spoken by Discrimination, concludes with praise for the undivided paramātama as the sākṣī, both immanent and transcendent (sabahī ke māhiṁ sabahī teṁ nyāro). The term is then used twice in the final act: once by the character portraying the Mīmāṃsāka Kumārila Bhaṭṭa, referred to as Kumārala in the Brajbhāṣā text and my translation (6.101), and immediately after, by Discrimination. Mind's son, enthralled by the philosopher's definition, praises the master of "the subtle points of Hermeneutics" (mīmāṃsā rahasya ko gāmī, 6.99), and further clarifies the meaning of Upaniṣad's statements (6.106).[34] While antarayāmī signifies pervasiveness in terms of mastery and control, sākṣī points to the presence of brahman in the individual self as an inactive consciousness or disinterested witness. This is conveyed by Discrimination through the classical simile of two birds on a tree:

31 The similes in PcN 5.99 and 6.87 touch upon the matter of agency, but it is not the primary issue. Sarasvatī and Discrimination aim to show Mind/Individual Self that he needs to alter his perspective of his own nature and realise his dependency on something higher. Rāmānuja was particularly vocal on the question of agency and defended the position that attributing an inner ruler to each embodied self did not negate the existence of "free will" for individuals. For further exploration, see Bartley 2002: 90–94.

32 RcM 1.104.3: sārada dārunāri sama svāmī | rāmu sūtradhara antarajāmī || jehi para kṛpā karahiṁ janu jānī | kabi ura ajira nacāvahiṁ bānī || (Sarasvatī is similar to a female puppet, o lord, and Rāma is the inner controller, who holds the strings. He greatly favours [whom] he knows [his] devotee, makes eloquence dance [in] the courtyard of a poet's heart.); RcM 4.10.4: umā dāru joṣita kī nāī | sabahi nacāvata rāmu gosāī || (O Umā, like wooden puppets – Rāma, the Master, makes everyone dance.)

33 It is used, for example, by Śaṅkara in the *Upadeśasāhasrī*, on which see Potter 1981: 233–235. A synthetic perspective on the views of several Advaita Vedānta thinkers can be found in Fort 1984. See also Gupta 1998: 17–32, 33–56. Madhusūdana Sarasvatī proposed a twofold understanding of the sākṣin, encompassing both metaphysical and epistemological aspects. See Gupta 2006: 92–98; Fort 1984: 283–285.

34 Halbfass 1983: 101n21.

yaha saṃsāra viṭapa kari bhākhā
  dvai khaga baiṭhe tākī śākhā
ika khaga vividhi bhāṁti phala khāī
  durabala rahata parantu sadāī
saṅkaṭa śoca vividhi so sahaī
  dukhita malīna dīna ati rahaī
tā samīpa khaga dūjo joī
  sākṣī rūpa lakhai saba soī
jo dukha sukha kachu bītata vāpai
  so saba lakhai na tāhi viyāpai
āpuna kachū khāya nahiṁ pīvai
  rahai tāsu ḍhiga tāhi na chīvai
yadyapi khāya piyai kachu nāhīṁ
  tadapi puṣṭa santuṣṭa sadāhīṁ
rahata prasanna parama sukha bhārā
  jānata hai jo jānanahāra

(6.106)

Talking [of] this cycle of existence [as] a tree, two birds sit on its branch. One bird eats many kinds of fruit, but it always stays weak. It bears several difficulties and griefs, sad, impure, miserable. The second bird, next to it, sees everything [with] the form [of] an observer. Sorrow and happiness happen to it [and] it observes them, [but they] do not affect [lit. "do not pervade"] it. It does not eat nor drink, stays next to it [i.e. the other bird], but without touching it. Even though it does not eat nor drink, it is always nourished and content. Delighted, full of supreme joy: one who knows, [just] knows!

The image of the two birds is one of the oldest and most reinterpreted motifs in Indic religious literature. Its earliest instance can be found in *Ṛgveda* 1.164.20,[35] followed by its appearance in the Upanishadic corpus, specifically *Śvetāśvatara Upaniṣad* (ŚvU) 4.6–5.7 and *Muṇḍaka Upaniṣad* (MU) 3.1–3.3, which use the same wording.[36] It is later mentioned in BhP 11.11.6–7. The description presented in the PcN aligns with the fundamental elements found in the aforementioned texts: two birds on the same tree or branch, one consuming and the other simply observing.

35 Scholarly discussions have extensively debated the possible meanings of these verses. See e.g. Norelius 2016; Johnson 1976. See also Goudriaan 1992.

36 For the text and translation of both Upaniṣads, see Olivelle 1998: 424–425, 448–449. For a slightly different translation, see Roebuck 2000. See also BĀU 3.4.2 for the concept of the seer (draṣṭā) who sees but is not seen.

There is a shared emphasis on the act of seeing or knowing ("sākṣī", "lakhai", and the final alliterative line, "jānata hai jo jānanahāra"). Additionally, similar to the BhP verses, Brajvāsīdās employs exclusive enclitic particles (-ī), adversative and concessive conjunctions (parantu, yadyapi . . . tadapi), and several negations (na tāhi viyāpai, kachū khāya nahiṁ pīvai, na chīvai, kachu nāhīṁ).[37]

However, there are several noteworthy differences and details to consider. First, it is important to clarify that the Sanskrit texts do not directly associate the term sākṣī/sākṣin with the image of the birds.[38] Second, in contrast to the Sanskrit texts, the birds in the PcN are depicted as perched on a specific tree branch, which can be understood to represent a particular embodied existence. There is no explicit mention in the PcN that the birds are friends. The only suggestion of a "similarity" is implied in the preceding verse. Verse 6.100 discusses two forms (anuhāra) of puruṣa, inactive and active, illustrated by the two birds. It is even possible that Brajvāsīdās, being familiar with the Sanskrit texts, replaced the term "sakhā" (friend) with "śākhā" (branch).[39] Verse 6.106 is somewhat uneven in its treatment, as more lines are dedicated to the sākṣī bird. Lastly, the topic of liberation or freedom from suffering is not addressed in this particular PcN verse.[40] This omission is likely due to the context in which Discrimination explains to Individual Self the meaning of Upaniṣad's declarations about brahman.

---

37 The BhP verses employ the terms "api", "na tu", and "tu" (twice): suparṇāv etau sadṛśau sakhāyau yadṛcchayaitau kṛtanīḍau ca vṛkṣe ekas tayoḥ khādati pippalānnam anyo niranno 'pi balena bhūyān; ātmānam anyaṁ ca sa veda vidvānapippalādo na tu pippalādaḥ yo'vidyayā yuk sa tu nityabaddho vidyāmayo yaḥ sa tu nityamuktaḥ. (Two birds, similar and friends, have made a nest on the same tree. One of them eats the fruit of the tree, the other does not eat, yet it is more powerful. The one not eating, omniscient, knows himself and the other as well, while the eater does not [know]. He is eternally trapped by ignorance, while the other, knowledgeable, is eternally free.)

38 The term or its compound form is present in the Upaniṣads, but it is challenging to determine what it refers to. See Gupta 1998: 19, 54. In Śaṅkara's commentary to MU 3.1.1, the term nityāsakṣin is used to refer to the Supreme Self as limited by māyā. See Gambhirananda 1965: 145.

39 I retain this as a conjecture since, despite the familiar imagery, I cannot find other parallels in the vocabulary used by Sanskrit texts and the PcN in this case. In his interpretation, Śaṅkara primarily associates the "same tree" mentioned in the Upaniṣads with the body due to "the identity of the place of their perception". See Gambhirananda 1965: 144. However, he also connects it with the upside-down banyan tree described in BhG 15.1, representing the phenomenal world. See Malinar 2007: 202–203.

40 Verses ŚvU 4.6–4.7 and MU 3.1–3.3 mention that when the deluded person, entangled in the cycle of existence, finally realises the greatness and contentment of the Lord, grief disappears. The BhP verses are part of a chapter where Kṛṣṇa explains to Uddhāva the characteristics of the liberated self and the bound self. In fact, BhP 11.11.7 explicitly mentions the "nitya-baddha" (eternally bound) and "nitya-mukta" (eternally liberated) selves.

Among these points, the notable absence of any mention of similarity between the two birds is significant. We have established that sākṣī represents the Supreme Self, in the sense of the inactive Self as a disinterested observer.[41] It is possible to establish a connection with the discussion about antarayāmī, which, as we have seen, qualifies Kṛṣṇa or the Absolute for Brajvāsī,[42] likening him to an immovable magnet propelling iron.[43] In both cases, the common denominator appears to be the potency of the Supreme Self and the dependence on it of the individual self. To my mind, this serves as part of the broader purpose of the teachings in the sixth act, setting the stage for the final instruction and realisation. These teachings seek to make Individual Self aware of who brahman truly is before realising his relationship with him.

The theme of the invisible presence and majesty of the Lord within embodied beings and the universe connects the concepts of antarayāmī and sākṣī to another crucial aspect that runs throughout the play, particularly in the sixth act. This thread revolves around another term used in the PcN to refer to the Ultimate Reality: Īśvara (sometimes spelled iśvara for metrical reasons). Interestingly, Īśvara only appears in the sixth act, in statements by Upaniṣad (6.139), Bhagavadgītā (6.133), Individual Self (6.136, 6.138), Discrimination (6.142), and Kumārala (6.99).

---

41 The concept of sākṣitva (witness consciousness) is a key feature of puruṣa in classical Sāṅkhya. See *Sāṅkhya Kārikā* 19 in Larson 1969: 261–262.

42 ŚvU 6.11–6.12 already refers to god (deva) as the inner self, observer, and controller of all beings. According to Gupta 2006: 92, the Advaita thinker Citsukha later explicitly identifies antaryāmin and sākṣin.

43 PcN 6.87: [Discrimination:] "The inactive Self is like a magnet, the active is like immovable iron, it does not move. Only if the magnet wills does it set it in motion. The power of the inactive is [present] in the active, that is why the active carries out actions. [. . .]." This simile is alluded to in BS 2.2.7, interpreted as illustrating the functioning of puruṣa (inactive consciousness) and pradhāna (active material cause). For Śaṅkara's critique of a Sāṅkhya interpretation of the sutra, see Śaṅkara 2009: 377–378. See also Larson 1969: 221–222. BhP 5.18.38 also uses this image to describe the way the Lord uses his power of material nature during creation. KPC 6.96 similarly employs the image in this context. See Kṛṣṇamiśra 2009: 257. For the Gauḍīya perspective on the BhP verse, see Dasa/Satyanarayana/Edelmann 2013. The YS *Bhāṣya* also alludes to this image when explaining the workings of the mind and the production of knowledge within it. The commentaries on YS 1.4 and 2.17 compare the citta to a magnet, as it functions simply by being present. Both commentaries emphasise that the citta is illuminated by the conscious puruṣa, indicating that the puruṣa is its master (svāmin). See Larson 2018: 124–128, 439–445. Conversely, in the commentary on YS 4.17, external objects are likened to a magnet since as soon as they are perceived, they affect the iron/citta. See Larson 2018: 914–915. By employing the image to describe the relationship between the inactive puruṣa and the active puruṣa, rather than between puruṣa and prakṛiti or pradhāna, the PcN thus appears to evoke its usage in the YS *Bhāṣya* and *Tattvavaiśāradī*. See also Ādiśeṣa's *Paramārthasāra* 1.11–1.12. Translation by Henry Danielson available in Ādiśeṣa 1980.

During their dialogue, Upaniṣad consistently addresses Individual Self as "deva".[44] Thus, when she concludes her narrative about her ordeal, Individual Self enquires about the identity of the Īśvara she has mentioned throughout (6.136). In response, Upaniṣad clarifies in verse 6.139:

jāni lehu aise yaha bāta
  puruṣa sanātana ke tuma tāta
tinateṁ tuma kachu nyāre nāhiṁ
  [. . .]
aise tuma Īśvara aho
  tāteṁ Īśvara tumako kaho

Understand this fact: you, dear, [belong] to the eternal Supreme Self. You are not separate from him [. . .]. In this way you are a god; hence I call you God!

It is unsurprising that we encounter these various appellatives side by side, as the BhP already presents such instances.[45] However, it is still worth reflecting on why Īśvara appears only in the play's final act. Similar to antarayāmī and sākṣī, I believe that the term Īśvara in the PcN signifies a specific concept, and the story provides some clues as to why it holds particular significance within the context established by Brajvāsīdās. In his exploration of the concept of a personal god in Indian thought, Jan Gonda illustrates that the Supreme Being is understood as possessing kingship and sovereignty. Īśvara emerged as a favoured term to denote a personal god, particularly starting from the Upanishadic corpus.[46] The concept was further elaborated in the BhG, where Kṛṣṇa's aiśvarya, or "lordship", lies in his role as a "*yogin*, who controls the powers of nature (*prakṛti*), yet remains detached from them".[47] The inclusion of Īśvara in Act VI of the PcN, in the verses where the identity with Supreme Self is unveiled to Individual Self "himself", may hold significance due to the specificities of the salvific path depicted in the drama. Although this will be discussed in greater detail later in the chapter, it is worth noting that in the PcN, Individual Self must recognise his true nature and es-

44 In fact, this is the term most frequently used by Upaniṣad when addressing jīvātama puruṣa (6.71, 6.75, 6.77, 6.79). Other appellatives include "bhagavāna" (6.71), "svāmī" (6.62), and "jagataguru" (6.71).

45 In these passages, another term employed is jagadīśa (6.86), meaning "Lord of the World", as noted above in footnote 7 of this chapter. This term was a favoured appellative for Kṛṣṇa in Jayadeva's twelfth-century poem *Gītagovinda*. In that context, the "dual divinity" of Kṛṣṇa paired with Rādhā was considered superior to Viṣṇu, and the term bhagavān, denoting "Ultimate Reality", was absent. See Stoler Miller 1982: 19.

46 Gonda 1968, 1969.

47 Malinar 2007: 146, referring to BhG 9.5.

tablish a relationship with Supreme Self. Furthermore, he must exercise control over his material nature and embody the qualities of a sovereign ruler: to exist within the world without being subject to its dominion but instead governing it (1.50).

Paramātama's Single Essence

One notable attribute of the paramātama puruṣa is that he is described as possessing or embodying eka rasa, which I have translated as "single essence".[48] Mind is encouraged to contemplate this single essence (5.108), while Peace is said to be absorbed in one essence (5.117). Science-of-Reasoning is criticised for being sapless, ignoring the taste (svāda) of the essence (rasa) (6.114). But what does this eka rasa, or single essence, refer to? If we examine the BhP, it does mention the rasa of brahman (4.4.15), Kṛṣṇa as an ocean of rasa of eternal bliss (7.7.45), as the bestower of rasa (10.42.1), and as possessing all rasa (10.87.34). The devotees are frequently addressed as knowers of rasa (rasajña, 3.15.48, 3.20.6, 4.31.21).[49] This aspect of brahman is also touched on in the RcM, but the significance of these references is less clear.[50] However, the PcN offers additional clues: eka rasa is samatā (literally "sameness", which I have interpreted as equanimity, 6.5) and ānanda (bliss, 6.93); there are no other sentiments (bhāva, 6.93). Furthermore, it is stated that Bond-with-Viṣṇu (viṣṇubhakti) is the rasa in the Blessed Lord (6.147) and consists of rasa (rasamaī, 6.158).

This, of course, is not a comprehensive elaboration of how our author perceives the connection between rasa and bhakti and between bhakti and brahman or bhagavān. However, these elements echo the concepts developed by the Advaitin Madhusūdana Sarasvatī and the Gauḍīya Gosvāmīs in their works during the sixteenth century. While it was not entirely novel to consider brahman as bliss in the Advaita milieu by that time,[51] Madhusūdana, a Bengali residing in Benares and a contemporary of Tulsīdās, aimed to establish bhakti as the highest goal of human life and the

48 See PcN 5.118–5.119, 6.22, 6.81, 6.97, and 6.101.

49 An earlier reference to brahman as rasa is found in *Taittirīya Upaniṣad* 2.7.

50 RcM 3.39a, 6.109.3, 7.29.5, and 7.77.3. The Gita Press English translation of these verses renders eka rasa as "unchangeable". While there might be a correlation between samatā and stability or immutability, I believe the samatā of the PcN more closely recalls its yogic soteriological path. Therefore, I find the translation "unchangeable" inappropriate in the case of the PcN.

51 On the concept of ānanda in the Vedas and Upaniṣads, see Buitenen 1979; Olivelle 1997. There has been extensive scholarly debate regarding Śaṅkara's consideration of ānanda as a constituent of the nature of brahman (along with sat and cit). Later Advaitins accepted this view. See Potter 1981: 74–76; Hacker 1995: 86–87. On Śaṅkara's position and contrasting opinions among scholars, see Myers 1998; Karmarkar 1919–2020: 108. See also Venkatkrishnan 2018.

supreme rasa. Like the Gosvāmīs, Madhusūdana initially identified bhakti with the highest form of ānanda and subsequently accorded it ontological status. According to Lance Nelson, Madhusūdana argued in his *Bhaktirasāyana* that since the object of bhakti is the Lord, and bhakti is a vṛtti (modification) of the mind, the vṛtti assumes the form of the Lord. Therefore, the form of the Lord serves as the basis for bhakti as rasa.[52] The Lord is the prototype for bhakti, which, in turn, is its reflection in the mind. Since the prototype and reflection are identical, bhakti is, at least implicitly, identified with bhagavān. Madhusūdana seemingly felt constrained by his Advaita commitment, which prevented him from identifying bhakti as a śakti (power) of the Absolute. However, this was not a limitation for the Gauḍīyas.[53]

Therefore, by defining brahman as a single rasa, with this sentiment or essence being bhakti or ānanda, Brajvāsīdās reflects the discussions surrounding aesthetic and religious experiences in Vaiṣṇava Vedānta contexts.[54] Brajvāsīdās's contemporary, the Gauḍīya Baladeva Vidyābhūṣaṇa (170?–1793), also defined the Supreme Brahman as eka rasa[55] in his *Govindabhāṣya* (1.1.1), one of his commentaries on the BS, as well as in his *Tattvadīpikā*.[56] The position presented in the PcN, which identifies bhakti with rasa and rasa with ānanda, the essence of the Lord, seems to align more closely with Madhusūdana's stance.[57] At the same time, it is plausible that the Gauḍīyas' perspective may have exerted some influence in this regard. In the PcN, there is also an implicit identification of bhakti with the Supreme Self, illustrated by Bond-with-Viṣṇu (viṣṇubhakti) manifesting to rescue Faith and Duty (śraddhā and dharma) from the assault of Bhairava's-Science (bhairavī vidyā), Skullman's daughter (4.13–4.15). This resonates with the description in verse 2.96, where parabrahman assumes a form and appears on earth to alleviate the burden of evil. The association of the single rasa with samatā (equanimity) will be further explored in the next section of this chapter since it is more pertinent to discuss it in the context of the PcN's perspective on salvation and bhakti.

---

52 On this point, see also Gupta 2006: 127–131.

53 For a comprehensive study that compares two positions, see Nelson 2004.

54 In the realm of literary theory, Bhoja had proposed in his eleventh-century *Śṛṅgāra Prakāśa* that a single rasa, śṛṅgāra, is paramount as it serves as the source of all other rasas. Here, śṛṅgāra should be understood not as one of the conventional eight rasas but, in Sheldon Pollock's words, as the "capacity for emotional intensity as such". See Pollock 1998: 126. Possibly inspired by Bhoja, the sixteenth-century Gauḍīya Kavikarṇapūra maintained in his work that a single primary rasa exists but, for him, it possesses the characteristic of bliss, from which the other rasas appear as variations. See Lutjeharms 2016: 192.

55 Translated by Kiyokazu Okita as "one nature". See Okita 2014: 146.

56 For references to the relevant verses, see Okita 2016: 161n21.

57 The Vallabhites were critical of Madhusūdana's perspective in this regard. See Nelson 2004: 371, 377–378.

Paramātama: Nirguṇa and Saguṇa

A crucial aspect revolves around the nature of puruṣa, where the PcN closely follows the RcM. While the BhP had already presented a similar viewpoint,[58] the PcN adaptively reuses the RcM in this instance as well:

[. . .] jo aguṇa anūpā bahu guṇa rūpā adbhuta rūpa apārā (PcN 1.36)

Who [is] without qualities, [and yet] a unique form with many qualities, a wonderful beauty, infinite.

[. . .] aguna anūpama guna nidhāna so (RcM 1.18.1)

Without qualities, [yet] incomparable treasury of qualities.[59]

The PcN verse is part of a longer chanda, with each pāda containing up to thirty syllabic instants, while the RcM verse is a caupāī with its usual sixteen instants. Consequently, Brajvāsīdās had more room to delve into the paradoxical nature of the Absolute. The PcN reuses the phrase "aguna anūpama" from the RcM but condenses it to "anūpā", amplifying the hyperbolic effect of the "guṇa" by adding "bahu". It replaces "nidhāna" with "rūpa", a key term in its metaphysical framework, to the extent that it is reiterated within the same verse, as seen in this case.

Another chanda establishes intertextual connections not only with the RcM (1.117.3–4) but also with the ŚvU (3.19)[60] and the *Viṣṇu Purāṇa* (5.1):

kou sakai na baranī adbhuta karanī binu pada pāna gauna grāhī
bina tvaka sparasī bina dṛga daraśī nāsā binā ghrāna cāhī
jo binu śruti sunaī bina mana gunaī binu jihva ko svāda grahai
jo binu bapu sumdara acaraja mamdira bina bānī bahu bacana kahai
(PcN 1.37)

58 This conceptualisation of nirguṇa/saguṇa is parallel to that found in the BhP, where, as Sheridan maintains, aguṇa or nirguṇa does not imply any deficiency in the Absolute but rather denotes his transcendence over the ordinary guṇas with which the created world is endowed. For references to the BhP and further explanations, see Sheridan 1986: 24–28.

59 See also RcM 7.12.1: jaya saguna nirguna rūpa rūpa anūpa bhūpa siromane | [. . .] (Victory to the crest-jewel [of] kings, whose beauty [is] incomparable, whose form [is] full of qualities and without qualities [at the same time].)

60 The theologically oriented perspective of the ŚvU renders it a significant text for Brajvāsīdās, who also uses other similes found within it, as we will see. The text portrays Rudra/Śiva as the sole deity and emphasises the significance of the god's prasāda as a means to attain salvation. For a summary of its content and a discussion on textual criticism and interpretation, see Cohen 2008: 213–245. According to Patrick Olivelle, its vocabulary resembles that of the BhG. See Olivelle 1998: 413. The Śaiva orientation of the text has been a subject of debate, as discussed in Cohen 2008: 244–245. For an initial exploration of the Viśiṣṭādvaita interpretation of certain verses in the ŚvU, see Stark 1992.

No one could describe [his] incredible actions! Without feet and hands, he moves and grasps.
Without skin, he touches; without eyes, he sees; without [a] nose, he smells. Without
ears, he listens; without mind, [he] ponders; without [a] tongue, [he] tastes.
Without a body, he is still beautiful, an abode of wonder; without a voice, he utters many words.

In the PcN verse quoted above, the First Actor responds to his wife's enquiry regarding the identity of the ākāśavāṇī who had instructed him to arrange the performance of the drama (1.33). The comparable RcM description constitutes a teaching by Śiva to Pārvatī, where the latter enquires why the nirguṇa brahman (unqualified brahman) descended to earth, acquiring qualities (1.109.2; becoming saguṇa). On the other hand, the third chapter of the ŚvU containing the similar passage identifies Rudra/Śiva with the cosmic puruṣa.[61] The PcN and the RcM construct the passage in a similar manner to the ŚvU, presenting their teachings as responses to questions.[62] However, while the ŚvU elaborates on the concept of puruṣa as a cosmic body containing other bodies, the PcN and the RcM focus on the enigmatic relationship between nirguṇa and saguṇa. At the conclusion of the passage, they each contribute their own perspective: in the PcN, puruṣa descends to earth as Kṛṣṇa, while in the RcM, it is Rāma (1.118).

On this theme, there is another example of adaptive reuse from Act V of the RcM, where Sarasvatī explains to Mind how to cultivate inner Peace by concentrating on brahman:

aguna saguna dui brahma sarūpā
akatha agādha anādi anūpā
(RcM 1.22.1)

Without properties, with properties – two [are] brahman's own forms,
inexpressible, unfathomable, beginningless, unique.

61 The ŚvU has: "He moves swiftly, but he has no feet; he grasps, but he has no hands; he sees, but he has no eyes; he hears, but he has no ears." Translation in Olivelle 1998: 423. In the verse in *Viṣṇu Purāṇa* 5.1.41, the context is still different as it is part of a lengthy praise to Viṣṇu/Hari by Brahmā: "Thou hearest without ears, and seest without eyes. Thou art one and multiform. Thou movest without feet; thou seizest without hands." The verse appears translated as 5.1.41 in Wilson 1868: 253.
62 In the ŚvU, it appears: "What is the cause of *brahman*? Why were we born? By what do we live? On what are we established? Governed by whom, O you who know *brahman*, do we live in pleasure and in pain, each in our respective situation? Should we regard it as time, as inherent nature, as necessity, as chance, as the elements, as the source of birth, or as the Person? Or is it a combination of these?" Translated by Olivelle 1998: 415.

Brajvāsīdās reuses Tulsīdās:

brahma ju vyāpaka eka rasa alakha anādi arūpa
aguna saguna dui rūpa so alakha agādhi anūpa
(PcN 5.119)

Brahman, pervading, a single essence, without distinguishing marks, beginningless, formless.
He [has] two forms: without qualities and with qualities – he [is] invisible, unfathomable, unique.

The PcN verse is a dohā, with each pāda consisting of twenty-four syllabic instants, while the RcM verse is a caupāī, with each pāda containing sixteen syllabic instants. Consequently, the pādas in the PcN verse contain more elements than those in the RcM. One notable observation is the reuse of the rhyme "rūpa . . . anūpa", and other terms reused by the PcN include "anādi" and "agādhi" (replacing "agādha" in the RcM). In the second pāda of the PcN verse, "aguna saguna dui rūpa so" is a reuse of "aguna saguna dui brahma sarūpā" in the RcM. The PcN omits "brahma" since it is mentioned in the first pāda. As the PcN verse is longer, it requires a syntactical connection between "aguna saguna dui rūpa" and the subsequent adjectives "alakha agādhi anūpa". It achieves this by using "rūpā" (form) instead of "sarūpā" (a tadbhava of "svarūpā" or "own form"), conveying the same meaning with less emphasis, and changing the prefix sa- to the pronoun "so". Both "aguna saguna dui rūpa so" in the PcN and "aguna saguna dui brahma sarūpā" (without the "brahma") in the RcM consist of thirteen syllabic instants.[63] Therefore, like the RcM, Brajvāsīdās conceives of the nirguṇa and saguṇa as two forms of brahman.

In explaining why the nirguṇa brahman assumes a form (rūpa dhar-), becoming saguṇa, the PcN presents two concurrent goals. The first one is found in Act II:

parabrahma jo hai jagata ādi kārana
dhare rūpa jaise mahī bhāra ṭārana
[. . .] kiye aurahū je carita bhāva jaise
bhayo pai kāraṇa tahāṁ nāma taise [. . .]
(PcN 2.96)

The Supreme Brahman, the primaeval cause of the universe, assumes a form in order to remove the weight [of evil people/wickedness] from the earth. [. . .] All the deeds he carried out according to the sentiment [evoked], indeed, [his] name was [acquired] in such a way, according to [their] cause.

63 In the PcN, Brajvāsīdās omits "akatha" and repeats "alakha". See footnote 110 of the annotated translation (Act V).

This first goal of assuming a qualified form aligns with the classical explanation for a god's manifestation, as we find it, for example, in BhG 4.7–4.8.[64] Rāmānuja, in his commentary on the introduction of the BhG, concurred with this account of the descent, along with the second account, which states that the Supreme Being embodies a qualified form to rescue the good. However, for Rāmānuja, and indeed for his own commentator Vedāntadeśika, relieving the earth of the burden of evildoers was a pretext for a more important goal. In their view, the accessibility of the Lord to his devotees was chief among his positive qualities.[65] And indeed, it seems that Brajvāsīdās expresses a similar sentiment regarding the second goal:

nirguna nirākāra nirdoṣa joī
  bhagata prema basa teṁ saguna hota soī
binā rūpa nirguna hiye meṁ na āvai
  su tāteṁ saguna ke carana citta lāvai
(PcN 5.120)

Unqualified, formless, faultless, he, indeed, out of the strength of the love [for/of] those bound to him (bhagata), is always qualified. He does not enter the heart without form, unqualified. Therefore, one (the bhagata) concentrates on the feet of the qualified [form].

The first and second pādas of the verse reuse the RcM:

aguna arupa alakha aja joī
  bhagata prema basa saguna so hoī
(RcM 1.115.1)

Unqualified, formless, invisible, unborn, he, indeed, [out of] the strength [of] love [for/of] those bound to him, is full of qualities.

The PcN verse maintains the rhyme "joī . . . hoī" as "joī . . . soī" with the initial series of negative adjectives in the first pāda. However, in this instance, Brajvāsīdās chooses to use the prefix nir- (nirguṇa) instead of the a- (aguṇa) favoured in

64 On how removing evil from the earth is integral to the figure of the king in the BhG, see Malinar 2007: 98.

65 See Clooney 2007: 332–335. For further philosophical and theological analysis of these passages, along with a comparison to Śaṅkara's commentary, see Ram-Prasad 2013: 60–66. Additionally, Carman 1974: 179–186 discusses how Rāmānuja developed the Pāñcarātra doctrine of vyūha and further expressed his views in his *Vedārthasaṅgraha,* which introduces the concept of the Lord's descent (usually called "vibhava") for his divine play (līlā). Furthermore, the Śrīvaiṣṇavas place importance on the manifestation of the Lord in a worshippable form (arcāvatāra). See Narayanan 1996.

the RcM. The bhujaṅgaprayāta is traditionally considered a varṇa vṛtta based on the number and order of syllables.[66] However, Brajvāsī deviates from the standard twelve-syllable scheme for each pāda. The pādas of this bhujaṅgaprayāta contain between twelve and sixteen syllabic instants. If we look at the order of syllables, the only sequence respected is at the end of each pāda, where we find a short syllable followed by two long syllables. Interestingly, if instead we count the syllabic instants, we find that all the pādas in this bhujaṅgaprayāta contain twenty syllabic instants, with the exception of the first one, which contains nineteen. The substitution of the a- with nir- enables Brajvāsī to give double syllabic value to the vowel of nir-, which is necessary to achieve the required nineteen syllabic instants. The bhujaṅgaprayāta also allows Brajvāsī to add the postposition "teṁ" in the second pāda. By including "teṁ", it becomes clear that "basa" is to be considered a noun. In contrast, in the RcM, the absence of "teṁ" leaves open the possibility of "basa" being an adjective.[67] Hence, the second pāda of the RcM could also be translated as "he, indeed, dependent [on] the love [for/of] those bound to him, is full of qualities".[68]

However, what remains unresolved in both texts is the connection between bhagata and prema: does the nirguṇa brahman appear in his saguṇa form out of his own love *for* the bhaktas or because of the love *of* the bhaktas towards him? Both the PcN 5.123 and RcM 1.198 emphasise the willingness of the Supreme Being to descend to earth for the sake of bhakti.[69]

Paramātama and Avatāra

Regarding embodiment, verse 2.96 of the PcN serves as the first and only statement of the canonical function of an avatāra as "removing the weight" of evil from the world.[70] This verse occurs in the context of the Materialist conveying Kaliyuga's message to Bewilderment, which introduces the character of Bond-with-Viṣṇu. The second statement of purpose, namely accessibility/knowability, is repeated

66 See subsection 1.2.4 and Table 4.

67 See HSS 4394.

68 Lutgendorf translates it as "by the power of the devotees' love". See Tulsīdās 2016–2023, vol. 2 (2016): 239.

69 PcN 1.39 states, in the words of First Actor, that puruṣa has taken an embodied form "[with] the desire [of] benefiting devotees" (janahita pūraṇa kāma). This is also present in the RcM in verses 1.23.1, 1.50.5, and 7.72.

70 It is the idea of the bhārāvatāraṇa, originally not exclusive to Viṣṇu, that, according to Paul Hacker, gradually came to be equated with the descent of a god. Thus, the term avatāra came to combine both the descent of a god and the removal of evil from the world. See Hacker 1960; Hardy 1983: 23–25; Biardeau 1978.

three times: in 5.120 (bhagata prema basa), in 5.123 (bhakti hita), and in 1.38 (jana hita). Consequently, it appears that Brajvāsī finds the second aim more interesting.[71]

Brajvāsī expresses embodiment through the conjunct verb rūpa dhar-, meaning "to assume, take up a form, a shape", and never employs avatāra or its derivatives. In contrast, the RcM uses the verb avatar- (1.50.5) in addition to tanu dhar-, and saguṇa ho-, "to assume qualities".[72] It is difficult to determine why the PcN does not explicitly use the term avatāra or its related forms. Freda Matchett astutely analyses the figure of Kṛṣṇa in three influential texts on Vaiṣṇava thought: the *Harivaṃśa*,[73] the *Viṣṇu Purāṇa*, and the BhP. In her chapter on the *Viṣṇu Purāṇa*, she notes that the text prefers the term rūpa over forms of the verb ava + √tṝ when describing Viṣṇu's manifestations in the cosmos. In contrast, it employs aṃśa to refer to his appearances in the human world, such as in the form of Kṛṣṇa, with whom the concept of avatāra is connected in the *Viṣṇu Purāṇa*. Matchett argues that rūpa relates to the act of seeing and serves as an appropriate way to express the fact that Viṣṇu's only action is to make himself "perceptible" to his devotees, who are the ones "acting", like Prahlāda and Dhruva. The term avatāra specifically pertains to Kṛṣṇa, who is Viṣṇu as the actor on the stage of the world.[74] However, this analysis does not seem to apply in the case of the PcN, where the use of rūpa dhar- is associated with parabrahman manifesting on earth as Kṛṣṇa or Rāma, two visible forms of the Supreme Being especially remembered and revered for their deeds (caritas).

If we examine the BhP, we find various overlapping reasons for the descent of the Supreme Being on earth.[75] Unlike the *Harivaṃśa* and the *Viṣṇu Purāṇa*, the BhP develops the idea that the god descends to liberate people from the cycle of rebirth (saṃsāra), offering them the possibility of mokṣa (freedom from rebirth, liberation) and emphasising bhakti as superior to mokṣa: "The idea that the *avatāra* comes to make liberation available is strong in the *Bhāgavata*."[76] This implies that even the enemies killed by Kṛṣṇa attain liberation. Rāmānuja also supported

---

71 The notion of accessibility through a saguṇa form for devotion or philosophical contemplation was also theorised by Vallabha, who emphasised above all the Lord's will to be perceptible. See Narain 2004: 82–83.

72 See RcM 1.23.1, 7.72, 1.115.1.

73 The *Harivaṃśa*, the appendix of the *Mahābhārata* that gives prominence to Kṛṣṇa's story, presents a complex introduction where two reasons for Viṣṇu's descent as Kṛṣṇa can be discerned. The first reason is his exploits as a child, adolescent, and young man in Braj. The second reason, which connects the *Harivaṃśa* with the epic, is his involvement in the conflict between the Pāṇḍavas and Kauravas, who aimed to free the earth from its burden of people. See Viethsen 2009.

74 For further insights and references to relative passages, see Matchett 2008: 89–93, 102–106. See also Corcoran 1995.

75 See Matchett 2008: 161.

76 Matchett 2008: 169.

the subordination of the purpose of removing the weight of evil to the presence of the deity for the bhaktas, although he appears to be unaware of the BhP.[77]

This conceptual shift, privileging the presence of the deity in the lives of its devotees, is further highlighted by Tulsī. RcM 1.50.5 states that Rāma descended for the sake of the bhaktas (avatareu apane bhagata hita). In the RcM, Rāma does indeed slay rākṣasas, most notably Rāvaṇa, but the framework of bhakti underlies these actions. Similarly, PcN 2.96, quoted above, is significantly framed by the presentation of Bond-with-Viṣṇu. Consequently, the BhP's perspective appears to influence both the RcM and the PcN. Returning to the distinction made by Matchett, we can say that the rūpa coincides with avatāra. There is no distinction here between a god who presents themselves to the bhakta for the bhakta to act and a god who acts in the world for and with the bhakta.

In this case, Hacker's contention that the terminology changed from rūpa to prādurbhāva to avatāra to differentiate the descent of the god from human birth does not hold. This is not to deny the validity of his thesis but to suggest that, from a historical point of view, since the idea of Viṣṇu/Kṛṣṇa's descent had become firmly established in early modern times, it was no longer a matter of terminology expressing the type of "birth". Instead, in the case of the PcN, a drama, it seems much more relevant that rūpa dhar- means "assuming a disguise" and highlights the aspect of the god as an "actor" (cf. BhP 10.18.11, 21.5–21.8, and 23.22) than the perspective of what kind of body the god assumes or whether or not he is subject to karman. In fact, these latter topics are not discussed in the PcN at all.[78]

But what rūpa does Brajvāsīdās favour? He shows favour towards Kṛṣṇa, referring to Rāma and the Rāma story in several passages.[79] He mentions Kṛṣṇa as Rādhikā's beloved in his prologue, before the proper beginning of the play (rādhikā pī, 1.10), and again at the very end with the term brajabāsīdāsa, that is, servants of the one who dwells in the Braj region, that being Kṛṣṇa, the brajabāsī par excellence (6.168):

jo cāhai ānanda sadā re brajabāsīdāsa[80]
  tau kīje hari bhakti ko choḍi viṣai kī āsa

77 Rāmānuja's commentaries and treatises encompass the BS, the BhG, and the *Viṣṇu Purāṇa*. See Carman 1974: 49–64. Additionally, other works are attributed to him, including three prose texts that explore the theme of seeking refuge in Nārāyaṇa and Śrī, as well as the *Nityam*, a manual outlining the daily practice of an individual who has surrendered to God. For further insights into these often overlooked works, see Clooney 2017–2018, 2021.

78 See Hacker 1960 for more information. References to God as an actor can be found in sources such as the BhP at 10.18.11, 21.5–21.8, and 23.22.

79 See section 3.3 for references to other characters and accidents in the Rāma story within the PcN.

80 Here, the interjection of "re" signals that Brajvāsī is calling the attention of his intended audience, the servants (dasa) of the Brajabasi (Krishna).

O servants of he who dwells in Braj (Kṛṣṇa), if you wish for eternal bliss,
then forsake desire for the objects of the senses and share a bond with Hari.

As we observed in Chapter 1, Brajvāsīdās plays with the meaning of brajabāsīdāsa, leading us to question whether he is addressing himself or his fellow bhaktas. In verse 6.167, I interpreted brajabāsīdāsa as his signature (bhaṇita), rendering "kaha brajabāsīdāsa" as "Brajvāsīdās says".[81] By contrast, in verse 6.168, I chose to translate brajabāsīdāsa as "the servants of he who dwells in Braj (Kṛṣṇa)". Both translations of the term brajabāsīdāsa, as the author of the PcN and the servants of Kṛṣṇa, are valid. I have privileged the "servants of Kṛṣṇa" in the translation of verse 6.168 because, to my mind, it makes sense to associate the ending of the drama with its opening and with its explicit appeal to the community of bhaktas.

As for further references to Kṛṣṇa, when the PcN's story begins, the First Actor praises the puruṣa and mentions that for the sake of people (or his devotees, jana), he assumes the saguṇa form of Nanda's son, the lover of fresh butter (navanīta piyāro), the flute player, and the rāsa dancer coupled with Rādhā, in his capital (rajadhānī) Vṛndavāna.[82] Then, in 1.52, he mentions him as Murāri. After that, in 1.126, two episodes from Kṛṣṇa's story are mentioned. Discrimination's first wife, Good Reason, comforts him by saying that wicked people do not hesitate to misjudge those who are innocent, as in the case of the god Kṛṣṇa and the jewel. I interpret this as a reference to the story of the Śyamantaka jewel. She further adds that wicked people do not pay attention to good advice or teachings, and Hari and Duryodhana are witnesses to this fact. This reference is more difficult to pinpoint precisely, but it most likely pertains to the narrative of the *Mahābhārata*.[83]

The other passages involving Kṛṣṇa are connected to contexts that in some way deal with bhakti, its practices, and bhakta yogis. In 2.21, a group of renunciants or bairāgī (Skt. vairāgin) is described as being absorbed in various activities, among them repeating Kṛṣṇa's names. In 3.5 as well, the sants are said to have affection for Rāma (and) Kṛṣṇa. In the previously mentioned 2.96, the Supreme Brahman assumes the form of either Rāma or Kṛṣṇa, and Bond-with-Viṣṇu creates a mental image of him and concentrates on it. The verses dedicated to Rāma mention his being Daśaratha's son (mahāraja dasaratha suvana), the breaking of the bow

81 PcN 6.167: bhakti hoya bhagavanta kī aura viveka prakāsa | bhakti binā tribhuvana dukhī kaha brajavāsīdāsa || [. . .] (There will be a bond with the Blessed Lord and the effulgence of discrimination. | Brajvāsīdās says: "Without a bond with god the three worlds are distressed || [. . .]".) See also subsection 1.2.3.

82 See PcN 1.35–1.38. Distinct from rasa (essence/sentiment), rāsa refers to a kind of circular dance.

83 See also footnotes 42 and 44 of the annotated translation (Act I).

(cāpa khaṇḍana),[84] his embodiment of the conduct prescribed by the Vedas (nigama nīti maṇḍana), the killing of Rāvaṇa (hanyo laṅka rāvana), and the liberation of Ahalyā (śilā śāpa mocana).[85] On the other hand, the lines dedicated to Kṛṣṇa evoke a slightly greater number of episodes from his life story, and the mention of Kṛṣṇa at the end seems to assign him a preferential role. These episodes are mostly recalled through various appellatives: Kṛṣṇa who severed Kaṃsa's head (kaṃsa ko śira nasāvana),[86] Ocean of Mercy (dayāsindhu), Lord of the Cows (govinda), Holder of the Mountain (giridhara),[87] Enemy of Mura (murāri),[88] Yaśodā and Nanda's son (yaśodā suana nanda), and the playful one in the courtyard (āṅgana bihārī).

Rāma, besides being mentioned together with Kṛṣṇa in 2.96 and 3.5, appears in two other instances: in a description of the activities of the silent sages (muni) who "sing the name of Rāma, untarnished"[89] and in a complex sentence uttered by Intention (saṅkalpa) and addressed to Mind. Intention wants Mind to recognise the transient nature of worldly relationships and accept that the only alternative is to take refuge in Hari. The verse states: "This is, indeed, the concern [of those] attached to a temporary existence, [but], in the end, who is your companion? King Rāma."[90]

Upon examining these instances, it becomes apparent that Brajvāsīdās leans towards favouring the saguṇa aspect of the Supreme Being, particularly in the form of Kṛṣṇa.[91] However, he also acknowledges Rāma as a potential manifestation. This alternative is presented not only through references to Rāma and the other characters/events from the story of King Daśaratha's son but also by skilfully incorporating elements from the RcM. In PcN 5.120–5.122, there is a depiction of Viṣṇu/Kṛṣṇa that closely resembles the portrayal of Rāma in RcM 1.146–1.147.1:

---

84 This incident takes place at Sītā's svayaṁvara (choice of groom through competition). See RcM 1.260.4.

85 This description of Rāma, highlighting his compassionate nature in addition to his prowess as a warrior, differs from the depiction in BhP 1.3.22, where his exploits as avatāra do not include this aspect of his personality.

86 Cf. BhP 10.44.

87 Cf. BhP 10.25.

88 Cf. BhP 10.59.

89 PcN 2.85: [. . .] | gāvata rāma nāma niradūṣaṇa ||.

90 PcN 5.86: [. . .] | ye tau hai hu taṅgī dinā cārahī ke raṅgī saba anta ke samaya ko tero saṅgī rāmarāya hai ||. On the various possible translations of this sentence, see footnote 106 in the annotated translation (Act V). John S. Hawley has briefly demonstrated that in later versions of a poem found in the Fatehpur Sikri manuscript of Kabīr's poetry, "rāma rāī" replaces "vanavārī" (typically referring to Kṛṣṇa) to "Ramacize" the lyrics. See Hawley 2005: 301, 390n44. For a more extensive discussion of the poem, see Hawley 2016: 156–159. To explore how Kabīr's poetry has been recontextualised and retextualised by different religious traditions, see Bangha 2019.

91 Another piece of substantial evidence is that his second work, the *Brajavilāsa*, is entirely dedicated to Kṛṣṇa.

<table>
<tr><th>PcN</th><th>RcM</th></tr>
<tr><td>a) sarada canda kete badana dekhi lājai |
b) paduma nīla maṇi ghana vapuṣa kānti rājai ||
c) adhara nāsikā rada cibuka cāru śobha |
d) hasana mādhurī meṁ su ko jo na lobhā ||
e) navala dala kamala lāla ambaka suhāye |
f) nirakhi cāru citavani kite ko vikāye || g) tilaka bhāla bhṛkuṭī ki śobhā suhāī | h) parama mādhurī chabi lakhai so lubhāī || i) kuṭila śyāma alakaiṁ mano nāganārī | l) apai viṣa bihīnī amīdhara piyārī || m) ḍasai draṣṭi jākī amara hota soī | n) kahai tāsu upamā kahā aura hoī || o) karana mukta kuṇḍala mukuṭa śīsa bhrājā | p) sakai ko barani cāru śobhā samājā || q) lasai kaṇṭha banamāla kaustubha suhāī | r) hiye cinha śrīvatsa śobhā mahāī || s) atīhī subhaga bāhu dīragha su doū | t) vibhūṣaṇa manina ke parama divya soū || 120 ||
a) pīta basana kaṭi kāchanī udara udāra sucāra | b) nābhi manohara bhaṁvāra janu śobhā sindhu apāra || 121 ||
a) sundara pada arabinda śubha sītala traitapa hara | b) bhare mukti makaranda hohu tāta mana bhṛṅga tahaṁ || 122 || [92]</td><td>a) nīla saroruha nīla mani nīla nīradhara syāma | b) lājahiṃ tana sobhā nirakhi koṭi koṭi sata kāma || 146 ||
a) sarada mayaṅka badana chabi sīṁvā | b) cāru kapola cibuka dara grīvā || c) adhara aruna rada sundara nāsā | d) bidhu kara nikara binindaka hāsā || 146.1 ||
a) nava abuṃja aṃbaka chabi nīkī | b) citavani lalita bhāvaṁtī jī kī || c) bhṛkuṭi manoja cāpa chabi hārī | d) tilaka lalāṭa paṭala dutikārī || 146.2 ||
a) kuṇḍala makara mukuṭa sira bhrājā | b) kuṭila kesa janu madhupa samājā || c) ura śrībatsa rucira banamālā | d) padika hāra bhūṣana manijālā || 146.3 ||
a) kehari kandhara cāru janeu | b) bāhu bibhūṣana sundara teū || c) kari kara sari subhaga bhujadaṇḍā | d) kaṭi niṣaṅga kara sara kodaṇḍā || 146.4 ||
a) taḍita binindaka pīta paṭa udara rekha bara tīni | b) nābhi manohara leti janu jamuna bhaṁvara chabi chīni || 147 ||
a) pada rājīva barani nahi jāhīṃ | b) muni mana madhupa basahiṃ jenha māhīṃ || 147.1 || [93]</td></tr>
</table>

92 In my translation: [Sarasvatī:] "How many autumn moons feel ashamed looking at his face? The body shines [like] a lotus, a blue sapphire, a cloud. The splendid beauty of the nose, lower lip, teeth, and chin. Who would not envy that sweet smile? The eyes attractive [as] fresh red lotus petals, the glance beautiful – at how much would it be sold? The auspicious mark [on] the forehead and the beauty of the eyebrows are charming. [When] one looks at this sweetest beauty, one desires [it]. Just as a snake-woman, he [has] curly, dark locks of hair. Devoid [of] the poison [of] the self, holding ambrosia, lovely. [When] his vision bites [someone], he becomes immortal. [If one] shall make a comparison with him – how can someone else [be similar]? The pearl earrings [at his] ears and the crown [on his] head sparkle. Who can describe the magnificence [in] a gathering? A garland of forest [flowers] shines on the neck, the kaustubha jewel is radiant [on the chest]. The great beauty [of] the śrivatsa mark on the heart. The two arms beautiful and strong, the adornment of jewels is perfect and divine" (5.120). "A yellow band [around] the waist, the belly [covered by] a garment, attractive and lovely. | The beautiful navel [is] like a swirl [on] a boundless ocean [of] beauty. ||" (5.121). "Beautiful lotus-feet, auspicious, cool, removing the threefold suffering, pearls filled [with] pollen – dear Mind, hover on them [like] a black bee! ||" (5.122).

93 Lutgendorf's translation of these verses in Tulsīdās 2016: 291–293 is well formulated: "His body was dark as a blue lotus, a sapphire, and a rain-bearing cloud, and the sight of his beauty put billions and billions of love gods to shame. The autumn moon of his face surpassed all splen-

Indic literature showcases a plethora of depictions featuring divine and human figures, both in Sanskrit and other regional and literary languages.[94] In the realm of bhakti literature, starting from the sixth or seventh century, the Āḻvārs and later the Śrīvaiṣṇavas composed poems that portrayed the physical appearance of the gods' bodies or temple images in Sanskrit, Tamil, and other regional languages.[95] In the domain of North Indian bhakti, Sūrdās's compositions serve as exemplary works within this genre.[96] While these representations often adhere to conventions, I will demonstrate how Brajvāsīdās adaptively reuses verses and reuses specific elements from the RcM's verses in his verbal visualisation.

However, there is a caveat regarding the metrical forms employed in the two passages examined. The PcN's passage in question includes a bhujaṅgaprayāta as the longest verse (5.120), followed by a dohā (5.121) and a soraṭhā (5.122). In contrast, the RcM presents two dohās (1.146, 1.147) and five caupāīs (1.146.1, 1.146.2, 1.146.3, 1.146.4, 1.147.1). Consequently, observations regarding the prosodical units are limited but still feasible. Furthermore, since the majority of the PcN passage falls under a single verse number (5.120), I have employed letters (starting again from "a" for each new pāda) to indicate the specific pādas I am referencing, aiming for greater clarity. I have also underlined the reused or re-used elements. First, I will compare the verses and then delve into the purpose behind Brajvāsī's selection of this particular passage from the RcM. As previously mentioned, although these descriptions may possess stylised elements, each author and work exhibit preferences for certain terms and imagery, and not all descriptions encompass precisely the same elements.[97] Despite the difference in metres, both de-

---

dour, with lovely cheeks and chin, conch-like neck, ruddy lips, charming teeth and nose, and a smile that mocked the moonbeams. His eyes eclipsed the beauty of new lotuses and the sweet glance delighted the soul. His brows stole the charm of the love god's bow and his forehead mark shimmered like lightning. Makara earrings hung below his radiant crown and his curly locks were like a swarm of black bees. On his breast was Shri's emblem, a wildflower garland, a nine-gem necklace set with diamonds. On his lion-like shoulders shone a sacred thread, and his armbands were beautiful, too, and lovely arms, stout as elephants' trunks, and he had a quiver at his waist and arrows and bow in hand. His yellow garment put lightning to shame, his belly bore three fair folds, and his deep navel seemed to steal the charme of blue Yamuna's eddles. There's no describing the lotuses of his feet, where sages' hearts hover like bees."

94 To enhance the erotic mood (śṛṅgāra rasa), courtly poetry often includes detailed descriptions of heroines from head to toe (śikha-nakha). For an example composed by the renowned court poet Keśavdās, see Busch 2011: 70–72.

95 On this topic, see e.g. Hopkins 2002.

96 See e.g. Hawley 2009a: 52, 71, 84, 90 (where Kṛṣṇa's body is compared to the Ganges).

97 Moreover, as Allison Busch has noted, a poet "cannot invent new body parts", so there is limited scope for "astounding" the reader. See Busch 2011: 72.

scriptions exhaustively cover the topic in the same number of lines – eleven[98] – focusing on identical physical details.[99]

Both texts begin with a metaphor, equating the resplendent dark beauty of God's body to a lotus, a sapphire, and a cloud. They employ hyperbole by asserting that the splendour of God's body surpasses that of the autumn moon (PcN) or the God of Love (RcM). In PcN 5.120a–b, which is shorter than RcM 146, Brajvāsīdās minimises the terms used. In 120b, he uses "nīla" only once, in contrast with the three occurrences of RcM 146a, and he also selects shorter words for lotus and cloud.[100] In 120a, he mentions only the face (badana) without elaborating on its beauty,[101] which he references in 120b. Furthermore, he maintains the double verb structure in the form of an absolutive and a predicative verb but employs two strategies to shorten them: substituting the verb indicating the act of seeing[102] and modifying the tense of the verb lāj-.[103] The hyperbolic effect conveyed by the repeated numerals "koṭi" and "sata" in RcM 146b is succinctly recreated by Brajvāsīdās by transforming the sentence into a rhetorical question using the pronominal adjective of quantity "kete" in 120a.[104] Interestingly, the phrase "sarada canda" and "badana" from PcN 120a correspond to "sarada mayaṅka badana" at the beginning of RcM 146.1a, suggesting that PcN 120a is constructed by combining elements from RcM 146b and 146.1a. In fact, PcN 120c appears to be a condensed version of RcM 146.1b–c, as the PcN retains the word order between "adhara" and "rada" from 146.1c and replaces the adjective "aruna" with the noun "nāsikā"

98 In the RcM, Rāma's depiction is followed by a brief portrayal of Sītā as primordial power (ādiśakti), which I am excluding from my comparison. This can be found in the second half of 1.147.1 to 1.147.2.

99 The descriptions in both texts follow a similar order: body, face (cheeks, chin, neck, lips, teeth, nose, smile, eyes/glance, brows, forehead, ears, hair), breast/chest, shoulders/arms, waist, belly, navel, feet. However, there is a slight adjustment in the PcN, where the hair is mentioned before the ears, and the cheeks are not included.

100 In the PcN, the terms "paduma" and "ghana" are used instead of "saroruha" and "nīradhara" in the RcM.

101 RcM mentions the splendour or beauty of the body (tana sobhā) here.

102 The word "dekhi" is used in the PcN instead of "nirakhi" in the RcM.

103 In the PcN, the subjunctive present tense "lājai" is used instead of the subjunctive future "lājahiṃ" in the RcM. These two tenses have overlapping usages. See Strnad 2013: 379–391.

104 Even if countless love gods or autumn moons existed, they would still pale compared to his dark-hued charm. This can be understood to correspond to the rhetorical device (alaṅkāra) known as "pratīpa" in Sanskrit literary theory. It involves a kind of "inverse comparison" (MW 674) where a habitual analogy is reversed: the moon or the god of love, typically considered embodiments of beauty, are humbled in the presence of God's beauty. I also see it as containing an element of exaggeration (atiśayokti). On this alaṅkāra and its slightly contrasting interpretations by the literary theorists, see Gerow 1971: 208–209.

(synonymous with "nāsā"), while retaining "cibuka" and "cāru" from 146.1b. Additionally, in PcN 120d, there is no reuse or re-use from RcM 146.1d, but both passages address the topic of the smile (hasana/hāsā). The subsequent lines focus on the eyes, and PcN 120e retains and expands upon RcM 146.2a: "nava ambuja ambaka" transforms into "navala dala kamala lāla ambaka", adding the syllable "la" to "nava", including the noun "dala" and the adjective "lāla", while substituting "ambuja" with "kamala" to create consonantal and vocalic alliteration. PcN 120f, g, and h re-use "citavani", "tilaka", "bhṛkuṭī", and "chabi" from RcM 146.2b, c, and d.

Following the previous lines, we encounter four pādas, labelled above as PcN 120i, l, m, and n, which delve into an expanded description of the curly hair (kuṭila alakaiṁ) by comparing them to the hair of a female nāga (nāganārī). This addition deviates from the description found in RcM, where the hair and the ears/earrings are paired in 146.3a–b. I perceive PcN 120i, l, m, and n as an inserted element that somewhat breaks the rhythm of the portrayal. However, this brief digression is not random; it serves as a reminder of the redemptive power of the envisioned form and aims to evoke, amid the depiction, the identity of the figure the text is depicting for us:

i) kuṭila śyāma alakaiṁ mano nāganārī
l) apai viṣa bihīnī amīdhara piyārī
m) ḍasai draṣṭi jākī amara hota soī
n) kahai tāsu upamā kahā aura hoī

The curly, dark hair, just like a snake-woman, empty of the poison of the self, full of nectar, beloved. Whose vision bites and one, indeed, becomes immortal. How could someone praise it further?

The snake holds a prominent place in bhakti compositions and is often associated with imagery surrounding Kṛṣṇa.[105] Rūpa Gosvāmī refers to Kṛṣṇa as bhujaṅga in his Sanskrit drama *Vidagdhamādhava* (2.52.6), depicting, in the words of Donna Wulff, "the irresistible nature of Kṛṣṇa's attraction, which is as inescapable and incapacitating as the intoxicating effect of a snake's bite".[106] Sūrdās also compares

105 There may already be a reference to Kṛṣṇa in the choice of metre used by Brajvāsī for the description, as mentioned in this section and subsection 1.2.4, known as bhujaṅgaprayāta, meaning "the movement of the snake (bhujaṅga)". Additionally, the adjective "śyāma", used to describe the curls, is a well-known epithet of Kṛṣṇa.

106 See Wulff 1984: 101–102. The snake/snakebite imagery in Bhakti literature is also associated with critiquing tantric beliefs, drawing on the symbolism of healing snakebites. For more on this theme, see Burchett 2013.

Kṛṣṇa to a snake in his poems.[107] In one instance, a young Kṛṣṇa asks his mother when his hair will grow like his brother's, and she responds by saying that one day his hair will resemble "a big black snake (*nāgini*)".[108] More generally, Sūrdās characterises Kṛṣṇa as the healer of the pain of separation that he himself inflicts on Rādhā and the other gopīs.[109] These references showcase both the negative and positive connotations associated with the snake imagery, which can also be discerned in the PcN verses. In these verses, Kṛṣṇa's lethal gaze, instead of causing death, bestows a precious elixir.[110] This seems to be the universe of inspiration these lines point to,[111] embedded in a portrayal that finds its model in the depiction of Rāma in the RcM.

Despite acknowledging the limitations of words in capturing such a portrait, Brajvāsī continues to use words to visually describe the rest of the saguṇa rūpa. In verse 120o, he proceeds to depict the ears/earrings and crown, almost quoting RcM 146.3a verbatim: "karana mukta kuṇḍala mukuṭa śīsa bhrājā", compared to "kuṇḍala makara mukuṭa sira bhrājā" in RcM. The PcN verse allows more space for the description, which motivates the addition of "karana" (ears).[112] "Mukta" in PcN indicates the type of earrings (pearl ones), as "makara" (makara-shaped) does in the RcM. The PcN favours alliteration, achieved by alternating terms with the same initial consonant (k-m-k-m) and using words that sound similar, such as mukta and mukuṭa (crown). "Śīsa" is synonymous with "sira", and "bhrājā" creates a rhyme with the following half-line, concluding with "samājā". Apart from this term, the two half-lines, PcN 120p and RcM 146.3b, being composed in distinct metres, have different quantities of syllabic instants but still share the same num-

---

107 See Hawley 2009a: 83.

108 In Hawley's translation, the poem is titled "Topknot". See Hawley 2009a: 57, 203n16.4. Rādhā's braid is also compared to a snake on page 79.

109 See Hawley 2005: ch. 7.

110 Hawley highlights a poem in the oldest manuscript of Kabīr's poetry, the Fatehpur Sikri manuscript, where Kabīr employs the snake and poison imagery differently. In this poem, the snake, representing māyā, is still female, with her poison representing the cycle of rebirth, while the poet invokes the snakebite-healer for the cure. The poem carries Haṭha yogic resonances, as the female snake can be interpreted as the kuṇḍalinī that needs to be raised from the basket of the body. See Hawley 2005: 298–300. Śabda 5 of Kabīr's *Bījaka* explores the power of māyā, likening it to a serpent-noose, while sākhī 38 depicts the serpent entwined around the sandalwood tree, a symbol of mukti (liberation). Sākhīs 97 and 99 mention the serpent of viraha (separation). For translations of these verses, see Singh/Hess 2002: 43, 93, and 100, respectively. For a slightly different translation of sākhī 38, see Dharwadker 2003. See also Vaudeville 2001: 169–172, 267–268, 306.

111 See also PcN 1.10 in Brajvāsī's narrative voice: "There is no greater joy than a bond with god, no other suffering – Rādhikā['s] beloved must be known."

112 Twenty syllabic instants to the sixteen of the RcM.

ber of syllables (thirteen), again highlighting the influence of the RcM verses as the model for the PcN. Subsequently, PcN 120q–t describe the neck/chest and shoulders by re-using terms from RcM 146.3c–d and RcM 146.4.[113]

PcN 121 and RcM 147 both take the form of a dohā and describe the waist/belly and navel. PcN 121a reuses the term "kaṭi" (waist) from RcM 146.4d and "pīta" (yellow) and "udara" (belly) from RcM 147a. However, the PcN again introduces an alliteration by adding "udāra" and "sucāra" after "udara". In PcN 121b, most of the terms are derived from RcM 147b. It begins similarly with "nābhi manohara" and employs "bhaṁvāra" as a comparative term, along with "janu" as a conjunction. The latter part of PcN 121b includes the location where the bhaṁvāra is found, the ocean, while the RcM mentions the Yamunā River. Finally, PcN 122 and RcM 147.1 focus on the feet. PcN 122 is a soraṭhā, with each pāda containing more syllabic instants (twenty-four) compared to the caupāī (sixteen) of the RcM verse. This is evident as PcN 122a expands the description of the feet by incorporating additional attributes: "sundara", "śubha", "sītala", and "traitapa hara". In PcN 122b, there is a reuse of "mana madhupa" from RcM 147.1b, which is transformed into "mana bhṛṅga", adding an additional layer of meaning. Here, mana refers to Mind, the character to whom this sequence in the PcN is addressed. The general description in RcM pertains to the minds of silent sages who concentrate on Rāma's feet, like bees on lotuses. In the PcN, the altered context turns the description into an exhortation for Mind to emulate a bee. The intertextual reference to the RcM adds another interpretive layer, suggesting that such behaviour is characteristic of the munis. Hence, Mind should emulate their conduct.

Those familiar with the context of the RcM verses, either through reading or listening to their adaptive reuse in the PcN, can establish deeper connections between the two texts and their stories. In the Rāma narrative, this particular segment revolves around King Manu and his wife, Satarūpā. After ruling for many years, the king feels the need for haribhagati (bhakti towards Hari; 1.142) and decides to abdicate in favour of his son. Together with his wife, he retreats to the Naimiṣa forest. They embark on a pilgrimage to various sacred sites, accompanied by sages, and then engage in ascetic practices, aspiring to behold the form of the unqualified Lord (1.143.2–1.145.4). After standing motionless on one leg for ten thousand years and declining the boons offered by other gods, the Lord eventually invites them to express their desire. Manu responds that their only wish is to witness the form beheld by Śiva in his heart (1.145.2). Subsequently, the Lord manifests in the form of Rāma. In this context, too, the RcM can serve as a conceptual model for the PcN, where Mind is prompted by Sarasvatī to relinquish kingship

113 The re-used words are "śrībatsa", "banamālā", "bahu", "bibhūṣana", and "subhaga".

in favour of his son, Discrimination. Mind is urged to detach from worldly attachments and engage solely in contemplating the Lord. As the text suggests, Mind has to become "like a bee", finding inspiration in figures like Manu, to whom the adaptive reuse of the RcM alludes.[114] Interestingly, Tulsī portrays Manu and his wife Satarūpā as embodiments of gyāna and bhagati (respectively, knowledge and devotion), two concepts that resonate with the themes explored in the PcN.

As anticipated above, in the PcN, Brajvāsī presents his favoured saguṇa manifestation, Kṛṣṇa. He establishes this identity by omitting the reference to the bow and arrow from RcM 146.4d and, it appears, by introducing the snake/snakebite imagery. The Manu episode also highlights that the Lord assumed a form for the sake of the bhaktas (143.4), and both the PcN and the RcM aim to portray the predominance of the deity's beauty and splendour. This is evident from the vocabulary employed in both passages: PcN uses terms such as kānti, rāj-, cāru, śobha, suhā-, mādhurī, chabi, bhrāj-, las-, manohara, sundara, śubha, and sītala, while RcM employs words like sobhā, chabi, sundara, lalita, bhrāj-, rucira, and so on. Through their descriptions of the captivating form of the deity, both authors, with Brajvāsī adaptively reusing Tulsī, seek to evoke a sense of appreciation for the god's beauty and cultivate attachment towards him.

Thus far, I have demonstrated that in the PcN, similar to the RcM, the nirguṇa and saguṇa forms are both manifestations of brahman, and the saguṇa form is essential for bhakti. Brajvāsī subtly reveals his preference for Kṛṣṇa while also acknowledging Rāma. This preference extends to the extent that the PcN adaptively reuses Rāma's description from the RcM to depict Kṛṣṇa through words. Another common aspect between the two texts is their consideration of the nirguṇa and saguṇa aspects of God as complementary, without relegating the saguṇa form to a mere aṃśa (partial manifestation).[115] While the KPC discusses the aṃśas of Nārāyaṇa descending to fulfil a mission (1.18),[116] there is no indication of aṃśas referring to the forms assumed by nirguṇa brahman in the PcN. In the RcM, Śiva responds to Pārvatī's questioning by sternly admonishing those who perceive the two forms of the Lord as separate.[117] Additionally, both texts seem to agree that the nirguṇa/saguṇa distinction does not imply a hierarchical relationship in terms of higher and lower brahman, with the former being ultimately real and the latter being provisionally real, as the conventional Advaita view sug-

---

114 In this sense, the reference is twofold: he serves as a model not only for King Mind but also for King Kīratibrahma, to whom the PcN is dedicated first.

115 See Pandey 1977: 112. It appears that also in the *Harivaṁśa*, Viṣṇu descends fully, unlike the other gods who descend partially. See Viethsen 2009: 231.

116 See Kapstein 2009: 11.

117 See RcM 1.115.1–1.118.4.

gests.[118] This viewpoint aligns with the perspective advocated by Madhusūdana Sarasvatī, even though he had a personal preference for Kṛṣṇa as the saguṇa form.[119]

The PcN upholds the view that the nirguṇa can only be known or perceived through the saguṇa (5.120) and that parabrahman assumes a form and name as either Kṛṣṇa or Rāma (2.96).[120] However, it does not contain any statements that diminish the reality of the saguṇa or present it as anything other than brahman manifesting on earth. Nor are there indications that the saguṇa form "disappears" once spiritual awakening is attained. While the PcN occasionally portrays brahman as the active puppeteer and inner controller, as we have discussed, and sometimes as the inactive magnet or disinterested witness, it does not explicitly discuss para/apara brahman. These aspects do not appear to be significant concerns for Brajvāsīdās in the PcN, nor do they seem to be specific areas of interest in the RcM.[121]

The influence of the RcM on the PcN can also be seen in the fact that Brajvāsī's drama does not assert Kṛṣṇa as the sole or primary saguṇa embodiment in an absolute sense. While the RcM considers Rāma to be both nirguṇa and saguṇa, representing the Highest Reality, the PcN depicts brahman as Hari/Viṣṇu. However, in this respect, the PcN's position differs from that of the Gauḍīyas and the Vallabhites. The BhP, while defining the supreme reality as both nirguṇa and saguṇa, champions Kṛṣṇa by considering him to be bhagavān svayam (Bhagavān himself, the Supreme Brahman).[122] The BhP holds a significant position as the primary scripture for Vaiṣṇava Vedānta schools of thought, leading to Kṛṣṇa being

---

118 See Potter 1981: 74–78. Śaṅkara's own distinction between nirguṇa and saguṇa brahman is not as straightforward as it may seem. First, his usage of the terminology is not employed consistently, as Hacker 1995 has demonstrated. Jacqueline Suthren Hirst further cautions against the hasty dismissal of the nirguṇa and saguṇa question in Śaṅkara's philosophy. She reviews various interpretations by scholars and suggests that it is more fruitful to examine Śaṅkara's underlying concerns than his usage of terminology. According to her analysis, it is clear that for him, it was possible to transcend the saguṇa conception and attain the ultimate nirguṇa. See Suthren Hirst 2005: 116–120.

119 Despite statements such as "I worship that *Brahman* [. . .]. Who was born in Vrindavana for the joy of all as a result of the austerity of Nanda, playing the flute, with [a] face as beautiful as the moon, having eyes like lotuses", Madhusūdana Sarasvatī maintains the view that saguṇa brahman only has a conditional existence. See Gupta 2006: 68. The quoted sentence is a translation from the opening verses of his *Saṅkṣepaśariraka Sārasaṅgraha*. For references and further reflection on the question, see Nelson 2007: 320–321.

120 We also find a reference to Viṣṇu in his four-armed form in verse 5.123.

121 R. K. Dasgupta points out that the reconciliation between nirguṇa and saguṇa was confusing for scholars at the beginning of the twentieth century. For further details, see Dasgupta 1977: 121.

122 See BhP 1.3.28 and 9.24.55. See Matchett 2008: 112–149; Sheridan 1986: 56–57. Of course, the BhG already gave unique precedence to Kṛṣṇa.

regarded as the supreme saguṇa brahman by both the Gauḍīyas and the Vallabhites.[123] Vallabha, in verse 3 of his *Siddhāntamuktavalī*, proclaims Kṛṣṇa as parabrahman. However, as Frederick Smith notes, commentators, such as Vallabha's son Viṭṭhalanāth and Lālūbhaṭṭha (seventeenth–eighteenth century), exercise caution in specifying that only Kṛṣṇa, and not Vāsudeva, Hari, or others, can be fully considered the Supreme Being.[124] Additionally, Vallabha presents a threefold distinction where the nirguṇa akṣara brahman, responsible for the manifestation of the material world, is accorded a secondary level of reality, recognising its limited possession of the ānanda nature of Kṛṣṇa.[125] In contrast, the position of the PcN does not seem to align with the Vallabhite perspective in this particular instance. This suggests that Brajvāsī, by adaptively reusing passages from the RcM, found actual inspiration from it in this philosophical and theological aspect, granting himself the liberty to portray Kṛṣṇa as a favoured personal deity.[126]

4.1.2 Māyā

After investigating the conception of the Highest Reality in the PcN, this subsection focuses on another personified concept within the play's cosmological framework: Māyā. The aim is to explore how Māyā corresponds to prakṛti and elucidate her ambiguous features. Furthermore, we will analyse her unique relationship with the concepts/characters of Misconception (avidyā) and Bond-with-Viṣṇu (viṣṇubhakti).

Although Māyā never appears in person throughout the play, she is discussed by other characters, particularly in Acts I and VI. In the passages of the first act, her identity and functioning are explored, albeit with only a hint towards salvation. As expected, the later parts of the drama provide more detailed insights into

123 For the Gosvāmīs, this applied to both Jīva and, later, Baladeva. See Gupta 2016a: 41–47, 2016b: 115.

124 See Smith 2016: 129–130.

125 See Barz 1976: 67; Narain 2004: 90–97.

126 I do not delve into the nirguṇa/saguṇa distinction as a symptomatic divide between two socially opposite tendencies, where believers in a nirguṇa Absolute traditionally oppose the varṇāśrama dharma, while saguṇa brahman believers support it. It can be imagined that Brajvāsīdās, due to his affiliation and source of inspiration (the RcM), supports the varṇāśrama order. However, this does not prevent him from reusing and reinterpreting images and expressions employed by nirguṇa exponents like Kabīr and Dādū. See footnotes 21, 52, 83, 110, 114, and 122 in the annotated translation. It is also significant in this context that the poetry attributed to Kabīr and other nirguṇa sants includes the names of Rāma or Kṛṣṇa, despite rejecting the avatāra doctrine. The presence of the avatāra names can certainly place them within the Vaiṣṇava fold for Brajvāsī. See Agrawal 2018; Callewaert 2018; Hawley 2005: 70–96. See also Callewaert 2000: 326–327.

her nature, focusing directly on salvation. While she is referred to as prakṛti (material nature) on a single occasion (1.96), she undeniably fulfils the role of prakṛti in relation to puruṣa. Numerous references are made to her qualities or guṇas (e.g. 1.140, 1.146, 6.16). In Act VI, when Discrimination teaches Individual Self how to understand his true nature through introspection, he instructs him as follows:

[. . .]
akāśa śabda maiṁ jāya bilāya
śabda tṛvidhi ahaṃkārahi milai
ahaṃkāra mahātatvahi ralai
bahurau mahattatva jo āhi
līna karau māyā maiṁ tāhi
māyā līna hoya jāmāhīṁ
so tū āhi aura puni nāhiṁ
(6.144)

Space will vanish into sound; sound will unite with the triple I-maker. I-maker mingles with the great element, then absorbs the great element in Material Nature. Material Nature will merge with nothing else but what you are.

Māyā here is clearly synonymous with prakṛti for Brajvāsīdās. This is evident as she is also described as dancing in front of puruṣa (6.86), thereby creating and destroying the world, as in Īśvarakṛṣṇa's *Sāṅkhya Kārikā* 42 (Verses about [the doctrine of] Enumeration; c. 350–400 CE).[127] For this reason, I have translated the character of māyā as both "Illusion" and "Material Nature". I have preferred "Illusion" when the verse addresses her role as a concealing power that obstructs Individual Self from recognising his true self, and "Material Nature" when the verse pertains to her projecting power or the cosmogonic process. The aforementioned quoted verse exemplifies the latter usage, while verse 1.52 represents the former sense:

yaha jīva sanātana hati ko
kachu mili māyā bhayo pharako
so māyā basa hai kaise
nalanī ko suka bhrama jaise
(1.52)

127 PcN 6.86: [Discrimination:] "Supreme Self, in the bending [of] whose eyebrows Material Nature creates the world. In one second destroys then creates, in this way she dances in front of him. [. . .]." The image is also present in RcM 7.71.1, and māyā is also identified with prakṛti in ŚvU 4.10. Moreover, in the PcN, māyā is also defined in one instance as mūla, "the root" (1.146), as it is in *Sāṅkhya Kārikā*. For a detailed discussion, see Larson 1969: 160–167.

The individual self [belongs] eternally to Hari [but], meeting with Illusion, became agitated. How is he [in] the power [of] Illusion? It is like the parrot's error about the lotuses.[128]

The dual aspect of māyā, both deluding and creative, is maintained in the RcM,[129] and this viewpoint also appears to be reflected in the BhP and the BhG, albeit with some variations.[130]

Māyā indeed represents the power (śakti) of the inactive puruṣa, through which he exercises control over the individual self, likened to a magnet attracting iron (6.87). Brajvāsīdās particularly emphasises this aspect of puruṣa as the controller of māyā or the possessor of māyā. He refers to brahman as "Illusion['s] husband" (māyāpati, 1.94), similar to RcM 1.50.5.[131] Furthermore, both texts associate her with being the servant (dāsī) of the Supreme Self and link her to ignorance (ajāna) and falseness (mithyā).[132] The KPC characterises her behaviour as that of a prostitute (veśyā),[133] as she feigns false emotions towards Puruṣa to manipulate him into doing her bidding. While the PcN shares this portrayal of her as deluding, and the term dāsī is not entirely devoid of the implications of veśyā,[134] I prefer to emphasise the connotation of service, as it appears that the dependency of Illusion/Material Nature on the Supreme Self is a central feature of her depiction in the PcN:

satya puruṣa te so bhaī ajāna dāsī bhāva
hai parantu aścaryamaya jāko amita prabhāva
(1.140)

---

128 See also footnote 21 in the annotated translation (Act I).

129 For example, RcM 3.14.2–3. On the concept of māyā in the RcM, see Bharadwaj 1979: 40–48.

130 In the BhP, the creative aspect of māyā is depicted positively. It is through māyā that the Supreme Self creates the world for the liberation of individual selves, and it is through māyā that he can embody in human form. See Sheridan 1986: 31–32. See also Bhattacarya 1960, vol. 1: 237–246. In the BhG, the term "prakṛti" signifies the creative power of a god, the nature through which he can manifest on earth with a material body. This body is apparitional in that it is not an ordinary body subject to karman, decay, and death. Prakṛti becomes māyā when controlled by a god, and māyā, in turn, represents both the activity and the result of the activity. See Malinar 2007: 95–102, 132, 235. On the concept of māyā in Vedic, epic, and Puranic sources, see Goudriaan 1978: ch. 1. See also Devanandan 1954.

131 See also RcM 1.116.4, 2.217.1, 7.62b, 7.77.3, and BĀU 1.4.3, where the primordial puruṣa divides his body into two halves: one masculine, called pati, and one feminine, patnī.

132 PcN 1.140 and RcM 7.71b.

133 In 1.91. See Kṛṣṇamiśra 2009: 36–37. Also Nambiar 1971: 14–15.

134 See HSS 2267.

Actually, she was [manifested] from Supreme Self, [she is] ignorance [with] the nature [of] a servant.[135] Whose infinite power is, however, wonderful.[136]

The depiction of māyā as a deceptive woman is common in vernacular poetry. In this regard, Brajvāsīdās follows not only in the footsteps of Tulsī but also Kabīr, the nirguṇa sant par excellence. Several of Kabīr's compositions portray her in such a manner.[137] Later on, Sundardās (1569–1689) of the Dādūpanthī tradition composed the *Pañca Prabhāva* (Five Categories [of Sants]). This allegorical work classifies the sants based on their relationship with personified Bhakti and Māyā. In this case, personified Bhakti is seen as the lawful wife and Māyā, her servant.[138]

In the PcN, this discussion is expanded by presenting all worldly women as manifestations of Māyā, a viewpoint shared by the RcM.[139] The RcM drew inspiration from the Rāma-related Upaniṣads and the *Adhyātma Rāmāyaṇa* in developing this concept further.[140] Furthermore, the PcN appears to have been influenced by the RcM's elaboration, which sees Bhakti as having an advantage over Māyā because Bhakti is also a woman (feminine according to the grammatical gender)[141] and, unlike men, will not be distracted by her beauty. While this idea is not explicitly theorised in the PcN, it can be discerned from the characters' behaviour and the overlapping terminology with the RcM.[142] Although a detailed exploration of Brajvāsīdās's understanding of bhakti will follow later, it is worth noting for the present purpose that he does not describe viṣṇubhakti as a power of brahman. Nonetheless, as mentioned earlier, he identifies Bond-with-Viṣṇu with the essence (rasa) of paramātama and describes her as the "beloved" of Supreme Self (1.173).

135 Alternatively, it can be translated as "with the nature of an ignorant servant".

136 I added the word "manifested" in this verse on the clue provided in the following verse, 1.141, which explicitly says that Illusion/Material nature was generated (jāta) from Puruṣa.

137 See Vaudeville 2001: 254–255, 267; also śabdas 44 and 62 and sākhī 147 in Singh/Hess 2002: 56–57, 61–62, 106.

138 For more on the typologies of sants and their relationships with the two personified concepts, see Rajpurohit 2019: 164–167. One of the most well-known episodes featuring the personification of Bhakti is found in the *Bhāgavata Māhātmya*, a text that celebrates the glory of the BhP. This text portrays Bhakti as moving from south to north India with her sons Jñāna and Vairāgya. See Hawley 2009b.

139 See BhP 3.31.37–40, which expresses a similar viewpoint.

140 For references to the verses from the Rāma-related Upaniṣads, the relevant passages from the *Adhyātma Rāmāyaṇa*, and secondary literature, see Caracchi 2010: 4n11, 13.

141 See RcM 7.115.2. However, in 7.114.8, it is mentioned that gyāna, birāga, joga, and bigyāna have a masculine nature and, as a result, are susceptible to the influence of women.

142 The behaviour of Bhakti's character follows the same line as in Kṛṣṇamiśra's PcN. However, I believe the reuse of terminology and parallel conceptions between the PcN and the RcM suggests a direct inspiration drawn from the latter text.

The term used, pyārī, is also employed in the RcM (piārī, 7.115.2) to refer to her. Additionally, the RcM contrasts her with Māyā, stating that Rāma holds a favourable disposition towards Bhakti in the same verse, while Māyā is merely a dancing girl. Māyā causes Individual Self to dance, whereas Bhakti liberates him.[143] In the PcN, the actions of Illusion/Material Nature and Bond-with-Viṣṇu are also diametrically opposed: the former keeps her son, Individual Self, separated from his father, Supreme Self, whereas the latter desires and endeavours to unite them (1.173).[144]

Regarding the ontological status of Māyā, the PcN aligns with the RcM in several instances. For instance, in 6.126, Māyā is described as being conformable (anurūpā) to Supreme Self. Similarly, Tulsī's work depicts Sītā, right after describing Rāma's physical appearance, as anukūlā, meaning conformable to him (1.147.1). The terms anurūpā and anukūlā can be considered synonymous.[145] Pinuccia Caracchi's interpretation emphasises that anukūlā signifies Sītā's complete agreement with, or "matching" of, Rāma's existence, implying that she does not exist independently from him.[146] This understanding seems to apply to Māyā in the PcN as well, where the verse in question warns against considering Māyā as an independent cause of creation apart from the Supreme Self.

Furthermore, Brajvāsīdās expresses the ontological relationship of Māyā with the Supreme Self and Individual Self in a manner similar to the RcM, re-using the noun "bīca" with a postpositional function:

tāte kabahūṁ cāhata nāhīṁ
  yaha vidhi bīca rahe pitu pāhīṁ
(PcN 1.173)

She [Illusion] never wants [Individual Self to wake], this way staying between [them], next to the father.

143 RcM 1.201.2: dekhā jīva nacāvai jāhī | dekhī bhagati jo chorai tāhī || (She saw the individual self, who is made to dance [by Māyā]; she saw Bhakti who frees him.)

144 Māyā is also metaphorically compared to a prostitute and described as being overcome through devotion in the *Mānasollāsa*, a Śaiva-oriented commentary attributed to the Vedāntin Sureśvara on a hymn that is itself attributed to Śaṅkara. See Potter 1981: 559.

145 See HSS 216. The portrait of Rāma in question is the same one analysed earlier in this chapter in comparison with the portrayal of Kṛṣṇa found in the PcN.

146 "La traduzione inglese della Gītā Press interpreta anukūlā come 'who is ever devoted to Him', ma l'aggettivo, etimologicamente, esprime un senso ancor più pregnante di conformità e di adesione, simile a quello dell'acqua che scende da un declivio adattandosi completamente alla conformazione del terreno. Come nel mito, anche nel culto Sītā viene sempre venerata in unione col suo sposo; a lei non viene neppure tributato un culto indipendente come, in diverse occasioni avviene, per esempio, per Lakṣmī o per Sarasvatī." See Caracchi 2010: 6.

ubhaya bīca siya sohati kaise
brahma jīva bica māyā jaise
(RcM 2.122.1)[147]

How magnificent was Sītā between the two [brothers]! As Illusion between brahman and Individual Self.

In the RcM, Sītā, as we have observed, is the embodiment of Rāma's power of deceptive creation, while Lakṣmaṇa, Rāma's brother, represents the jīva in this case. Moreover, Sītā's remarkable beauty chimes with the PcN's characterisation of Māyā, for example in PcN 1.141: "She is from beginning to end very beautiful, generated [by] Supreme Self. [. . .]."

In agreement with the RcM, the PcN affirms that puruṣa is the master of māyā and that, since he is all-powerful, she is too.[148] A pāda from the PcN echoes the sentiment found in another pāda from the RcM regarding māyā:

vividhi vicitra caritra amita guṇa bharī nikāyā
kyoṁ na bhramāve sabahi puruṣa kī hai yaha māyā
(PcN 1.140)

A collection full [of] infinite qualities, [of] multiple, wondrous actions. Why would she not mislead everyone? She is Supreme Self's Illusion!

prabhu māyā balavanta bhavānī
jāhi na moha kavana asa gyānī
(RcM 1.61.5)

O Bhavānī, the Lord's Illusion is strong: who [is] so wise [that] he is not bewildered?

Brajvāsī re-uses the rhyme of "nikāyā" and "māyā", often employed by Tulsī in contexts distinct from that of the PcN.[149] Moreover, both verses conclude with a rhetorical question that further emphasises their admiration for māyā. Brajvāsī emphasises the immense (adbhuta) and wonderful (vicitra) aspects of māyā, portraying her as a mysterious force[150] that defies description.[151] By exploring her nature and dedicating

147 A similar half-line is found in RcM 3.6.2: ubhaya bīca śrī sohai kaisī | brahma jīva bica māyā jaisī ||.

148 Both texts refer to the infinite power (prabhāva) of māyā ("hari māyā kara amiti prabhāvā" in RcM 7.59.2 and "amita prabhāva" in PcN 1.140). See also RcM 7.59.1 and 7.61.5.

149 See RcM 1.182.2, 1.224.2, 3.12.3, 3.14.1, 5.20.2, and 6.78.2.

150 "andhero" in PcN 1.138, "aścaryamaya" in PcN 1.140, and "adbhuta" in PcN 1.146.

151 PcN 1.146: yaha māyā kehūṁ barani na jāī | [. . .].

space to contemplation about her, he aligns himself clearly with the tradition of Vedantic reflection. Similar verses extolling māyā's incredible nature can be found, for example, in the *Pañcadaśī*, a compendium composed by the Advaita Vedānta philosopher Bhāratītīrtha/Vidyāraṇya (thirteenth–fourteenth century):

Without in any way affecting the real nature of ātman, māyā creates the world. It makes the impossible look possible. How powerful māyā is! [. . .] But since the nature of māyā itself is a wonder, it is only to be expected that its power too is marvellous. [. . .] Māyā is materialised marvelousness; the wise must make efforts to abolish it. If you want to know the nature of māyā before trying to eradicate it, alright – do so! Māyā being indefinable does not lend itself to any logical definition.[152]

Another significant element of the Vedantic ontological and epistemological framework in the PcN is the introduction of a distinct character called Avidyā, which I have translated as Misconception, the twin sister of I-maker (ahaṃkāra, or ego), the first-born children of Mind and his first wife Engagement-with-Action:[153]

mana ke prathamahiṁ bhava pravṛtti te ahaṃkāra suta
  prakaṭo take saṅga avidyā kanyā saṃyuta
tina so mana kī prīti adhika china karata na nyāro
  [. . .]

(1.152)

Mind's first-born from Engagement-with-Action [was] a son, I-maker. A daughter was delivered with him, Misconception, [his] twin. The affection of Mind towards them was excessive; he was not away [from them for] an instant.

Although Misconception never actively participates or speaks in the narrative, she is briefly mentioned in Act VI as vanishing from Mind due to the manifestation of her counterpart Vidyā, which I have translated as Knowledge (6.154). Knowledge is the twin sister of Wisdom Moon (prabodhacandra), and Brajvāsīdās may have deliberately established a symmetry between the two opposing pairs: Misconception/I-maker and Knowledge/Wisdom Moon. It is no coincidence that in the final verses of the play, Brajvāsī states that reading, listening, understanding, and contemplating on the PcN leads to unravelling the knot of ahaṃkāra and avidyā

---

152 Vidyāraṇya 1954: vv. 128–142. See also Mahadevan 1969: 72.

153 In terms of genealogy, Bewilderment (moha) should be considered the elder son, but his birth is not discussed together with that of I-maker and Misconception in the PcN. Kṛṣṇamiśra's PC also does not present them together; moreover, it only mentions the birth of I-maker from Mind and does not introduce a character named Avidyā. See Kapstein 2009: 39.

(6.166).[154] According to our author, egoism and ignorance are undoubtedly the primary adversaries on the path to self-realisation and awareness of one's true form.[155]

Why did Brajvāsī introduce avidyā as a character, and what does it represent? While the play does not explicitly provide an explanation, we can gather some insights. Previously, in verse 1.140, māyā was referred to as ajāna (ignorance or ignorant), and in the final act, there is a connection between avidyā (ignorance, misconception concerning one's self) and māyā. When the character of Knowledge appears, she radiates intense luminosity, dispelling the obscurity of Misconception. As a result, the extent of Illusion's manifestation (māyā prapañca) – that is, the phenomenal world – also disappears (6.154).[156] By introducing Misconception as a character, Brajvāsī may have intended to highlight the existence of an individual dimension of cosmic māyā, which he refers to as avidyā, although this concept is not fully developed.

In doing so, Brajvāsī engages in a discussion that has spanned several centuries, focusing on the relationship between avidyā and māyā, particularly the underlying substratum (āśraya) of avidyā/māyā. In the Vaiṣṇava Vedānta tradition, both Rāmānuja and Vallabha consider avidyā to pertain to the individual self (jīva), but they differ in explaining its causality. Rāmānuja firmly rejects the notion of māyā and avidyā as being related to illusion or misconception/ignorance, as he believes that all knowledge is valid (satkhyāti). This stems from his view that the world and individual selves are real and integral parts of the body of God.[157] He interprets māyā as "the conjunction of souls with material bodies".[158] Furthermore, he sees avidyā not as an uncaused principle but as directly arising from karma, whereby past actions cause individual selves to overlook their true nature and relationship with the Lord.[159]

Vallabha considers māyā to be one of the śaktis (powers) of the Supreme Kṛṣṇa, and in essence, identical with him. Through māyā, Kṛṣṇa conceals his nature as cit (consciousness) and ānanda (bliss) to manifest the world. In this context, Kṛṣṇa, while concealing part of his ānanda, is seen as the impersonal akṣara brahman (imperishable brahman), whereas māyā is identified with the prakṛti (material nature)

---

154 See MU 2.1.10; *Kaṭha Upaniṣad* 6.15; CU 7.26.2; BhP 1.2.21, 5.5.9.

155 See e.g. PcN 5.10.

156 "Knowledge arrived, with shining radiance. That mass of obscurity Misconception drifted apart from Mind. Knowledge showed her form and destroyed it immediately. She annihilated the whole extent of Illusion's manifestation. Then, together with Mind, she became absorbed in the Self. Like wood burned by fire, even the own self will merge [into the Self]."

157 See Dasgupta 1922, vol. 3: 179–188, 198–201. See also Lipner 2018.

158 See Carman 1974: 184.

159 See Carman 1974: 56.

issued by this lower form of Kṛṣṇa. Avidyā and vidyā are also considered śaktis connected to māyā, activated by the will of Kṛṣṇa, who, for the sake of his līlā (divine play), renders individual selves ignorant.[160] According to K. Narain, Vallabha emphasised that avidyā is a power of brahman, which produces the misunderstanding of saṃsāra (cycle of rebirth) and the perception of a separate individuality or ego within the embodied self.[161] In contrast, the PcN does not reference avidyā as a cause of ahaṃkāra. Instead, in the play, avidyā and ahaṃkāra are depicted as being born or originating together from Mind and Engagement-with-Action as Misconception and I-maker.[162] Differently, the Gauḍīyas present a variety of śaktis whose significance depends on their proximity to brahman, specifically Kṛṣṇa. These śaktis are inconceivable (acintya), thereby safeguarding the Lord's supreme position.[163] Among these śaktis, māyā is considered an inferior type, belonging to the external potencies of the Lord. Through māyā, the world is manifested, and she obscures and confuses the jīva about its true nature.[164]

The Advaita Vedantic perspective on this matter is quite intricate, often called the "hard problem" within this school of thought.[165] Paul Hacker has shown that Śaṅkara, widely regarded as the most authoritative philosopher of Advaita Vedānta, does not extensively develop a comprehensive doctrine of māyā. It is his followers who delve deeper into conceptualising māyā, while Śaṅkara, in terms of terminology, prefers to use the term avidyā. In his commentary on the BhG, māyā is described as "deceit, magic", representing something that illusory phe-

160 See Barz 1976: 66–67, 70, 78. Marfatia 1967: 56–65; Dasgupta 1922, vol. 4: 330–331; Smith 2016: 135.
161 Narain 2004: 304–305.
162 Also absent from the PcN are other features of the Vallabhite doctrine of māyā and avidyā. For example, it does not refer to māyā as "yogamāyā" nor to avidyā as "vyāmohika māyā". It does not mention the five "knots" of avidyā but instead portrays her with a single knot associated with ahaṃkāra. See Narain 2004: 267–274, 305–306. Yogamāyā is a term used in the BhP to indicate a concept that scholars such as Sheridan, Bryant, and Bhattacarya have interpreted as distinct from māyā. According to this interpretation, yogamāyā would pertain to Kṛṣṇa's līlās, while māyā would refer to the realm of saṃsāra. However, Gopal Gupta has recently argued against this distinction. Based on his analysis of the occurrences of the words māyā and yogamāyā in the BhP and their respective meanings, Gupta suggests that these concepts often overlap in their usage and significance. He proposes that māyā is frequently a contracted or abbreviated form of yogamāyā due to metrical requirements. For references to the works of these three scholars and Gupta's argument, see Gupta 2020: 34–36.
163 For a concise comparison of the ideas of Śrīdhara and Jīva Gosvāmī on māyā, see Gupta 2007: 70–71.
164 It is māyāśakti who possesses the power of avidyā and vidyā in relation to the individual jīva. See Gupta 2007: 41–42, 173–174, 205.
165 See Kaplan 2018.

nomena can be compared to.[166] On the other hand, avidyā is understood as adhyāsa (superimposition) and mithyājñāna (false cognition).[167]

According to the Kevalādvaita philosopher, avidyā is not the cause of the phenomenal world but rather a kleśa, an affliction of the mind. It is neither a jaḍa (insentient) nor an anādi (beginningless) metaphysical entity.[168] In Śaṅkara's commentary on the BhG, māyā is sometimes referred to as the illusory creative power of God.[169] However, Śaṅkara is not primarily concerned with providing a definitive theory regarding the ontological status or location of avidyā.[170] Among Śaṅkara's followers, Padmapāda (c. 820 CE) equates māyā, prakṛti, and śakti with avidyā, treating them as synonymous. Padmapāda appears to be the first to attribute some ontological reality to avidyā and to describe it as material (jaḍa).[171]

Regarding the foundation of avidyā, the thinkers of non-dual Vedānta can be divided into two distinct schools: the Vivaraṇa school and the Bhāmatī school.[172] Supporters of the Vivaraṇa school believed that brahman was the underlying substratum of avidyā. In contrast, proponents of the Bhāmatī school argued that the jīva, the individual embodied self, bore avidyā, and there were as many avidyās as there were multiple jīvas.[173] The second perspective, which does not challenge the supremacy of brahman, aligns well with Vaiṣṇava Vedāntins and can be easily adapted to their doctrines.[174]

---

166 Hacker 1995: 79–80.

167 Hacker 1995: 58–59.

168 Hacker 1995: 64–65.

169 Hacker 1995: 80–81.

170 Śaṅkara's emphasis on the epistemological aspect is relatively less developed, but it would later become central to Advaita through his followers. His primary concern is mukti (liberation), in view of which misconception must be eradicated. Hacker 1995: 65.

171 See Dasgupta 1922, vol. 2: 104–105.

172 These two doctrines are associated with two contemporaries of Śaṅkara in the eighth century: Maṇḍana Miśra and Sureśvara. Sureśvara is credited as the first proponent of the viewpoint that would be upheld by the Vivaraṇa school, which derives its name from the homonymous work by Prakāśātman (eleventh–twelfth century). Prominent proponents of this school of thought include Padmapāda, Citsukha, Vidyāraṇya, and Sarvajñātman. On the other hand, Maṇḍana Miśra originated the Bhāmatī view in his *Brahmasiddhi*, which would later be further elaborated by Vācaspati Miśra (ninth century), the author of the seminal work that gives the branch of thought its name. See Dasgupta 1922, vol. 2: 108–111.

173 See Potter 1981: 78–80.

174 Starting with Madhusūdana Sarasvatī in the sixteenth century, attempts were made to reconcile these two positions. Gianni Pellegrini identifies this tendency in works such as Appaya Dīkṣita's *Siddhāntaleśa* (sixteenth century). See Pellegrini 2018: 605. Pellegrini also provides two useful tables comparing the broad positions of the two schools on pp. 606–607. See also Potter 1999: 181–182. For Madhusūdana's own approach to reconciliation, see Gupta 2006: 14–28.

Brajvāsī follows a similar interpretative approach as reflected in the RcM and the sensibility of local languages when it comes to terminology and conceptualisation. He synthesises the theistic view of māyā as an extraordinary power inherent in brahman while acknowledging her inexplicable, material, and overall "ambiguous" nature in line with Advaita reflections.[175] Moreover, the author of the PcN distinguishes māyā from avidyā, emphasising the purely epistemological interpretation of avidyā as primarily a product of the mind.[176] Although avidyā's character is not extensively developed in the play, its mere presence indicates Brajvāsī's awareness of the debates that animated the Vedantic landscape from a historical perspective.

4.1.3 The Phenomenal World

While the treatment of māyā/avidyā in the PcN takes a middle path among the various Vedantic perspectives, the portrayal of the world (jagata) aligns mainly with the Advaita tradition. The PcN emphasises the illusory and unreal nature of the world, which has been brought into existence through māyā. Although Hari is referred to as "jagata pitā" (father of the universe) once (PcN 1.173), the predominant focus is on the transient and impermanent aspects of the world. This theme is particularly highlighted in the play's fifth act, where Sarasvatī and Dispassion persuade Mind to renounce worldly attachments.[177]

jo vicāri kari dekhiyata tau kachu sthira nāhi
jagata budabudā nīra ko upajata aru vinasāhi
(PcN 5.9)

If it is observed through reason, then nothing is permanent.
The world is produced and destroyed [like] a water drop.

The central idea conveyed is that nothing in the world is permanent, a point reiterated throughout the act. The world is compared to fleeting phenomena, such as the false appearance of the sun's rays, a dream, an abode filled with smoke, a

175 In the PcN, māyā is not portrayed as the consort of a personal embodiment of God. In fact, she is characterised rather negatively. For instance, in verse 5.40, she advises Mind to pursue whatever pleases him, in this case, acquiring supernatural powers through the practice of yoga.
176 Advaita Vedānta reflection came to talk about a radical ignorance (mūlājñāna) that is related to brahman itself, as well as a derivative ignorance (tūlājñāna) associated with the embodied self. See Potter 1981: 79; Pellegrini 2018: 608n40. Also Pellegrini 2015.
177 However, we do also find earlier mentions in verses 1.157 and 1.78. In 1.50, the world is identified as the "seat" of desires.

well of falsehood, error, and darkness.[178] Such descriptions emphasise the transient nature of the world and underscore its illusory and deceptive qualities.

There are two instances in the play where the term prapañca appears: in the fifth act, "jagata prapañca" (5.143), and in the sixth, "māyā ko prapañca" (6.154). In both cases, prapañca signifies the "expansion", "extension", or "manifestation"[179] of the world or the material power of brahman. The contexts in which these expressions are used are also similar. In the first case, it describes someone who has recognised their true nature and is no longer affected by the multiplicity of the visible world. In the second case, in Act VI, with the manifestation of Knowledge, Misconception and a portion of Illusion's creation are said to vanish from Individual Self.

Notably, the PcN does not seem to distinguish between jagata and prapañca, as Vallabha had. In his theory of perception (khyāti), prapañca represents the real universe as the Supreme Kṛṣṇa with his concealed ānanda and cit. In contrast, jagata is a deception experienced by the jīva due to māyā. According to Vallabha, indeterminate cognition precedes determinate cognition, so when the individual self first perceives prapañca, māyā obstructs its proper understanding. Māyā increases the tamas (darkness, confusion) in the intellect (buddhi) of the individual self, leading to an incorrect perception of the actual object. Instead of grasping prapañca, the jīva apprehends a distorted intellectual object called jagata.[180] While jagata represents an epistemological error, prapañca signifies the expansion of the Lord's power.

The analogies employed by Brajvāsī to discuss the ontological status of the world are in line with classical Vedantic discourse:

[. . .]
jagatau rabi marīcikā jala jyoṁ bhrama soṁ bhayo amithyā
jāko udaya nāśa lakhiyata hai satya kauna vidhi kahiye
jaba lagi ātma bodha nahiṁ hoī tabahīṁ lai yaha lahiye
mālā visaya sarppa bhrama jaise sīpa madhyā jimi rūpā
bhrama miṭi gaye sarpa nahiṁ rūpā lakhiye śuddha sarūpā
(PcN 6.126)

178 "jaga saba jhūṭha hai marīcaka ko yota jaise" (5.106); "jaga svapana" (5.138); "hai jaga jhūṭha dhuvāṁ ko so dhāma" (5.141); "yaha jaga mithyā bhrama tama kūpa" (6.144). See also verse 5.90: "what originates is destroyed".

179 See MW 681.

180 Rāmānuja's theory of perception, technically known as anyakhyāti, is distinct from other philosophical viewpoints such as those of Advaita Vedānta, Madhva, Nyāya, and Rāmānuja's own satkhyātivāda. Narain 2004: 224–228 provides detailed explanations and comparisons. Additionally, a brief mention of satkhyātivāda can be found on pp. 178–179. However, in contrast, Barz does not perceive this terminological distinction and suggests that jagata and prapañca are seemingly used interchangeably. Refer to Barz 1976: 66 for further analysis.

The universe is like the mirage of water [due to the sun, which], by mistake, appears as not false. If you can see the origin and destruction of something, then how can this be called real? Until the individual self is awakened, this [universe] is [what is] perceived. Like the error [of mistaking] a rope for a serpent or [of perceiving] silver in a seashell. The confusion dispelled, [there is] no serpent nor silver, [only] the pure true being is perceived.

These words are spoken by Discrimination, who discredits Science-of-Reasoning. This didactic verse employs similes as examples, followed by assertive explanations. The mention of the sun/water/mirage example is followed by philosophical questioning and reflection.[181] The jagata appears to exist due to a mistake (bhrama) and seems to be amithyā, that is, "not false", in the sense of "not unreal", like a mirage. The use of amithyā is significant as it emphasises a stark contrast between the impermanent world and the imperishable reality. Satya, meaning "true" as in "real" or "reality", which appears on the following line in an interrogative sense ("How can this be called real?"), could have been chosen instead of the negating mithyā. However, the choice of amithyā ("not false") suggests an even stronger distinction between the two. Thus, satya and the jagata are not juxtaposed. The world's unreality is justified by its temporary nature, having a beginning and an end ("you can see the origin and destruction of something").

Next, in the same verse, there is the dual simile of mistaking a rope for a serpent and a seashell for silver.[182] When correct perception arises, the epistemic error is dispelled, and the snake and silver vanish. These instances prompt the reader/listener to contemplate the following questions: What is eternal, that which is not originated and destroyed? What is the true and inherent form underlying the illusion of the world? What is the real form of the world, the rope to its serpent?

Despite the PcN's overall emphasis on the deceptive nature of the world, the second part of verse 6.126 opens up an additional dimension of interpretation:

māyā aru paramāṇa puruṣa saba hai vāke anurūpā
  jala taraṅga aru kaṅcana bhūṣaṇa ekahi bastu anūpā
nitya prakāśarūpa hai joī udai bināśa na jāhī
  jaga karatā bharatā haratā so nahiṁ vikāra kachu tāhī
jyoṁ ambara meṁ jaladhara śreṇī vividhi bhāṁti daraśāvai
  vinaśai bahuri milai tāhī meṁ nahiṁ vikāra kachu āvai

181 These similes were commonly employed in non-shastric literature, including Sanskrit dramas. For insights into their usage in such dramas, see Goldman 1986.

182 Śaṅkara mainly employs the rope-snake analogy to illustrate superimposition (adhyāsa) and the doctrine that the effect pre-exists in the cause (satkāryavāda) as superimposition. See Suthren Hirst 2005: 86, 105–109; Rao 1998.

nṛpati viveka samīpa puruṣa ke yoṁ jaba bhāṣi sunāyo
harṣavanta hvai puruṣa kahyo taba yaha mata moko bhāyo
(PcN 6.126)

Material Nature and the atoms are all conformable to Supreme Self: like waves and water, ornaments and gold, [they] are just one thing, incomparable. Always a brilliant form, unborn, imperishable, creating, sustaining, and withdrawing the universe – no change occurs to him! Different masses of rain-bearing [clouds] can be seen as they disperse, [then] they gather [again, yet] there is no modification in the sky!

The first part of verse 126, with its similes paired with explanations, is embedded in a broader elucidation: the world is unreal, and so are its presumed constituents, be it māyā or atoms. However, the significance conveyed by the imagery is that their unreality is established when they are perceived as different from the puruṣa. When one forgets that the essence of the world is nothing but the Supreme Self, the ability to discern between the imperishable and eternal and that which is not becomes obscured.

It should also be noted that this verse reinforces the perspective that the relationship between māyā and puruṣa does not affect puruṣa, just as the clouds do not truly modify the sky.[183] In terms of causality, the PcN suggests the involvement of brahman in the creation of the universe, as without his impetus, his radiant light, nothing would have been possible (1.142). However, the text carefully emphasises that it is māyā that undergoes change, not brahman (1.143). In Act VI, a concise teaching is presented regarding the relationship between puruṣa, māyā, and the phenomenal world: māyā is the ineffable power of the Lord, who allowed the world to manifest through her (6.21).

Examining the RcM, Tulsī's text does not extensively delve into a metaphysical reflection on causality, but it does depict the world as founded on Rāma (āśrita, 1.117.1). It warns against superimposition in 1.116.1–4, as it leads one to believe that everything possesses an independent reality, while it is Hari's light that truly animates the worlds and their inhabitants.[184] Rāma is praised as the basis of the world (ādhāra, 1.197) and as the father, or father and mother, of the world (jagata pitu mātā, 1.199.1; jagata pitā, 1.201.4), similar to PcN 1.173. Tulsī employs the analogies mentioned earlier of the sun/water and the silver/seashell to convey the perceptual error of regarding the world as real when it is false (mṛṣā). While I cannot establish a complete dynamic of adaptive reuse between the PcN and the RcM in this instance, it is still noteworthy that both PcN 6.126 and RcM

183 See *Paramārthasāra* 35.
184 See Bharadwaj 1979: 43.

1.117 convey superimposition through variations of the same locative postposition: PcN "madhyā" and RcM "mahuṁ".[185]

The PcN presents a tension between the negative characterisation of the world as an illusion and, simultaneously, the belief in the existence of a unique, all-encompassing, variegated reality. This tension remains unresolved, as the only indications that the world may indeed possess a certain degree of reality are found in the verse mentioned above (6.126) and another earlier in the same act:

jāko anta na pāiye pāravāra madhi eka
  alakha arūpa anādi jala laharaiṁ jagata aneka
laharaiṁ jagata aneka uṭhata jo jala ke māhīṁ
  līna hota puni tahīṁ chūṭa bāhira nahiṁ jāhīṁ
niramala athala agādha eka rasa rasa haiṁ tāko
  brahmānanda samudra veda gāvata yaśa jāko
(PcN 6.97)

Endless, one in this and the other shore. Not visible, without a form, without a beginning, the waves [of] the ocean [are] the multiple worlds. The multiple worlds, [like] the waves of the ocean, rise. Then, they are absorbed again in [the ocean]; they do not separate [from it]. Pure, unfluctuating, profound, his essence a single essence. The ocean of bliss of brahman, whose glory the Vedas sing.

Even so, this verse does not directly address the world itself but rather brahman. The imagery employed here appears to suggest the ultimate identity between brahman and the world(s). It may also indicate shared characteristics or nature between brahman and the world, but this aspect is largely left to interpretation. Still, these verses provide a glimpse of a possible affinity between Brajvāsī's thought and those outside the realm of Advaita concerning the phenomenal world. Notably, while the simile of the sky and clouds does not imply a similarity or identity in nature between the two, the similes of the ocean and its waves and jewels and gold do suggest a common or similar nature. The first instance implies separation, while the second conveys both difference and identity. There is, therefore, an observable inclination towards the conceptions of Bhāskara's Bhedābheda Vedānta. In his commentary on BS 2.1.14, Bhāskara employs the example of the ocean and its waves to illustrate his Vedantic philosophy.[186] However, unlike Bhāskara, the PcN does not seem to support the pariṇamavāda in the sense of an

185 PcN: mālā visaya sarppa bhrama jaise sīpa madhyā jimi rūpā; RcM: rajata sīpa mahuṁ māsa jimi jathā bhānu kara bāri.
186 See Dasgupta 1922, vol. 4: 329; Ingalls 1967: 66.

actual transformation of brahman into the world. Instead, through its frequent use of illusory/dream imagery when discussing the world, the PcN appears to conceive of the modifications as apparent (vivartavāda).

Of course, by positing a degree of identity with the Supreme Self, Brajvāsī aligns himself with Vaiṣṇava Vedāntins such as Vallabha and Rāmānuja. The BhP, the primary text for these philosophers, also regards the Supreme Self as the support (ādhāra) of prakṛti and thus the ultimate cause of the world.[187] However, the BhP is much more inclined to affirm the reality of the world and the identity between bhagavān and the world compared to the PcN.[188] As we have already observed, Brajvāsī does not adopt the specific terms and threefold ontological distinctions developed by Vallabha,[189] nor does he assert that the jagata is the body of God like Rāmānuja does.

When examining later developments in the Vedantic tradition, we find parallels on the conceptual level, even if exact correspondences in terminology are not evident. Vijñānabhikṣu, an important Bhedābheda thinker of the sixteenth century, in his commentary on the BS, the *Vijñānāmṛtabhāṣya*, employs a combination of Advaita terminology (such as mithyā, "not false") and affirmations of the reality of the world.[190] In his interpretation, material cause (upadāna kāraṇa) has two meanings: the primary meaning is the locus cause (adiṣṭhāna kāraṇa), which remains unchanged and is identified as brahman; the secondary meaning is the changing cause, whose underlying substratum is brahman.[191] By presenting this elucidation, Vijñānabhikṣu demonstrates that he aligns with neither the pariṇamavāda nor the vivartavāda perspectives. Andrew Nicholson reminds us that the distinction between these two perspectives was not always well-defined. He also suggests that Vijñānabhikṣu likely considered himself not to be a pariṇamavādin like Bhāskara since "he understands *pariṇamavādin* to mean one who believes that Brahman is a changeable material cause". Nicholson further notes that in this regard, Vijñānabhikṣu is "much closer to his 16th century Advaita contemporaries than to early bhedābhedavādins such as Bhāskara".[192] Notably, Vijñānabhikṣu's position bears some similarity to that of his contemporary, the Advaitin Prakāśānanda, who also posited the idea of a locus cause and clearly identified the changing cause with māyā. As an adherent of Advaita, Prakāśānanda considered the world an unreal transforma-

187 See BhP 11.24.19.

188 See Sheridan 1986: 26–27, 31.

189 According to Vallabha, brahman is the inherent cause (sāmavayikāraṇa). See Dasgupta 1922, vol. 4: 327–328.

190 See Nicholson 2007: 385.

191 See Nicholson 2007: 394–395 for an extensive account with examples.

192 Both quotations from Nicholson 2007: 396.

tion and did not concede provisional reality to it. According to him, the only two ontological categories were reality and unreality/falseness.[193]

Rather than attempting to precisely assign a specific viewpoint to the PcN regarding the phenomenal world, it is more accurate to say that the text, despite its emphasis on the falseness of the world, allows for multiple interpretations. The emphasis on a deceptive world and the limited exploration of the complete or partial identity of the world with brahman may be attributed to the primary purpose of the PcN, which is to cultivate detachment from the world through the discernment of one's true nature.

4.1.4 Jīvātama Puruṣa/Mana

The final aspect within the metaphysical realm of the PcN to be examined is the concept and character of Individual Self, the jīvātama puruṣa (Skt. jīvātman). We will delve into three interconnected questions concerning Individual Self: his origin, his relationship with Mind (mana) and other terms used in the drama to refer to both entities, and his relationship with Supreme Self. As we have previously encountered in PcN 1.150, it states that "from the Supreme Self [comes] the Individual Self, from the Individual Self [comes] Mind, from Mind the brilliance [in] the universe".[194] Regarding the origin of Individual Self, the PcN upholds the perspective that it is as though the Lord deceives himself through his own power of māyā, through which the jīvātama is manifested:[195]

> [. . .] prakaṭa bhayo sundara suta tāte jīva nāma kahi gayo
> vividhi bhāṁti ke mandira sundara pāṅca tattva meṁ kīnha
> nava nava dvāra savāṁre tina meṁ bahu vidhi śobhā dīnhā
> lakṣaṇahuṁ te jite māyā meṁ so saba sutahi sikhāye
> kīnhoṁ rāja lalaki tribhuvana ko sukha dukha bhoga suhāye
> (PcN 1.146)

A lovely son was born, called Individual Self, by her [Illusion]. He built multiple kinds of beautiful temples in the five elements. He gave them a variegated beauty

193 "The existence of the objects is nothing more than their perception (dṛṣṭi)." See Dasgupta 1922, vol. 2: 221, 224.

194 See also subsection 4.1.1.

195 See PcN 1.136, 1.144, 6.17–6.18, and 6.21. The KPC also contains similar elements, as seen in verse 1.89 in Kapstein 2009: 35–37. These elements resonate with texts such as Gauḍapāda's *Kārikās* on the *Māṇḍukya Upaniṣad* (2.19) and the *Paramārthasāra* (32). Sheridan supports the perspective that even in the BhP, the Supreme Self is influenced by his own power. See Sheridan 1986: 43.

[by] decorating [them with] nine new doors. He [Individual Self] taught everything to [his] son [Mind], including the many characteristics of Illusion. He [Individual Self] desired to be king of the three worlds, to be beautiful [with] the enjoyment [of] happiness and sorrow.

When Individual Self manifests from Māyā, he becomes connected to the physical world. This connection is expressed through the metaphors of the body composed of the five elements, a building with nine doors,[196] and the reference to mana (Skt. manas). In this context, the negative implications of being bound to the phenomenal world are mitigated by describing the bodies as beautiful temples. Similar to the PcN's overall perspective on materiality, which leans towards a negative view, there is a subtle hint that the body can also be seen positively. This echoes the play's opening scene, where the First Actor tells his wife that the human body (nara tana) is valuable as it enables one to pursue higher goals.[197]

The introduction of Mind, not as the son of Supreme Self and Illusion/Material Nature but as the son of Individual Self, represents an innovation. The PcN explicitly defines the relationship between the two entities in the following manner:

jāte prakṛti garbha ko pāyo
  suta mana nāma prakaṭa tina jāyo
kahata ātamāhū manahī ko
  mana hai viṣaya ātamāhi ko
ubhaya kachū ātama mana nāhīṁ
  ekai haiṁ jimi tana parachāhīṁ
(PcN 1.96)

From that [movement], Material Nature became pregnant. She gave birth [to] a son, visible, named Mind. Mind, indeed, is also called Self; Mind is the field of Self. [But] Self and Mind are not distinct; they are just one – like body and shadow.

The visible (prakaṭa) manifestation of Material Nature is Mind, which is also material. Mind serves as the object (viṣaya)[198] of pure consciousness, of Individual Self. Here, I have translated viṣaya as "field" in the sense of "field of action", em-

196 PcN 3.20: [Jaina Ascetic:] "To what extent can I describe him [Individual Self]? All listen carefully: this body is [made of] the five elements, [and] inside is filled [with] dross and dust. It is manifested with nine doors; it [has] impurities like urine, mucus, etc." See also *Sāṅkhya Kārikā* 35; BhG 5.13; BhP 4.29.2–16.
197 PcN 1.52: "When Murāri gives a human body, then one is entitled [to] pure knowledge." This viewpoint appears to align with the perspective supported by the RcM. See Bharadwaj 1979: 117–118.
198 See HSS 4557.

phasising its meaning as a "sphere of influence or activity" – akin to such terms as dominion, kingdom, territory, region, district, country, abode, or plurals like lands or possessions – as a synonym for lakṣya, kṣetra, vistāra, or vibhāga.[199] This translation choice aims to underscore the notion that it is through Mind that Individual Self can function as a knower, a doer, and an enjoyer.

Individual Self is dependent on Mind and vice versa. This mutual dependence is why they are described as being one rather than two separate entities. Through the illumination of Individual Self, Mind also possesses luminosity, meaning he is conscious and capable of perceiving sensory data from the sense organs.[200] Examining the structure of the lines "ubhaya kachū ātama mana nāhīṁ | ekai haiṁ jimi tana parachāhīṁ" above, we can understand that the Individual Self (ātama) corresponds to the "body" (tana), while Mind (mana) corresponds to the "shadow" (parachāhīṁ).[201] The correlation of Individual Self with the term "tana" does not imply materiality, and likewise, the parallel between Mind and "parachāhīṁ" does not diminish his nature as a physical entity. Beyond the level of identification, the imagery of the body-shadow also signifies the primacy of Individual Self over Mind. This is further demonstrated by passages highlighting Individual Self as Mind's father (1.149) and describing him as Mind's prathama rūpa (first, original form; 1.173).[202]

However, it would be extreme to completely distinguish between the two, as the text allows for an overlap of the concepts/characters. Both Individual Self and Mind share the characteristics of being powerful and blissful, as they are both connected to Supreme Self (1.100, 5.31, 6.102). The sequence of verses that present these two concepts/characters reinforces this overlap: in 1.127, Mind is said to be manifested by brahman and partake in its greatness in 1.128–1.129. Then, in verses

---

199 See MW 997; BHK 1073.

200 In CU 8.12.5, the manas is referred to as the divine eye of the ātman, enabling perception of objects through it.

201 The term "ātama" or "ātma" is consistently used throughout the play as a shorter form of "jīvātama".

202 While my translation of viṣaya as "field" may evoke Sāṅkhya conceptions of kṣetra (field) and kṣetrajña (knower of the field), these terms are not found in the PcN. Nonetheless, the passages being analysed may allude to the idea that an individual comprises the conscious jīvātama as the kṣetrajña and the cognitive faculty, mana, as its immediate kṣetra, symbolising the broader kṣetra of the body. Among Vedantic thinkers, Śaṅkara introduced the distinction between kṣetra and kṣetrajña in his commentary on the MU verses concerning the simile of the two birds on the same tree. According to him, the inactive bird represents Īśvara, while the bird eating the fruits represents the kṣetrajña, the individual self. In the PcN, the latter bird also represents the individual, embodied self, but it cannot be said that the former bird represents Īśvara in the sense of an "inferior" form of brahman. For more details, see "Paramātama's Pervasiveness" in subsection 4.1.1. Śaṅkara's commentary on the MU verses can be found in Sastri 1905: 153–156.

1.130–1.131, Discrimination depicts Mind as being subjugated by Bewilderment, followed by verse 1.132, which portrays Individual Self as being deceived by Bewilderment. Further examples of this relative interchangeability between the two can be found in verse 1.96, where Mind is described as Material Nature's (Māyā) first son, and in verse 1.146, where Individual Self is mentioned as Illusion's (Māyā) offspring.[203] It is important to note that while the alternation between the concepts/characters of Individual Self and Mind can be fluid at times, there are no other terms substituting or paraphrasing Mana when it is specifically intended as the character in the play from both cosmogonical and soteriological perspectives.

There is only one instance where neither the term jīvātama nor mana appear when we would expect them to do so. In verse 6.144, Discrimination teaches Individual Self how to interpret the famous Vedic statement (beda bacana) of "neti neti".[204] His instruction revolves around reversing the tattvas, starting from the gross principles to the more subtle ones. After dissolving the "last" of the mahābhūtas (ākāśa or space) into the corresponding tanmātra (śabda or sound), śabda unites (milai) with the triple ahaṃkāra. Subsequently, the latter mingles (ralai) with the mahātatva, and the mahātatva is absorbed (līna) into material nature (māyā). Finally, material nature is immersed in the Highest Reality.[205]

We have observed that Individual Self and Mind are the first to manifest from the union between Supreme Self and Illusion/Material Nature and that I-maker is born from Mana and his first wife Engagement-with-Action. However, in this instance, although māyā is mentioned, the term mahātatva or "great/large constituent" – not mana or jīvātama puruṣa – is used.[206] The concept of mahattattva is generally seen as a crystallised form of the mahān/mahān ātmā from Upanishadic speculation and is more closely associated with classical Sāṅkhya theory.[207] I contend that in the PcN, the term mahātatva represents the "compound" of jīvātama and mana and exhibits some characteristics of the mahān that are not integrated into the Sāṅkhya worldview while also not entirely aligning with the Upanishadic account of it.

---

203 From the point of view of the flow of the narrative: in 1.96, Desire provides a glimpse of the family's genealogy, which will be further expanded by Discrimination in 1.146–1.150. Desire's explanation lacks the extended critique of Mind as the cause of suffering for Individual Self articulated by Discrimination at 1.118–1.119 and 1.151–1.152.

204 For more on this topic, see section 4.2.

205 It is worth noting that the KPC does not precisely explain "neti neti" in the same way. In the KPC, Viveka does refer to the dissolution of the tattvas but does not specify what these tattvas are or the process involved. See Kṛṣṇamiśra 2009: 269–271.

206 The Brajbhāṣā term "mahātatva" is a tadbhava of the Sanskrit term "mahattattva". Interestingly, "mahattattva" also appears immediately after in the same verse.

207 Buitenen 1964: 114.

Individual Self, unlike the mahān of the BĀU, is not portrayed as the Highest Principle, and there is no explicit description of him as vast or complete. Furthermore, the PcN does not explicitly state that Individual Self recognises himself prior to embodiment by declaring, "I alone am this creation, for I have created all this."[208] However, verse 1.146 of the PcN does report that Individual Self creates multiple temples/bodies out of Material Nature's guṇas because he desires (lalaki) to experience embodied existence.[209] This echoes BĀU 1.4.17, where the mahān ātmā desires a wife, progeny, and wealth to perform sacrifices. Additionally, the mahān ātmā is identified with puruṣa in BĀU 2.5.18, and it is worth noting that Jīvātama's complete appellation in the PcN is Jīvātama Puruṣa.

These thoughts and actions attributed more to Jīvātama Puruṣa than to Paramātama Puruṣa seem to reflect the shift in emphasis noted by Johannes van Buitenen in the *Kaṭha Upaniṣad* and other texts, where the manifest mahān ātmā is conceptualised as the "first-born", subordinate to an unmanifest principle (avyakta).[210] While van Buitenen observes that the Upaniṣads do not delve into whether this unmanifest principle is material or "spiritual",[211] the PcN presents its own perspective through Discrimination's explanation to his wife, Good Reason:

aho dūri daraśī sumati puruṣa rūpa kī rāśa
tāte hai yaha prakaṭa jiya tiya te jagata prakāśa
(PcN 1.148)

O farsighted Good Reason, [Supreme] Self is a receptacle of form.
By him, the Individual Self was manifested through the woman, [he is] the light in the universe.

Jīvātama Puruṣa, as embodied, is the first-born progeny of Māyā, the unmanifest tattva into which he, as mahātatva, will be absorbed, and Puruṣa, whose nature he shares (6.139). The quasi-identity between Jīvātama and Mana, as discussed earlier, further reinforces the composite nature of their origin, with the embodiment of Individual Self emphasised when addressed as Mind. Viewing the PcN's mahātatva as Jīvātama/Mana helps make sense of the fact that in verse 6.144, ahaṃkāra is sublated by mahātatva, aligning with the PcN's genealogy that positions I-maker (ahaṃkāra) as the son of Mind (mana).

208 Buitenen 1964: 104, 107–108.
209 As above in this chapter. "He built multiple kinds of beautiful temples in the five elements" (PcN 1.146).
210 Buitenen 1964: 106–107.
211 The first viewpoint would be adopted by classical Sāṅkhya and the second by Vedānta.

The last relevant issue raised by van Buitenen is that the Upanishadic mahān transformed into the mahat or mahattattva of classical Sāṅkhya, "which made innocuous a term that in its original development of conception reflected a fundamental position opposed to the dualism of *puruṣa* and *prakṛti*, that of a supreme being, creating itself in the universe".[212] However, in the context of the PcN's Sāṅkhya framework, the term mahātatva reveals that Brajvāsī's text assimilated the Upanishadic mahān as Jīvātama, a being who creates himself out of the desire to multiply and experience enjoyment, within a broader structure of duality encompassing Paramātama and Māyā/Prakṛti. Nevertheless, as we explored in relation to the nature of the phenomenal world, this duality is ultimately a non-duality.

In the PcN, the concept of buddhi does not explicitly arise in this level of discussion, as it appears that mana serves as an umbrella term encompassing the functions of both buddhi and citta. In the case of citta, this equivalence is substantiated by Brajvāsī employing "mana" in place of the "citta" or "cittavṛtti" used in the YS when defining yoga.[213] As for buddhi, it is not specifically addressed as an independent entity, but rather, mana, functioning as the mind, encompasses not only the organisation of sensory data but also the "faculty of intellect that produces refined decisions, resolutions, and determinations".[214]

However, in contexts that are not directly connected to the identity of the characters, the terms buddhi and citta/cita do make appearances. Buddhi is primarily used to denote the capacity for reflection and secondarily refers to the content of that capacity, encompassing thoughts and ideas.[215] On the other hand, citta/cita is more frequently utilised than buddhi[216] and can often be translated as "heart". Thus, there are instances where citta/cita is employed to refer to the seat of emotions and feelings.[217] In fact, throughout the play, citta/cita is prominently featured in dialogues related to the process of liberation, often associated with the verbs dhar-, lā-, and de-, which convey the meaning of "to concentrate".[218]

---

212 See Buitenen 1964: 114.

213 In the context of Passion's description of Discrimination's fourth minister, Breath Control. PcN 1.76: [. . .] | mana ko kare nigraha apara abhyāsa yoga bakhāniye ||. In the commentary to the YS and Vācaspati Miśra's own commentary to the YS, the terms "citta", "buddhi", and "manas" appear to be used interchangeably. See Whicher 1997: 39n20.

214 In the words of James Fitzgerald, who has written two informative and insightful articles on the meanings of buddhi. See Fitzgerald 2017a, 2017b.

215 See e.g. PcN 3.49 (intellect), 6.60 (intelligence), 6.67 (insight), and 6.126 (thought).

216 I have counted eleven instances of the term "buddhi" (excluding words formed from the same root) and thirty-nine occurrences of "citta" or "cita".

217 See e.g. PcN 1.42, 3.27, 3.5, 4.4, and 4.77.

218 See PcN 2.21 (dhar-), 5.118 (lā-), and 2.113 (de-).

Furthermore, there is another direct echo of YS 1.2, where the term "cittavṛtti" is specifically broken down as "citta kī vṛtti" in PcN 2.96.

Jīvātama Puruṣa and Paramātama Puruṣa

The relationship between Jīvātama Puruṣa and Paramātama Puruṣa is a crucial aspect that has been briefly touched upon in the previous subsections. As discussed earlier, Jīvātama is associated both with Māyā and Paramātama as the first-born. Furthermore, Jīvātama is referred to as pauruṣa (6.139), indicating his participation in Paramātama's nature or his belongingness to Paramātama.[219]

Previously, Brajvāsī employed a simile using the reflection of sunlight on different surfaces to illustrate the relations among Paramātama, Jīvātama, and Mana.[220] Verse 1.150 stresses a shared quality or nature (light) among these categories of beings. It also highlights the dependence of the latter two on the former, as light cannot exist separately from the sun. However, the verse remains somewhat cryptic. The imagery of sunlight reflected on water surfaces may suggest that Brajvāsī attempts to explain their relationship using the Advaita theory of reflection (pratibimba). Still, it should be recalled that the PcN also mentions the Supreme Self residing in every body[221] (saba ghaṭa bāsī, 1.96), which recalls the opposing Advaita theory of delimitation (avaccheda).[222] As we saw earlier in the case of avidyā,[223] it is challenging to determine with certainty which position on individuation the PcN favours, as Brajvāsī's text does not explicitly align with a specific school of thought. Additionally, it is worth noting that the distinctions between the

---

219 See also PcN 6.17, where Faith reports Paramātama Puruṣa as saying that individual selves "come" or "are" from his own self and remain non-conscious; that is, bound to the materiality of the world by Māyā: jaga jīva jite mama ātama teṁ | jaḍa hoya rahe saba yā kara teṁ ||. (As many individual selves are [in] the world from my own self, all stay unconscious by her [Material Nature's] hand). In verse 1.117, instead, Mana is said to belong to Self (puruṣa), which I chose to interpret as Individual Self. The ambiguity of puruṣa, deprived of a qualification as Jīvātama or Paramātama, could still point to the fact that ultimately Mana and Jīvātama are identified in the text, so in this verse it is implied that both belong to Paramātama Puruṣa.

220 See subsection 4.1.1, especially "Paramātama's Light".

221 Or, metaphorically, every "pot". See "Paramātama's Pervasiveness" in subsection 4.1.1 and footnotes 24–26 in this chapter.

222 In the theory of pratibimba, the individual self is conceptualised as an appearance or reflection of brahman, with brahman being the sole reality and the "prototype" (bimba) of the reflection. Conversely, in the theory of avaccheda, the individual self is viewed as a limitation of brahman, as space is confined by pots. See Potter 1999: 172–182.

223 See subsection 4.1.2.

two schools started to become less rigid during the early modern period, so applying these terms in this context may be anachronistic.[224]

These analogies coexist with another perspective on the relationship between Paramātama and Jīvātama/Mana. In two instances, Mana and Jīvātama are likened to a drop in the ocean, symbolising Paramātama.[225] Initially, this may seem like a variation of the sun/light analogy. However, verse 1.129 intriguingly juxtaposes the ocean-drop metaphor with the expression "a particle of such mass" (dānā aiso rāśi ko), hinting at a different kind of relational possibility. Another verse, 1.146, further supports the understanding that Jīvātama/Mana is to be seen as a "fragment" of Paramātama, as it states that Māyā obtains a part (pāyo aṃśa) of Paramātama Puruṣa.

The conceptualisation of the individual self as an aṃśa of the Supreme Self is also present in the BhG and the BhP.[226] This idea has been part of a philosophical debate dating back to BS 2.3.42 (or 2.3.43 depending on the edition). The challenge for interpreters of Vedānta lies in deciding whether to interpret the sutra text literally or metaphorically.[227] Śaṅkara, opposing the notion of the Supreme Self having parts, rejected a literal interpretation that would undermine its changelessness and imply duality.[228] Conversely, Rāmānuja and other Vaiṣṇavas accepted the literal interpretation of the individual self being a part, although each explained it from their own unique perspective. For instance, Rāmānuja placed it within his ontological framework by asserting that "part" refers to an aspect or quality of something.[229] Vallabha also considered the jīva as a part but one manifested from the akṣara brahman rather than the parabrahman.[230]

In this regard, the PcN's overarching metaphor, where Paramātama is the father and Jīvātama is the son (and Mana the grandson), adds another layer of significance. Unlike the previously discussed metaphors, this image not only represents an ontological relationship but also sheds light on the nature of liberation and the path towards it. While salvation in the PcN will be explored in greater detail in the subsequent part of this chapter, the focus here will be on the metaphysical dimension, with only a brief mention of the redemptive aspect. The grandfather/father-son

224 See Minkowski 2011: 212–213.

225 See PcN 1.129 and 6.144.

226 See BhG 15.7 and BhP 11.11.4.

227 The hermeneutical question of meaning and levels of meaning lies at the heart of much of Indian intellectual history. For a comprehensive synthesis of these aspects in philosophy, see Keating 2020.

228 See Śaṅkara 2009: 506–520.

229 See Vireswarananda/Adidevananda 2003: 296–298.

230 See Smith 2016: 125; Narain 2004: 276, 279–284; Barz 1976: 68–69.

imagery was already present in the KPC and can be found in such later Vedantic texts as Bhāratītīrtha/Vidyāraṇya's *Pañcadaśī*, mentioned earlier in this chapter.[231] Far from implying a part/whole relationship between brahman and ātman, Bhāratītīrtha/Vidyāraṇya employs the familial analogy to explain that when māyā is superimposed on brahman, it is referred to as īśvara, and when avidyā is superimposed on brahman, it is referred to as jīva. Thus, it is like a man who is a father to a son (īśvara) and a grandfather to a grandson (jīva). However, if māyā and avidyā are neutralised, brahman is not truly connected to īśvara or jīva, just as a man is not a father or a grandfather in and of himself, without the association *with* a son or a grandson.[232]

Two centuries later, the Bhedābheda philosopher Vijñānabhikṣu offered criticism of the theories of reflection and limitation in his *Vijñānāmṛtabhāṣya* on the BS. He deconstructed the examples of the sun's reflection on water surfaces and of the space enclosed in pots to demonstrate their limited validity.[233] Instead, Vijñānabhikṣu advocated for the concept that the individual self is an aṃśa of the Supreme Self, understood in the technical sense of it sharing properties and belonging to the same class as the whole.[234] Vijñānabhikṣu employed the father-son analogy to emphasise this point further: "And at the time of the creation of the universe, just due to the Lord's own desire, the souls, after attaining effective consciousness, become manifest just as the sons of the father become manifest. Therefore, souls can be called the parts (aṃśa) of Brahman."[235] Furthermore, he posited that the whole can exist independently from its parts, just as a father can exist separately from his son, unaffected and unaltered by the son's absence. However, when it comes to the dissolution of the universe, Vijñānabhikṣu's father-son analogy falls short in explaining how a son could be reabsorbed into the father at the time of death, as noted by Andrew Nicholson.[236]

Returning to the PcN, we have observed that Brajvāsī's clearest statement is that the jīva is an aṃśa of brahman, and like Vijñānabhikṣu, he emphasises the

---

231 See subsection 4.1.2.

232 My paraphrase of *Pañcadaśī* 3.37–3.42. See Vidyāraṇya 1954: 69–70.

233 See Nicholson 2007: 386–388.

234 According to Vijñānabhikṣu, brahman and jīva have selfhood, existence, and consciousness in common. See Nicholson 2007: 389.

235 Quoted and translated by Nicholson 2007: 391.

236 Nicholson 2007: 391: "Either Vijñānabhikṣu means that the father inherits his son's possessions (*vetanāḥ*) when the son dies, or his consciousness (*cetanāḥ*) is absorbed into his father's at death. The first seems irrelevant, while the second is simply false. The basic problem with this metaphor is that although the father can certainly be said to be responsible for the manifestation of his son, it is hard to see how he might re-absorb his sons at the time of his son's death."

shared nature between the jīva and brahman.[237] However, what sets Brajvāsī apart from Vijñānabhikṣu is his inclusive approach to the various analogies used. Brajvāsī integrates the sun-reflection and space-pot images into the broader father-son model of the PcN rather than viewing them as competing or less valid. Additionally, the PcN specifically addresses the topic of "reabsorption" within the context of the father-son discourse. The following verse captures this idea:[238]

[. . .] putra pitā kī baiṭhe godī
  hoya parama ānanda binodī
pitā putra mili hoya bhalāī
  hoya daśahu diśi ananda bhadhāī
pitā putra doū mili rahahīṁ
  nityānanda parama sukha lahahīṁ
yoṁ ātama paramātama bheṁṭai
  pitā putra ko kaṅṭha lapeṭai
(PcN 1.173)

The son will sit on his father's lap, they will be joyful and there will be supreme bliss. By the meeting of father and son, there will be goodness, there will be a surge of bliss in the ten directions. Father and son will stay together, they will experience supreme joy and eternal bliss. In this way Supreme and Individual Self meet, the father embraces the son.

The meeting of Paramātama and Jīvātama depicted in this verse signifies their reunion after being previously separated by Māyā. This analogy emphasises both difference and non-difference,[239] illustrating that liberation is characterised by a sense of renewed belongingness.[240] However, in this case, too, we should exercise caution before applying labels to Brajvāsī and his PcN, as the father-son analogy

---

237 It has also been suggested that Brajvāsīdās may have been drawn to the concept of brahman as the locus cause, a view shared by sixteenth-century Vedāntins from different lineages, including Vijñānabhikṣu. See subsection 4.1.3.

238 It should be noted that Vijñānabhikṣu did not show an interest in bhakti despite the times in which he lived. Therefore, it is likely that the manner in which Brajvāsī expanded and thought about the father-son image and its ontological and soteriological significance was not an option for the Bhedābheda thinker. For further insights into Vijñānabhikṣu's perspective on the path to liberation, see Ram 1995: 127–156.

239 Or separation and non-separation, as Vijñānabhikṣu would interpret it, where the natural state of things is the non-separation of the Supreme Self and Individual Self. See Nicholson 2010: 44–50.

240 In PcN 6.139, Upaniṣad tells Individual Self: jāni lehu aise yaha bāta | puruṣa sanātana ke tuma tata || (Understand this fact: you, dear, [belong] to the eternal Supreme Self.).

had not disappeared from Advaita speculation either. In the seventeenth century, Nārāyaṇa Tīrtha, an Advaitin scholar closely associated with Madhusūdana Sarasvatī, utilised the father-son metaphor to explain that once the individual self's ignorance about its true nature is dispelled, it attains unity with the Supreme Self.[241] Nārāyaṇa Tīrtha's intellectual project harmoniously merged Advaita Vedānta, bhakti, and yoga, much like Vijñānabhikṣu's approach.[242]

The father-son relationship, after all, is compatible not only with Vedantic ontology but also bhakti. In BhG 11.44, Arjuna, upon realising that Kṛṣṇa is the Supreme God, entreats him to forgive his previous casual manner of address and embody the roles of a father to a son, a friend to a friend, or a lover to his beloved. On this point, Angelika Malinar observes that the BhG integrates various types of relationships, including familial ones, into the fabric of bhakti.[243] Similarly, in the RcM, the jīva is indeed a part of brahman. In verses 7.86.1–4, Rāma teaches the crow Bhuśuṇḍi that he loves all beings in the universe and compares himself to a father with multiple sons, each with their unique way of building a relationship with him.[244] Brajvāsī's PcN appears to engage with both classical Vedantic discourses and adapt the PC story to his own bhakti models and sensibilities. By incorporating these diverse elements, Brajvāsī crafts a narrative that resonates with both traditional Vedānta and his devotional perspective.

Having explored the cosmological framework of the PcN and established its connections not only with classical sources of Vedantic philosophies but also with the RcM, the final section of this chapter delves into the path to liberation. By examining the PcN's proposed soteriological path, we will observe a similar pattern of adaptively reusing elements from the RcM. Moreover, this analysis will shed light on the PcN's unique approach to "bringing the son closer to the father" by synthesising yogic bhakti and Vedānta.

4.2 A Bhakta Yogi and the Path to Freedom

As discussed in Chapter 1, the depiction of Kṛṣṇamiśra in the early stages of the PcN presents him as a compassionate teacher.[245] In this particular instance, Kṛṣṇamiśra, or rather Kṛṣṇadās, decides to liberate his disciple from the cycle of saṃsāra

241 See Venkatkrishnan 2015: 216.

242 Nārāyaṇa Tīrtha's blending of Advaita, bhakti, and yoga shares similarities with the PcN's understanding, which will be further explored in the next part of the chapter.

243 See Malinar 2007: 184–185; also Malinar 2015: 386–408.

244 See RcM 7.116.1.

245 See subsection 1.2.2.

and concludes that composing a play is the most effective solution. The student, who has a deep passion for literature, resists traditional modes of instruction:

[. . .] guru upadeśa na mana meṁ āvai
  jimi jvara grasita sunāja na bhāvai
jau māṅgai tau kupatha khaṭāī
  jāke khāta tāpa adhikāī
tāhi vaida samarattha sujānā
  cūraṇa deta khaṭāī sānā
jāke khāta hoya jvara nāśā
  auguna amala na karai prakāśā
(PcN 1.13)

The teachings [of] the teacher did not enter his heart; like [someone] taken [by] fever, he did not enjoy a healthy meal. When he begged, he [obtained] unhealthy food, sour; [as soon as] he ate it, his temperature increased. To him, a physician, experienced and wise, gave a powder mixed [with] sour [food]. [When] he ate it, the fever was eradicated, [and] there was no way [for him] to distinguish between impure and pure.

Analogies between teaching/philosophy and healing/medicine are prevalent in Indian philosophical traditions.[246] By considering some of the key themes explored by Clare Carlisle and Jonardon Ganeri regarding the relationship between these two domains, we find the PcN addressing their questions:

Is the medicine a curative, a tonic, or a prophylactic? Is there a tension between philosophy as treatment and the autonomy of the subject? Can those who suffer cure themselves? To what extent must a philosophical "treatment" be adapted to fit the particular individual's needs? Is there a risk that, in locating the cause of suffering in all that individualises human beings, the cure deprives individuals of their individuality?[247]

In the PcN, philosophical medicine serves as a remedy for an existing unhealthy condition. The treatment is tailored to the specific disposition of the individual.[248] This approach seems to lead to a state where the healed individual can continue

246 For an overview, see Halbfass 1991.
247 Carlisle/Ganeri 2010: 3. An additional question is, does health correspond to liberation/freedom (mukti)? This question will be addressed at the end of this chapter.
248 This notion also applies to the understanding of brahman and its personal manifestations in PcN 5.38–5.40, which confirms the existence of various forms of brahman, allowing individuals to choose their preferred form to worship.

their life with a renewed understanding of themselves and the world.[249] Health is restored by contemplating and correctly discerning Vedānta.[250]

Regarding Carlisle and Ganeri's question of whether identifying individuality as the cause of suffering ultimately results in the loss of individuality itself, the PcN offers a distinct perspective. The mentioned verse highlights the potentiality of individual, albeit circumscribed, paths to healing. The aim is not to negate or erase individuality but to transcend toxic individuality. Relationships, impossible without the individual and envisioned in various forms, are foundational to any expression of bhakti.[251] From this standpoint, the misguided individuality that needs to be discarded is when a person fails to recognise their inherent connection to the divine and acts solely for selfish motives and objectives.[252]

Regarding the likely direct sources of inspiration for the teaching/healing analogy in the PcN, we find parallels in the *Bāla Kāṇḍa* of the RcM, where Tulsī also employs the medical simile in various passages. While we cannot establish a complete dynamic of adaptive reuse, there is a clear re-use of terms from the RcM in the PcN. Similarly, in the RcM, an unwell person desires to consume unhealthy food (kupatha māga), which the skilled physician (baida) does not provide in itself but mixed with the medicine.[253] What further connects these two texts is the concept in the RcM that the healing powder, referred to as cūraṇa or the dust of the guru's feet,[254] can remove impairments to sight, allowing one to perceive and think with discernment (bibeka bilocana).[255] Furthermore, the RcM describes the absence of discrimination (abibeka) as a kind of fever (jvara).[256] Therefore, apart from sharing terminology, both texts view the actions of the guru, whether

249 As seen in PcN 1.16, where Brajvāsī's extradiegetic narration communicates the phala: "[There will be] the destruction [of] ego, [the] sense of possession, bewilderment, and the obtainment [of] the prosperity of knowledge. [For one who] listens, understands, [and] reads [this work] with delight, the adversity [of] the world will vanish."

250 See PcN 1.13–1.15.

251 See Malinar 2015: 396: "In the context of *bhakti* the individual is recognised in an unprecedented way as being ontologically connected to god and entitled to both earthly well-being (*bhukti*) and liberation (*mukti*)." See also Fuchs 2019.

252 In pursuing an ideal state, one aims to elevate their existence from the particular to the universal level, embracing the all-encompassing reality of brahman. This aligns with the teachings of many Upaniṣads, which encourage a kind of "spiritual exercise", as described by Jonardon Ganeri, drawing upon Pierre Hadot's concept. Refer to Ganeri 2010 for further insights. Hadot's conception of "exercices spirituels" has also been explored by Bouthillette 2020: 18–20.

253 See RcM 1.32.1, where the doctor is Rāma.

254 See RcM 1.7.1.

255 See RcM 1.1.1.

256 See RcM 7.120.19. In *Upadeśasāhasrī* 19.1, Śaṅkara also employs the imagery of someone consumed by the fever of desire. See Potter 1981: 254.

through the dust of his feet or the composition of the PC story by Kṛṣṇadās, as redemptive.[257] According to Bhuśuṇḍi, the Rāma kathā reinforces dispassion, discrimination, and bhakti – just as Kṛṣṇadās aimed to consolidate bhakti, knowledge, and dispassion, according to Brajvāsī.[258]

In Chapter 3, the concept of viveka was highlighted as a means to bridge the gap between bhakti and dharma, with dharma being subservient to bhakti and upheld rightly because of bhakti. This interplay was primarily embodied in the character of Bharata, Rāma's brother. Viveka emerged as his distinctive quality as a devotee, and the chapter noted the frequent comparisons of Bharata to a haṃsa (gander) and a yogi.[259] His bhakti is infused with elements of yoga and asceticism, and his identity as a bhakta is characterised by nema (observance, discipline) and prema (love) (2.311.1). I argue that Brajvāsīdās drew inspiration from the RcM, as Tulsī's portrayal of Bharata closely resembles Brajvāsī's own conception of the ideal bhakta. In Act I of the PcN, the First Actor informs his wife that through the PC story, King Kīratibrahma will gain insight into his true nature, relinquish worldly desires, and live a life that combines discipline and love.[260] At this juncture, it is worth exploring exactly how the PcN depicts its bhakta yogis, yoga itself, and the personification of bhakti.

4.2.1 Bhakta Yogis, Yoga, and Bhakti

In the previous section, I used the term "bhakta yogis" in the plural form because the PcN introduces multiple figures that can be seen as variations on the theme of disciplined individuals devoted to God. Act II opens with a scene set in the holy city of Vārāṇasī, where Bewilderment aims to establish his capital (v. 2.4). To accomplish this, he dispatches I-maker and Deceit to corrupt those referred to collectively as vivekī (Skt. vivekin), translated as "discerning people" (2.10), or siddhi sādhaka, "accomplishers of perfection" (2.4).[261] Furthermore, various specific

257 See PcN 1.13 quoted above. See also PcN 1.14–1.15.

258 See RcM 7.14.4 and PcN 1.13.

259 See subsection 3.2.2.

260 PcN 1.50: [Actress:] "He shall win the organs of sense [thanks] to his own strength, shall leave [behind] any desire linked to [this] world. He shall combine love [for god] and observance, he shall live with the teacher, thirsty for the sentiment of beholding [god]. Destroying [his] own self, he shall reduce [it] to dust; laying down his life, he shall watch the spectacle [of the world]. When he acts in such a way, he will obtain peace – [the world will be but] a sugar-cake [in] milk!".

261 Not to be confused with those practising austerities and forms of yoga to obtain siddhis (supernatural powers). As is generally expected in a bhakti text, the PcN criticises the pursuit of

terms are used to describe these individuals. They are referred to as bairāgī/vairagina (Skt. vairāgin), "[Vaiṣṇava] renunciants" (2.8), muni, "silent sages" (2.9), and tapasī (Skt. tapasvin) or ascetics, "those who practise austerities" (2.25). In verses 2.7–2.9, the First Actor vividly portrays the appearances and behaviours of these individuals: paṇḍitas (scholars) holding the BhG (2.7); a bairāgī wearing a necklace of prayer beads, overcome by "divine love [. . . singing] the praises of Hari's qualities" (2.8); and munis with "gleaming matted hair" praying silently with the help of beads, keeping their arms raised, gazing at the sun in meditation, or sitting between five fires (pāñcagni ko tapai, "practis[ing] the fivefold heat austerity]") (2.9).

Brajvāsīdās views all of them as both bhaktas and yogis. Not only are they collectively referred to as vivekī, but the comparison to the haṃsa is also used.[262] This analogy becomes evident when Deceit is depicted in the form of a mahant: his beauty is likened to a "[pair of] scissors covered in honey" (lapeṭī manahuṁ kataranī), and "his external appearance is that of a gander (haṃsa)", although his actions resemble those of a crow (2.13). Additionally, in 2.20, I-maker encounters "a group of renunciants (vairaginas), like a big flock of ganders" while walking along the banks of the Ganges, "their aspect pristine, their words clear".

In the next verse, the bairāgīs are portrayed as practising bhakti (bhakti karai), or building a bond with Hari:

koū hari guṇa varṇana karai
  koū suni suni ura meṁ dharai
koū baiṭhi kiratana karai
  premānanda sindhu ura bharai
koū kṛṣṇa nāma ko rarai
  koū arcana bandana karai
koū caraṇa kamala cita dharai
  vividhi bhāṁti sevā bistarai
koū dāsa bhāva ko dharai
  prabhu kī icchā soī śirai
koū sakhā bhāva ko gahai
  antargati hari sana gahi rahai

power. The character of Dambha or Deceit serves as an example of a deceitful yogi, friendly to tantric practices, who exploits his powers to exert control over others. See PcN 2.13. On the other hand, the bhaktas disdained displays of powers for their own sake. Numerous hagiographical accounts reflect this sentiment, including the example of Tulsīdās, who reportedly refused Emperor Akbar's request to perform a miracle in his presence. See Pinch 2006: 218–219 for further insights. Additional relevant information can also be found in Callewaert/Snell 1994.

262 In PcN 1.124, Discrimination and his wife Good Reason are likened to a pair of haṃsas.

koū tana mana dhana kī arapai
  hohiṁ nichāvara radhābara pai
yahi vidhi bhakti karai
  bairāgī dekhī tinhai āgisī lāgī
(PcN 2.21)

Someone described Hari's qualities, someone else – listening – treasured them in their heart. Someone, sitting, sang devotional songs and filled his heart with an ocean of love and bliss. Someone was repeating Kṛṣṇa's names, another was paying homage, someone else was meditating on the lotus-feet – the worship was of various kinds. One was inclined to the attitude of service and puts on his head [i.e. obeyed] the Lord's will. Someone else had turned to the attitude of friendship, with Hari in his heart. Someone else had dedicated his body, mind, and wealth as a sacrifice to Rādhā's groom. In these ways, he [I-maker] saw the renunciants sharing a bond with God in these ways and became purple with rage.

Even though these descriptions may seem conventional, Brajvāsīdās strategically introduces them in the plot to create conceptual space for his bhakta yogi, which will be further developed in Act II with the introduction of the character Bond-with-Viṣṇu. This intention is evident in the speeches of I-maker, who becomes angry and condemns them by mocking the Vedas they believe in and asserting that mortifying the body through ascetic practices is futile (2.22–2.25). Interestingly, in verse 2.26, I-maker refers to them as fools since yoga renders the body useless for worldly enjoyment and, in that sense, resembles a bodily disease.[263] The narrating voice concludes the scene in 2.27 by commenting that I-maker was criticising "the actors [playing the devotees of] Hari".

The connection between bhakti and yoga is consolidated in the portrayal of Bond-with-Viṣṇu described by Materialist in 2.96. While this verse has already been analysed in relation to the avatāra issue,[264] its primary purpose is to outline the essential features of bhakti in the PcN.[265] It begins by narrating how Bond-with-Viṣṇu always remains united with her allies, Firmness (dṛṛhatā) and Love (prema). It then mentions the descent of the Supreme Brahman to earth, assuming various names and forms, before returning to focus on her character and actions:

263 The verse in question, e mūṛha yāhī joga hai | mānahu sadehi roga hai ||, seems to sarcastically reverse the notion of physical benefits resulting from the practice of yoga.

264 See "Paramātama and Avatāra" in subsection 4.1.1.

265 For how bhakti is defined as the singular essence of brahman, see "Paramātama's Single Essence" in subsection 4.1.1.

[. . .] racai tāsu pratimā dharai bhava setī
  karai arcanā bandanā bhāṁti ketī
karai kīratana nāma antara na rākhai
  dṛgana mitra kī mādhurī rūpa cākhai
karai citta kī vṛtti ko līna tāmeṁ
  chapai capala dāminī śyāma ghana meṁ
caraṇa bāndhi nūpura karai nirtta āge
  visārai dasā deha kī prema dhāge
ahara niśi sadā rīta aisī nibāhai
  viṣaya bhoga kabahūṁ na svapane meṁ cāhai
yadyapi karoṁ maiṁ vidhannahu ghanere
  tadyapi baṛhe sūra phirate na phere

(2.96)

[Bond-with-Viṣṇu] creates a mental image of him [brahman] and concentrates [on it] with emotion – in how many ways she honours and pays homage [to brahman]. She sings and does not hold distinction [among brahman's] names; she relishes the sweet form of [her] friend [with] her eyes. She makes the fluctuations of the mind absorbed in [the mental image], just as swift lightning is hidden by dark clouds. She ties the anklets [to her] feet and dances in front [of the image]. She forgets the condition of the body [through] the threads [of] Love. Day and night, always, she carries on in this way; she never desires the enjoyment of sense objects, [even] in dreams. Even if I create many obstacles [against her], nonetheless, the mighty warriors do not come back, [and they are won by her].

The profile of bhakti in the PcN extends beyond its conventional activities such as arcana (worship), vandana (prayer, reverence towards God), and kīrtana (chanting about God), encompassing distinct yogic practices as well. One such practice involves creating a mental object, in this case, the preferred form of brahman, upon which concentration is focused.[266] The simile of the dark cloud (śyāma ghana) may allude to Kṛṣṇa in this context. Furthermore, the language used is specifically related to yoga, with "citta kī vṛtti" evoking YS 1.2. Another notable element that emerges is the complete absorption in yogic worship, which entails detachment from the world and a mind fully turned towards brahman. While Māyā/Prakṛti manifests the world through dancing,[267] Bond-with-Viṣṇu (Bhakti)

266 PcN 5.120 provides a description of a saguṇa form of brahman, likely inserted into the text for this specific purpose. For a demonstration of how this verse adaptively reuses the RcM, see "Paramātama, Nirguṇa, and Saguṇa" in subsection 4.1.1.
267 See PcN 6.86.

remains unbound by materiality (her body or sensory objects) but in her dancing is instead connected to brahman through prema.

Up to this point, yoga and yogic practices have been mentioned in a broad sense, except for occasional references to the YS. It is important to note that the PcN offers its own bhaktified version of the eight limbs of Pātañjala Yoga, which intersects with accounts found in the BhG and BhP and with Haṭha Yoga. While I have partially investigated this topic elsewhere,[268] this section will summarise previous findings and present additional novel observations.

In Act I, Desire and his wife Passion discuss the conflict between the two parts of their family. Passion advises her husband not to underestimate his adversaries and enumerates the ministers of Discrimination, explaining their qualities in detail (1.71–1.78). Considerable space is given to Desire's speech, in contrast to the relatively brief account of the ministers of Bewilderment that follows.[269] Discrimination's "eight strong ministers" are listed as Moral Principle (yama), Observance (nema), Balance (sama), Breath Control (prāṇāyāma), Withdrawal (pratyāhāra), Concentration (dhāraṇa), Meditation (dhyāna), and Absorption (samādhi). Notably, Moral Principle and Observance in the PcN present more than the five rules of conduct associated with them in YS 2.30–2.32. Moral Principle is expanded to ten principles, while Observance contains ten or more (1.74). These augmented principles and observances integrate with those found in the YS, as well as with the modes of conduct aligned with bhakti that partly correspond to instances in the BhG and BhP.[270] The fact that Brajvāsī has framed them within an aṣṭāṅga (eight-limbed) structure highlights that, as in the case of the descriptions of the vivekīs, he may have seen these traditions as interconnected and sought to draw authority from the Pātañjala paradigm as well.[271]

---

268 See Pastore (forthcoming a).

269 See 1.81–1.87. Although the number of verses for each party may be equivalent, it is worth noting that two verses are solely dedicated to Kāma's glorification of his power, while the characters of Violence, Greed, and Intoxication remain underdeveloped.

270 Similar to the belief in the teacher and the reverence for the sants. See Pastore (forthcoming a). See also Horstmann/Mishra 2013: 174n23. It is important to note that the yamas and niyamas outlined in the YS were intended for renunciants. However, the PcN, despite the framing tale of Kṛṣṇadās and his twice-born disciple, addresses the devotional community (satsaṅga), which includes householders. While Brajvāsī acknowledges ascetics (tapasī) in the descriptions of the bhakta yogis discussed above, tapas is notably absent among the niyamas, unlike in YS 2.32. Hence, Brajvāsī likely adapted the yamas and niyamas to align with his own bhakti mindset and audience. See Birch/Hargreaves 2016a: 32–35, 2016b; Mallinson/Singleton 2017: 55–85.

271 Other Vaiṣṇavas did not always find it necessary to rely on aṣṭāṅga formulations when it came to ethical matters. See O'Connell 2016: 135–162.

There are a couple of variations worth exploring in relation to the YS: the treatment of concentration (dhāraṇa) and balance (sama). Inner concentration and worship (mānasī pūjā) are underscored within the context of observance, as well as in the case of viṣṇubhakti. Concentration is personified as the sixth minister (or limb) of dhāraṇa, but there is a difference if one considers Bond-with-Viṣṇu's portrayal. While Bond-with-Viṣṇu "relishes the sweet form" of the mental icon of brahman (2.96), Concentration places the mahābhūtas (gross elements) and tanmātras (subtle elements) along the backbone and holds his breath at each of them (1.77). It is uncertain whether the practices described in Bond-with-Viṣṇu's case align entirely with the depiction of Discrimination's ministers. Still, the use of yogic terminology inevitably invites a comparison between the passages.[272] Although further details on the relationship between these two modalities of concentration are lacking – whether they are to be conceived as separate, complementary, or subordinate – it is evident that the PcN envisions diverse focal points for concentration.

Regarding the second variation, sama occupies the position of āsana in the eightfold yoga system of the YS. It appears immediately after yama and nema, and considering the context of enumerating the limbs, I have chosen to translate it as "balance" because the verse mentions the term "āsana" and refers to sitting postures and concentrating on the tip of the nose.[273] In this regard, it appears that Brajvāsīdās drew inspiration not directly from the YS themselves but from the description of yogic practices in the ŚvU. In ŚvU 2.8, for instance, the yogi keeps their body straight (saṃa).[274]

272 The verb translated as "concentrates" in 2.96 is dharai, to which dhāraṇa among the ministers is connected etymologically.

273 PcN 1.75: [Passion:] sama nāma mantrī tṛtiya tāke cinha aise jāniye | mana ko kare nigraha apara abhyāsa yoga bakhāniye || kari siddhi padmādikana āsana dṛṣṭi trikuṭī mahaṁ dharaiṁ | puni tattva pāñcoṁ sādhi nīkai roga kāyā ke haraiṁ || (The third minister is called Balance, his features are known as such: yoga is described [as] an incomparable practice [that] restrains the mind. Mastering the postures such as the lotus and so on, he fixes the view on the three points [at the base of the nose]. Afterwards, meditating on the five elements, he removes diseases from his body for good.). Brajvāsīdās mentions only the Lotus Posture as an example but alludes to the presence of other postures (padmādikana āsana). This reference could be interpreted in two ways. First, it may allude to the commentary of YS 2.46 by the same author-redactor, where the Lotus Posture is listed as the first example among others. Second, it could refer to the plethora of postures available in Haṭha Yoga texts and practices. On the varied interpretations of YS 2.46 and the possible identification of the āsanas intended, which in the YS are all characterised by bodily stillness (sitting, supine, kneeling), see Maas 2018. On the multiplication of the postures, starting with non-seated positions, in Haṭha Yoga texts, see Birch 2018.

274 For the Sanskrit and translation, see Olivelle 1998: 418–419. Cf. also BhG 6.13.

However, there is a textual issue as the term sama overlaps with another usage, namely that of sama (tranquillity), usually mentioned in conjunction with dama (self-restraint). These two characters seldom appear in the play, but when they do, they are often paired with nema and yama or just yama.[275] Brajvāsī does not elaborate on their qualities, but it is evident that he considers them to be the children of Mind (5.143) and ministers of Discrimination, placed on the same level as yama, nema, and the other members of the eightfold enunciation (4.91, 6.25).

One might consider that the verses in the PcN that start with sama and dama but stop at yama do not necessarily imply that the other limbs of classical yoga follow. They may be interrupted not only for the sake of brevity but also to leave them open to further elaboration by whoever is reading the text. The relative fluidity and inclusive tendency in such enumerations, which may have led the author to name the third member of his aṣṭāṅga as sama instead of āsana, can be observed in the RcM. For example, RcM 1.36.7 presents "sama jama" and "niyama", as does 6.79.5, while 7.37.4 slightly differs by including "sama dama niyama" with the addition of "nīti". Tellingly, two other instances are found in the portrayal of Bharata's practices and dedication as a bhakta:[276] the first verse, 2.324.2, shows a sequence of "sama dama sañjama niyama upāsā"; the following verse, 2.325.5, inverts the order as "jama niyama sama dama". Thus, the RcM provides a precedent where sama follows jama and niyama, which, considering the adaptive reuse made of this text by the PcN, may have served as the foundation for the PcN's presentation of sama as the third limb of the aṣṭāṅga path, in addition to the Upanishadic reference. It is within the context of Sama's profile that we encounter Brajvāsī's own definition of yoga:

[. . .] mana ko kare nigraha apara abhyāsa yoga bakhāniye (1.75)

Yoga is described [as] an incomparable practice [that] restrains the mind.

In this case of adaptive reuse,[277] the conceptual backdrop remains yoga, but the goal of the PcN's yoga is not referred to as kaivalya (aloneness) as in the YS. Instead, here yoga is seen primarily as a means to an end.[278] The PcN is still preoccupied with explaining liberation in yogic terms, as its samādhi (absorption) is a

275 "sama dama nema yamādika" (2.92), "sama dama nema yamādi" (4.91), "aura putra sama dama yamādi" (5.143), and "sama damādi" (6.25).

276 For an extended discussion of the topic, see section 3.2.

277 In the PcN, "mana" is equivalent to "citta" in the PcN. See subsection 4.1.4.

278 Refer to Mallinson/Singleton 2017: 4–6 for textual references regarding the expression "yoga abhyāsa" in Haṭha Yoga sources.

state where there is "no difference between the servant and the Lord" (1.78).[279] As in the YS, this state entails a dissolution of distinctions between the contemplating subject and the object of contemplation.[280] However, the terminology of sevaka (servant) and svāmī (master) present in the PcN reveals the bhakti perspective of the text in this dimension, which is absent from the YS.[281]

At the same time, to fully understand this yogic account, it must be integrated with PcN 6.157, which explains the final state of liberation using the term "yoga". In this instance, the term yoga should be translated as "union" with God, which aligns with Bharata's yoga described in the RcM.[282] This interpretation is further supported by two elements of the metaphysical structure of the PcN story. First, there is a strong emphasis on the excellence and superiority of the Supreme Self. Second, there is a depiction of the relationship between the individual self and the Supreme Self, where the jīva is the son or a part (aṃśa) of the Supreme Self. Additionally, in the synthesis of the story provided by Discrimination in 1.173, the liberation of Mind/Individual Self is envisioned as a reunion, an encounter with the Supreme Self, akin to a son meeting his long-forgotten father. Therefore, the concept of yoga as union in the PcN implies a degree of identity or unity while preserving the overall bhakti orientation of the text, as will be further explored in the next subsection.

Returning to the concepts of sama and dama, as previously discussed, they are combined with the aṣṭāṅga in both the PcN and the RcM. These composite series allow the texts to bring together classical yoga and Vedantic reflection. Sama, understood as "balance" or the bodily state achieved through postures, is a derived form of śama, meaning "peace" or "tranquillity". Dama means "self-restraint". Both are considered prerequisites for engaging in the inquiry into brahman according to Śaṅkara in his commentary on the first aphorism of the BS.[283] Barbara Stoler

279 aṣṭama saciva samādhi ahai | jagata upādhi meṭi jo rahai || āpāhū miṭi jāta jahāṁ | svāmī sevaka bheda na tahāṁ ||.

280 See Stoler Miller 1996: 61.

281 When discussing the highest stage of bhakti, Sundardās of the Dādūpanthī tradition employs the imagery of the "sevaka" (one who serves) and the "sevya" (one who is to be served) to represent the bhakta/devotee and the Lord. In *Jñānasamudra* 2.52–2.53, he states that the sevaka merges into a single body with the sevya. Both the Lord (Hari) and the devotee partake in the eternal divine play (līlā). See Thiel-Horstmann 1984: 276–277.

282 The PcN's expression is "viṣṇu hona ko yoga bhayo". See subsection 3.2.2 for the comparison between Bharata and a yogi and his yogic bhakti.

283 Their past participle forms (śānta and dānta) are employed in the commentary. See Gambhirananda 2009: 9. The other essential conditions are discrimination (viveka) between what is eternal and what is not eternal, indifference (virāga) towards the fruits of one's actions, and a desire for liberation (mumukṣutva). See Suthren Hirst 2005: 41–45. See also *Upadeśasāhastrī* 1.1.2

Miller, in her commentary on the limb of āsana in the YS, notes that a state of equanimity can be attained through physical equilibrium, where "pairs of opposites, such as heat and cold, pleasure and pain, self and other, cease to shape the yogi's awareness of existence".[284]

Connected to this, the concept of samatā (lit. "sameness"), which is characteristic of the yogi, holds a significant place in the PcN and can be interpreted as equanimity. As discussed earlier in this chapter, brahman in the PcN is portrayed as a single essence (eka rasa). This essence is identified with Viṣṇubhakti, who in turn is described as embodying both samatā and ānanda (bliss).[285] Moreover, as we have observed, the identification of bhakti as rasa with ānanda demonstrates parallels with the ideas of Madhusūdana Sarasvatī and the Gauḍīyas.[286] As for samatā, the PcN explicitly mentions it in both the opening and closing acts. In the first act, a verse sets the premise for what is to be attained on the path to salvation:

samatā gahaiṁ sacca ko jānai duḥkha suḥkha sama āṛā hai [. . .] (PcN 1.49)

[The king] shall obtain equanimity [because] he knows the truth: pleasure and pain are the same, interposed [between the true path and himself].

In the final act, the perspective is distinct because Mind has already achieved samatā:

[. . .]
taji viṣai bhoga vilāsa āśā meṭikai āpā dayo
bhayo asthita capalatā taji eka rasa samatā gahī
saba nīca ūṅca miṭāya duvidhā prīti ātama kī sahī
(PcN 6.5)

Forsaking the enjoyment of worldly objects, removing [any] hope for pleasure, he [Mind] has given up egoism. He has become stable, abandoning fickleness, and has found a single essence, equanimity. All opposites vanishing, he felt love again for Individual Self.

in Potter 1981: 218. BĀU 4.4.23 also presents a set of qualities that define the knower of brahman: śānta, dānta, uparati, titikṣu, samāhita.

284 Commenting on YS 2.46–2.48. See Stoler Miller 1996: 57. Although see Maas's novel interpretation of sutra 2.46 combined with sutra 2.47, following which, a stable posture is not the cause of but "the result either of the meditative practices of merging meditatively into infinity or of a slackening of effort in practice". Maas 2018: 51.

285 See also PcN 5.141.

286 See "Paramātama's Single Essence" in subsection 4.1.1.

Sameness (samatā) here implies indifference towards pleasure and pain connected to embodied existence: once true knowledge of the self has arisen, everything is perceived as the same (sama), where all oppositions (saba nīca ūṅca) due to the perception of duality (duvidhā) fade. These verses inevitably evoke the BhG, where yoga is defined as equanimity (2.4) and peace (śānti) arises from renunciation (12.12). In BhG 12, the ascetic is depicted as a bhakta, and among the various kinds of ascetics dear to Kṛṣṇa, the first kind is a yogi devoted to Kṛṣṇa.[287] Angelika Malinar suggests that one of the two aims of liberation outlined in BhG 6 is that, through the yogi's power of discrimination, they can "see the 'same' in everything because it is itself 'indifferent', 'same' (samabuddhi; 6.9), being now connected with its cosmological dimension".[288] This state of sameness brings bliss to the yogi (5.21). Similarly, samatā in the PcN signifies a state where the mind is pacified, having relinquished its usual volatility, and has become asthita (stable).[289] This term evokes the sthitaprajña of BhG 2.72, which refers to "one whose insight is firm" due to meditative practice. Evenness of mind and detachment from worldly enjoyment are features of this state of equanimity.[290]

In the PcN, meditative practices are not only presented from a bhakti perspective but also integrated into a Vedantic framework. The following subsection will further explore the interaction between the yogic bhakti depicted in the PcN and the text's Vedantic viewpoint on salvation.

4.2.2 Yogic Bhakti and Wisdom

In terms of the history of the ideas, Brajvāsī's work falls into the traditions of both Sanskrit and vernacular texts, each presenting their unique combination of bhakti, yoga, and Vedānta. Michael Comans has analysed how the term samādhi enters and is understood within Advaita Vedantic reflection.[291] The first chapter of the *Pañcadaśī*, titled "Tattvaviveka", delineates the practices of śravaṇa (listening), manana (reflection), and nididhyāsana (deep concentration) and subsequently sa-

---

287 Malinar 2007: 191.

288 Malinar 2007: 122.

289 The prefix a- should not be confused with the privative a-, as the Brajbhāṣā "asthita" is attested as a form of the Sanskrit "sthita". See HSS 384.

290 See Malinar 2007: 75–79. Sameness as a way of perceiving other beings in the world, resulting from the knowledge of brahman, is also praised in regional-language compositions, such as Kabīr's sākhīs. See Vaudeville 1974: 253.

291 The doctrines presented in texts that blend Advaita and yoga are referred to as "Yogic Advaita" by Andrew Fort. See Fort 1998.

mādhi (absorption) as the experiential outcome triggered by the teaching of the Upaniṣads.[292] Another pivotal non-commentarial work within this tradition is the *Vivekacūḍāmaṇi*, which extols the faculty of discrimination and considers samādhi as a necessary step towards liberation.[293] Sadānanda's *Vedāntasāra* (fifteenth century) not only integrates samādhi with listening, reflection, and deep concentration but also quotes and reinterprets the eight limbs of Pātañjala Yoga in a Vedantic key.[294] In the sixteenth century, Vijñānabhikṣu, mentioned in the previous section, sought to reconcile Bhedābheda Vedānta, Pātañjala Yoga, and Sāṅkhya.[295]

During the seventeenth century, yoga flourished in the Advaita milieu, exemplified by the corpus of the 108 Upaniṣads, established between the end of the century and the first half of the eighteenth century.[296] The Advaitin Nārāyaṇa Tīrtha regarded Pātañjala Yoga as ancillary to Advaita Vedānta, linking samādhi with nididhyāsana in his commentary on the YS.[297] Within the realm of regional-language compositions, the Dādūpanthī Sundardās had formulated a hierarchy of yoga paths in his *Sarvāṅgayogapradīpikā* in the seventeenth century, considering Advaita Yoga the highest.[298] In the eighteenth century, the Rāma devotee Jayatarām infused Haṭha Yoga with bhakti by envisioning the final stage of the practice as the contemplation of the divine couple of Rāma and Sītā.[299]

The interactions among the various schools of Vedantic exegesis (whether Advaita or not), bhakti, and yoga were multiple, and each author forged a unique synergy based on their own understanding of these traditions.[300] But the paths of

---

292 See Mahadevan 1969: 9–11.

293 This text, attributed to Śaṅkara, incorporates both savikalpa samādhi and nirvikalpa samādhi in its explanation of nididhyāsana. See Menezes 2017: 141–142; Sundaresan 2003.

294 Comans begins his analysis by emphasising the significance of samādhi as a direct experience for modern Vedāntins like Vivekananda. He then traces the usage and interpretation of the term, primarily focusing on the Upaniṣads and the works of Śaṅkara while also touching upon some medieval and early modern texts. I have confined my discussion to these texts since they are closer in chronological proximity to the PcN. For further details, see Comans 1993: 31–32.

295 See Nicholson 2005.

296 By a Vedāntin author who selected passages mainly from Haṭha Yoga texts. See Bouy 1994: 81–110.

297 See Venkatkrishnan 2015: 222.

298 The other paths being Bhakti Yoga, Haṭha Yoga, and Sāṅkhya Yoga. See Burger 2014a: 683–708.

299 See Burger 2014b. In the southern regions of India, too, some poets merged yoga and bhakti into a singular practice. Shulman 1991 provides examples.

300 The *Haṭhayogapradīpikā*, authored by Svātmārāma in the fifteenth century, draws verses primarily from texts originating in the Vedantic milieu. These include the thirteenth-century *Vivekamārtaṇḍa*, the *Gorakṣaśataka*, and the *Dattātreyayogaśāstra*. The last text also exhibits Vaiṣṇava inclinations. See Mallinson 2014.

yoga and knowledge were not always embraced without contention. In his *Bhramargīta*, the Vallabhite Nanddās depicted the gopīs firmly rejecting the yoga and jñāna (knowledge) proposed by Uddhāva, Kṛṣṇa's friend and messenger.[301] Nanddās, following Vallabha, upheld bhakti – in this case, the bond between bhakta and Kṛṣṇa – as supreme. Vallabha argued that only bhakti could foster true knowledge of Kṛṣṇa and grant participation in his divine play, considering the paths of jñāna and yoga to be of lesser significance.[302] The term "yoga" itself underwent constant reinterpretation. For instance, Rāmānuja understood the jñānin (knower of God) to be the best bhakta and considered the path of jñanayoga (yoga of knowledge) to be the most effective. He interpreted yoga as the intuitive realisation by the individual self (ātman) of being a separated being (viśiṣṭa) from brahman.[303]

Regarding the PcN, I have demonstrated that the portrayal of Discrimination's ministers combines the qualities of an ideal Vaiṣṇava with the eight limbs (aṣṭāṅga) of the YS, and the inclusion of sama and dama expands their scope to encompass Vedantic reflection as well. Although constitutive of bhakti and comprising bhakti,[304] yoga as a meditative practice appears to be insufficient on its own and in need of support. Indeed, Discrimination's advisers are described as gaining strength from the birth of Vidyā on two occasions.[305] Vidyā, or Knowledge, is the daughter of Discrimination and Upaniṣad and precedes her twin brother Wisdom Moon (prabodhacandra). Thus, she represents the initial outcome of Discrimination and Upaniṣad's collective effort to awaken Individual Self (jīvātama puruṣa) to his true nature in Act VI.

Before delving into the type of knowledge generated through this collective endeavour, it is necessary to take a step back. Act V portrays a crucial stage on the path to liberation, where Mind comprehends the impermanent nature of the world and worldly relations. His teachers, Sarasvatī and Dispassion, are both sent by

---

301 See Nanddās 1973: 47–54.

302 According to Vallabha, Pātañjala Yoga and the knowledge derived from the study of the Upaniṣads served as a valuable means of purification, encompassed within his conceptualisation of maryādā bhakti. However, he regarded this path as exceedingly challenging but leading to a secondary form of liberation: union with the akṣara brahman. For further insights, see Narain 2004: 367, 396.

303 The awareness of being a separated being from the Supreme Brahman does not compromise non-dualism since a viśiṣṭa is still considered a part of the body of brahman. See Lipner 1986, 2018: 111–116; Freschi 2017–2018.

304 Building upon the elements discussed in the previous section, we can observe the influence of bhakti in the PcN's aṣṭāṅga, the integration of yogic practices within the portrayal of Bond-with-Viṣṇu, and the depiction of the bhakta yogis in Act II.

305 PcN 1.106, 1.173.

Bond-with-Viṣṇu. Sarasvatī helps Mind realise that by meditating on a personal form (i.e. with qualities) of brahman in the nine ways,[306] he can discover inner peace.[307] Peace serves as the antidote to attachment to or care (cintā) for the family and the world, as it engenders detatchment (acintā) (5.142).[308] In 5.143, Sarasvatī advises Mind to adopt the ways of his lost son Dispassion, in the sense of an internalised vow of renunciation (saṃnyāsa). This entails relinquishing the responsibility of kingship to his son, Discrimination.[309] Before suggesting this, she instructs him to stay in the company of his other sons, beginning with the ministers Balance, Self-Restraint, and Observance. At this point, we witness a further integration of yoga into the Vedantic framework, as the goddess tells Mind to have his sons "stay next to you, reading the Vedas" (5.143). Through the constant and uninterrupted exercise of viveka and its associated positive concepts (sama, dama, nema), Mind will be able to live in the world even after recognising his true nature without succumbing once again to the entanglements of Bewilderment.[310]

However, the actual reading of the Vedas, explicitly interpreted as the understanding of the Upaniṣads, occurs in Act VI. The final act aligns entirely with the Vedantic mould as Discrimination applies his critical discernment to the words of Upaniṣad in a process of listening, reflection, and deep concentration. While these three activities are commonly regarded as central to the inquiry into brahman in Advaita Vedānta,[311] inspired by BĀU 2.4.5, their mechanics are unclear. This is particularly true regarding nididhyāsana, as even Śaṅkara's perspectives

306 The concept of navadhā bhakti, or "ninefold bhakti", initially presented in BhP 7.5.23–24, holds central importance in various religious traditions across North India. These practices include śravaṇa (listening to stories about god), kīrtana (singing about the deity), smaraṇa (remembering the divine activities), pādasevana (serving the feet of the deity), arcanā (worshipping a sacred image of the deity), vandanā (reverence towards the deity), dāsya (serving the deity), sākhya (friendship), and ātmanivedana (dedicating oneself to the deity).

307 "Peace comes [to your] heart when you concentrate on brahman" (5.118); "By worship of the form of brahman with qualities in nine ways, | the [form] without qualities, incomparable, then naturally will enter your heart ||" (5.124). See PcN 5.117–5.130 for the full passage.

308 "The moon [of] peace has risen inside [me], the obscurity [of] care has faded. The [moon]light [of] indifference [has] appeared, bestowing calmness and joy, destroyer of the threefold affliction."

309 In courtly culture, ministers had a crucial responsibility to prevent the king from committing misdeeds influenced by his selfish tendencies. Furthermore, the delegation of tasks was perceived as a symbol of power. For further insights, see Ali 2004: 56–59.

310 The PcN briefly illustrates the consequences of deviating from the prescribed path when Mind succumbs to the allure of Honeyed Knowledge, which represents the temptations and powers derived from the practice of yoga. See PcN 6.34–6.41. The importance of viveka, or discrimination, as a means to attain freedom is emphasised, even after achieving gnosis, in texts such as Vidyāraṇya's *Jivanmuktiviveka*. See Madaio 2018.

311 Dasgupta 1922, vol. 1: 490.

on it have been debated.[312] The initial part of the exegetical process of śravaṇa and manana between Upaniṣad and Discrimination is dedicated to refuting Sacrificial Science, Hermeneutics, and Science-of-Reasoning, in addition to correctly interpreting the words of the Upaniṣads.[313] Upaniṣad's speeches in the presence of Individual Self primarily focus on her conception of brahman by juxtaposing both its negative descriptions and its pervasiveness and essence as bliss,[314] represented by the paradox of satkāryavāda, which posits that everything already exists in its cause. In this case, the universe pre-exists in brahman, and any change in brahman is illusory (6.122).[315] As we have already observed, māyā/prakṛti is the material source of the universe, while brahman is its unchanging substratum.[316] After hearing about Upaniṣad's misadventures among the incorrect views, Individual Self starts to unravel his confusion about brahman and develops a desire to know "his own true form" (6.135).

The subsequent verses of the act explore the true nature of Individual Self, portraying his realisation of his true relationship with his own father, Supreme Self. This revelation is primarily conveyed through the exegesis of the Upanishadic statements "so tū āhi" and "naita neta", which are "Brajified" forms of "tat tvam asi" and "neti neti".[317] Individual Self inquires about the identity of the god mentioned by Upaniṣad in her speeches, to which she responds that he himself is such a god. However, Individual Self finds it hard to believe and demands further explanations, to which Upaniṣad replies:

---

312 See Suthren Hirst 1996; Dalal 2014. A recent article examining the interactions and incorporation of yoga techniques and terminology in Vedantic discourse, including the equating of nididhyāsana and samādhi, is Schwarz 2017.

313 The three female characters, as in the KPC, are Yajñavidyā, Mīmāṃsā, and Tarkavidyā. Yajñavidyā personifies Vedic rites, while Mīmāṃsā embodies the homonymous philosophical school devoted to the interpretation of the statements of the Vedas. Neither value the knowledge of brahman derived from the Upaniṣads in view of mukti. Tarkavidyā is also called Nyāyaśāstra and represents the school of thought which refuses to uphold the Vedas and the Upaniṣads as the ultimate and incontestable source of knowledge. For these reasons, they are portrayed as misunderstanding Upaniṣad's words and disqualified by Viveka's refutation.

314 See PcN 6.81, 6.97–6.98.

315 "[When] the name of the tree exists, it is said that in that flowers and fruits [exist already]. [But] in the absence of a seed, [how can there be] branches, leaves, roots and a big tree? The tree exists, indeed, in the seed; the seed, in fact, is the true being [of] the tree. I always praise the form of this all-pervading brahman."

316 See PcN 6.126, where prakṛti and the atoms are said to be conformable to the Supreme Self, being equal to him as ornaments made of gold are to raw gold. See also subsection 4.1.2.

317 Respectively, in CU 6 and BĀU 2.3.6. "Naita neta" can be deconstructed as the negation "na" plus the adverb "ita", where the latter corresponds to the Sanskrit "iti". See HSS 508.

jāni lehu aise yaha bāta
  puruṣa sanātana ke tuma tāta
tinateṁ tuma kachu nyāre nāhiṁ
  māyā kari dvai jāne jāhiṁ
jaise sūraja ekahiṁ hoya
  jala hima upala nāhiṁ kachu doya
aise tuma īśvara aho
  tāteṁ īśvara tumako kaho
(PcN 6.139)

Understand this fact: you, dear, [belong] to the eternal Supreme Self. You are not separate from him, who is known to be two due to Illusion. As the sun is but one, water, snow, and hailstones are not distinct either. In this way you are a god; hence, I call you God.

Nevertheless, Individual Self becomes even more perplexed than before and questions Discrimination:

tuma sunata hau kachu tāta
  yaha upaniṣada kī bāta
mopai na samajhī jāya
  ye kahata kahā banāya
parabrahman baranata jāhi
  muhiṁ kathati so tū āhi
sata citānanda anūpa
  so ahai tero rūpa
(PcN 6.141)

Are you listening, dear, to what Upaniṣad has been saying? I cannot understand; what is she saying, [is she] making it up? What is described as [the] Supreme Brahman, she tells me, "That you are. Being, consciousness, and bliss, incomparable, that is your form."

Verses 6.139 and 6.141 mark the beginning of the paraphrase of "so tū āhi" and "tat tvam asi". In verse 6.139, Upaniṣad's words put into perspective how the formula is to be understood: the sense of belonging that Individual Self must regain towards the Supreme Self in order to recover. The term "ke" in the possessive postposition indicates "belonging", implying non-separateness (tinateṁ tuma kachu nyāre nāhiṁ) because Māyā creates the illusion of separateness. The subsequent sentences, connected by the correlatives "jaise" and "aise", illustrate the concept of non-duality (nāhiṁ kachu doya): multiple forms that ultimately trace back to a single form, just as snow and hailstones are fundamentally composed of water. Individual Self can be equated with Īśvara because Īśvara is essentially the only reality, as Discrimination's earlier statements in the act emphasise.[318]

Turning to verse 6.141 and the expression "so tū āhi", we can draw a parallel with "tat tvam asi". "So" functions as a distal (or far) demonstrative pronoun,[319] corresponding to "tat", and, similarly to the relationship between "tat" and "tvam", there is a grammatical apposition between "so" and "tū".[320] The paraphrase partially elucidates the meaning of "so tū āhi": "tū" more precisely refers to the rūpa, the form, of the individual self. The ultimate objective of the play is for Mind/Individual Self to recognise his original form (prathamo rūpa, 1.173).[321] "That you are" means "that is your form", representing existence, consciousness, and bliss. The account presented so far aligns closely with the Advaita perspective, where the statement "tat tvam asi" signifies identity between brahman and the individual self, with "tat" representing brahman and "tvam" representing ātman, from which empirical limitations have to be abstracted.[322]

Looking at Vaṣṇava Vedānta interpretations of the great statement, Rāmānuja's differs significantly. According to him, the sentence cannot indicate identity between brahman and the individual self but rather unity. In his view, "tat" and "tvam" refer to brahman in two distinct modalities. The Supreme Self and the individual self cannot be equated. The first modality pertains to brahman as the cause of the universe – infinite, flawless, and encompassing all its greatness and glory. The second modality, indicated by "tvam", represents brahman as the inner controller, enlivening the individual self as its body.[323] Later Vaiṣṇava schools, such as the Gauḍīyas, shared a similar perspective. They acknowledged the statements of differentiation found in the scriptures and considered them alongside the concept of unity.[324] Vallabha also concurred with this view, seeing "tat tvam asi" as a means to prevent the bhakta from dying due to separation from the Lord. However, he could not accept it as a statement with ontological value.[325]

---

318 See PcN 6.126.

319 See Strnad 2016: 287–288.

320 See Dalal 2020: 53; Suthren Hirst 2005: 156–157.

321 See also PcN 4.31.

322 See Suthren Hirst 2005: 152; Dalal 2020: 54, 56–77, quoting Śaṅkara's commentary on BĀU.

323 See Bartley 2002: 95–55.

324 Caitanya himself was said to have assigned limited authority to the "tat tvam asi" in the *Caitanyacaritāmṛta*. See Gupta 2007: 47. The Gauḍīya thinkers strategically interpreted the mahāvākyas as statements that exclusively point to Kṛṣna as the sole Reality upheld by the Vedas. See Uskokov 2018.

325 See Narain 2004: 291, 315–316, 392.

The passage in the PcN makes it challenging to draw a clear distinction between unity and identity. Unity amid differences is a recurring theme in the final acts, extending beyond the path to liberation. It is evident in the dialogue between Peace and Faith that opens Act V. While recounting the events from the battlefield, they discuss the interconnectedness and relative equivalence of all viewpoints derived from the Vedas, emphasising that they all worship a form of brahman. Furthermore, the father and son(s) analogy reinforces the harmonising intent. The Vedas are likened to the father of all doctrines, with other doctrines like sons who unite to protect their father's authority.[326]

It is noteworthy that despite the mention of māyā in verse 6.141, the question of the complete unreality of embodied existence is not raised. The understanding of identity or unity is further elucidated in Discrimination's subsequent speech in 6.144. He employs the technique of naita neta/neti neti to guide Individual Self in recognising his true and fundamental form. In this final act, naita neta/neti neti does not pertain to brahman, as in previous verses of the PcN.[327] Instead, it is used to determine the nature of the jīva:

prathama apanapau rūpa nihāra
  tako nīke karau vicāra
bolata sunata su to maiṁ kauna
  dekhata samujhata samajhau tauna
veda vacana puni karau vicāra
  naita neta jo kahata pukāra
nahiṁ tuma pṛthvī nahiṁ tuma nīra
  nahiṁ tuma teja akāśa samīra
nahiṁ haṃkāra nahīṁ mahatatva
  ina meṁ tero nā kachu satva
aba ina tatvan lai kari dehu
  jaise bhaye ulaṭi tyauṁ lehu
pṛthvī gandha mahiṁ lai karau
  gandha bahuti lai jala meṁ dharau
jala rasa meṁ teja milāya
  teja rūpa meṁ dehu samāya
rūpa vāyu meṁ lai kari dehu
  vāyu sparasa eka kari lehu
sparasahi dehu avakāśa milāya
  akāśa śabda meṁ jāya bilāya

326 See PcN 5.37–5.41.
327 See PcN 1.4, 1.127.

śabda tṛvidhi ahaṃkārahi milai
ahaṃkāra mahātatvahi ralai
bahurau mahattatva jo āhi
līna karau māyā meṁ tāhi
māyā līna hoya jāmāhiṁ
so tū āhi aura puni nāhiṁ
tāhī sāgara kī tū chīṭa
nahiṁ tū sura nara āsura kīṭa
yaha jaga mithyā bhrama tama kūpa
tāhi miṭāya dekha nija rūpa
svaprakāśa sukha nidhi jo koya
so tū veda vacana yoṁ hoya
(PcN 6.144)

First, distinguish your "I" from your own self, ponder it deeply. One talks and hears, but who am I? Realise who is seeing and thinking. Then reflect on the words of the Vedas, its appeal "Neither this, nor that." You are not soil, water, fire, space, and wind. You are not the I-maker, nor the great element – there is no reality of yours in them. Now, taking those elements, place them: take them as if they were reversed. Put the earth in the smell, then place the smell in the water; the water in taste, then bring together taste with fire; unite fire with form, then form in wind. Afterwards, place together wind and touch and, then, touch and space. Space will vanish into sound; sound will unite with the triple I-maker. I-maker mingles with the great element, then absorbs the great element in Material Nature. Material Nature will merge with nothing else but what you are. You are a droplet of that ocean – not a god nor a man, not an evil creature nor an insect. This world is false, a well of darkness – by eradicating it, see your own nature! Someone self-luminous, a treasury of joy, "That you [are]", thus the Vedas affirm.

In this verse, Discrimination teaches Individual Self how to recognise his true form by making him reverse the process through which he was manifested. We have noted how this return to the source encompasses both Sāṅkhya and monistic elements. On the one hand, the Sāṅkhya aspect is evident in the recognition of the two main principles of manifestation and the role of Māyā/Prakṛti in the embodiment of Individual Self, despite his inherent belonging to the Supreme Self. On the other hand, the dualistic structure ultimately merges into a monistic framework where Paramātama Puruṣa is the sole reality.[328]

328 See subsection 4.1.3 for an exploration of how the material causes of the universe are ultimately conceived as non-different from brahman. Another notable difference from classical Sāṅ-

The concluding lines of this verse do not propose a complete identity between Individual Self and the Supreme Self but rather emphasise their shared nature and unity. Individual Self's own form (nija rūpa) is likened to a drop (chīṭa) in the ocean that is the Supreme Self. This supports the argument presented in the earlier part of the chapter, establishing that the jīva is a part (amśa) of the Supreme Self. Notably, this notion is further reinforced by verse 6.157, where Jīvātama Puruṣa declares the attainment of union (yoga) with Paramātama Puruṣa.

After the stages of śravaṇa and manana in the Upaniṣads, nididhyāsana is the final segment of the Vedantic path to liberation. Although not extensively elaborated upon in the PcN, it appears to involve deep and repeated contemplation of Upaniṣad's teachings, correctly discerned by Individual Self.[329] However, from the narrative's point of view, Upaniṣad's role is also to reincorporate a bhakti dimension into the story. She states that she has been sent by Bond-with-Viṣṇu and is meant to inhabit Individual Self's heart alongside love (prema). Upaniṣad relays a message from Bond-with-Viṣṇu, instructing her to give birth to Knowledge and Wisdom Moon. As she embodies the vidyā derived from the study of scriptures, Knowledge will reside within the now tranquil Mind. Wisdom, as a higher level of insight, will be bestowed upon Individual Self. Bond-with-Viṣṇu's elevated position is further emphasised by her authority over both Discrimination and Upaniṣad, who are commanded to live by her guidance. Moreover, the encounter between Individual Self and Supreme Self is described as facilitated by Bond-with-Viṣṇu, as noted not only by Discrimination in 1.173 but also later by Individual Self himself:

paryo rahyo duṣṭana ke saṅgata apano śuddha rūpa bisarāya
  so prasāda śrī viṣṇubhakti ke karunā karikai liyo bacāya
aba kāhū soṁ prīti karauṁ nahiṁ eka sanātana brahma bihāya
  karoṁ pratīti na duṣṭa janana kī tinake saṅga bahuta dukha pāya
diśi aru bidiśa nahīṁ kahuṁ heroṁ śānta bhayo bhava bharama naśāya
  viṣṇu hona ko yoga bhayo aba viṣṇubhakti kī kṛpā sahāya

(PcN 6.157)

I had fallen and stayed in the company of evil people, forgetting my pure form. This is the favour of Bond-with-Viṣṇu [who], compassionate, saved [me]. Now I

khya is that in the PcN, the ahaṃkāra is considered a hierarchically lower principle than mana/manas. This is because mana/manas in the PcN also encompasses the functions of the higher intellect or buddhi. It is also worth noting that the mahatatva can be seen as representing both the Individual Self and Mind simultaneously. The alignment of ahaṃkāra and manas can also be observed in the BhP. See Sheridan 1986: 44.

329 See PcN 6.145–6.146.

will not love anyone else, I will never forget the single eternal brahman. I will not trust evil people [for] I have experienced a lot of pain from them. I cannot see the different regions, I have become pacified, having obliterated the confusion of worldly existence. There was the union with the being of Viṣṇu, with the help of Bond-with-Viṣṇu.

The state of liberation in the PcN is defined in terms of renewed awareness, which is derived from wisdom gained from scriptural knowledge, peace (śānta) and union (yoga), and love (prīti). The combination of insights gained from the Upaniṣads and the equanimity achieved through yoga creates a synergistic effect with bhakti personified as Bond-with-Viṣṇu. She has determined the course of the war between Discrimination and Bewilderment,[330] directed Mind towards his father, Individual Self, and led Individual Self to his own father, Supreme Self.

How do we understand, then, the relationship between jñāna (the knowledge gained through the Upaniṣads) and yogic bhakti in the PcN? Barbara Holdrege refers to the Gauḍīyas' incorporation of Pātañjala Yoga and Advaita in their soteriological discourse as a "theology of superordination".[331] For the Gauḍīyas, both Pātañjala Yoga and Advaita led to lesser forms of realisation since the former posited a dualistic ontology, while the latter proposed a monistic one. The Gauḍīyas, as supporters of a Bhedābheda position, could not endorse either approach without raising objections, although they did regard Pātañjala Yoga as superior to Advaita.[332]

In the case of the PcN, there is no explicit critique of Pātañjala Yoga as distinct from bhakti. While the text makes references to the YS, its description of yoga also draws from other traditions, even while using Patañjali's conception as a model. For this reason, I have referred to "yogic bhakti" and "bhakta yogis" as broad categories encompassing diverse types of yoga-inflected devotion and bhaktas who are both yogis and ascetics. Unlike the Gauḍīyas, in Brajvāsī's work, yoga abhyāsa (yogic practice) and Vedantic knowledge are not treated as separate or lesser ways but rather integrated into the discourse of bhakti (devotion). The bhakta yogi attains the condition whereby his equanimity and knowledge enable him to dedicate himself solely to devotion. This devotion is understood as encompassing both samatā (sameness, equanimity) and ānanda (bliss), which are identified as the single rasa (essence) of the Supreme Self.[333]

---

330 See e.g. PcN 4.20, 4.98, 4.100, 6.7, and 6.9.

331 Holdrege 2013: 170. See also Holdrege 2014.

332 See Holdrege 2013: 169.

333 See PcN 6.5. Additionally, in PcN 6.93, Sacrificial Science equates a state of equanimity with bliss while reasoning on Upaniṣad's words, before rejecting them since they would make her (i.e. sacrifice) useless. See also 6.161.

In addition, while the Gauḍīyas rejected mukti (liberation from the cycle of rebirth), particularly the Advaita conception of it which involved jñāna as the salvific means, and envisaged bhakti alone as the ultimate goal for a bhakta,[334] the PcN acknowledges the significance of mukti. While bhakti is still considered the ultimate purpose of an individual's life in Brajvāsī's drama,[335] mukti, connected to wisdom or intuitive realisation (prabodha) and knowledge, is not rejected altogether. Instead, the PcN's path encompasses *both* mukti and bhakti. In terms of progression, it appears that mukti *precedes* bhakti, as one must first discern one's true nature and brahman's nature before engaging in unfettered devotion.

The emphasis on bhakti as the supreme path and goal, while rejecting the sole importance of knowledge, is not unique to the Gauḍīyas but a characteristic of Vaiṣṇava Vedānta schools that considered the BhP as their primary scripture. These schools aimed to establish distinct worldviews separate from Advaita Vedānta.[336] However, during the early modern period, a tendency emerged to reconcile seemingly opposing positions, leading Advaitins to confront the challenge of bhakti,[337] resulting in varied responses. Madhusūdana Sarasvatī viewed jñāna and bhakti as two equivalent yet distinct paths, while Nārāyaṇa Tīrtha, on the other hand, subordinated jñāna to bhakti. In his commentary on the *Bhakti Sūtras* by Śāṇḍilya, the *Bhakticandrikā*, Nārāyaṇa Tīrtha asserts that knowledge of brahman, which entails the eradication of ignorance, is authentically mokṣa only when characterised by prīti/prema (love).[338]

While his interpretation is reminiscent of PcN 6.157 quoted above, which highlights particularly the power of Viṣṇubhakti, one should be careful not to underestimate the role of knowledge in Brajvāsī's text. After Individual Self recognises his true nature and is prepared to love only the Lord, Bond-with-Viṣṇu appears and grants him a boon (6.158). He requests her assistance in helping other jīvas to cross the ocean of existence and realise their true nature – a path which entails jñāna – to which she willingly agrees (6.161). As the conflict between Discrimination and Bewilderment reaches its conclusion with the former's triumph over the latter, the other two narrative frames of the PcN resume. The story of Kṛṣṇadās Bhaṭṭ and his disciple follows, revealing that the pupil has attained discrimination and detachment from the phenomenal world through his

334 See Holdrege 2013: 171.

335 See PcN 6.160, where Individual Self praises Bond-with-Viṣṇu: "What is difficult to obtain if one has your favour? You are the fulfilment of all desires."

336 See Dasgupta 1922, vols 4 and 5.

337 See Pellegrini 2018: 603–605.

338 My paraphrase of the translation and text in Venkatkrishnan 2015: 214–215. In the preceding pages, Venkatkrishnan examines Madhusūdana's thoughts on knowledge and bhakti.

guru's narration of the PcN story (6.165). Then Brajvāsīdās speaks in the first-person voice, explaining the phala (fruit) of the book as the dissolution of the knot of ego and misconception. Thus, he assigns the primary purpose of the PcN as the generation of knowledge (6.166). He then extols the greatness of discrimination and pairs it with bhakti, emphasising their indispensability for those bound to Hari (Kṛṣṇa):

bhakti hoya bhagavanta kī aura viveka prakāsa
bhakti binā tribhuvana dukhī kaha brajavāsīdāsa
(PcN 6.167)

jo cāhai ānanda sadā re brajbāsīdāsa
tau kīje hari bhakti ko choṛi viṣai kī āsa
(PcN 6.168)

There will be a bond with the Blessed Lord and the effulgence of discrimination.
Brajvāsīdās says: "Without a bond with God the three worlds are distressed.

O servants of he who dwells in Braj (Kṛṣṇa), if you wish for eternal bliss,
then forsake desire for the objects of the senses and share a bond with Hari."

In this fusion of knowledge derived from discrimination and bhakti, we can discern the influence of the RcM. In the RcM, the crow Bhuśuṇḍi proclaims that faith cannot exist without knowledge, and without faith, the love that nourishes bhakti is not possible.[339] While the RcM upholds the primacy of bhakti as the most accessible means in the age of Kali,[340] it also provides a starting point within the realm of the local language for those seeking to reconcile jñāna and bhakti. Besides the emphasis on the power of discrimination, as exemplified by Bharata's character, in the *Uttara Kāṇḍa* Bhuśuṇḍi condemns those who do not walk the path of bhakti to Hari together with detachment (birati) and discrimination (bibeka).[341] Furthermore, there is a lengthy passage praising "the lamp of knowledge", where Bhuśuṇḍi teaches Garuḍa that the most radiant flame of this lamp is sohamasmi, "I am he", quoting *Īśā Upaniṣad* 16 in verse 7.117.1. Thus, regarding jñāna, the RcM refers to Upanishadic wisdom, which is also central to the PcN. In Brajvāsī's

339 RcM 7.88.4: jāneṁ binu na hoi paratītī | binu paratīti hoi nahiṁ prītī || prīti binā nahiṁ bhagati diṛhāī | jimi khagapati jala kai cikanāī || (Without knowing there is no faith; without faith, there is no love. In the absence of love, bhakti is weak. O king of birds, [love is] like the silkiness of [bhakti's] water.) See also Tripathi 1977: 132–133.

340 See e.g. RcM 7.118.1–5.

341 See RcM 7.100b. In PcN 6.7, the path of bhakti is also mentioned as being cleared of the thorns and mires of delusion and fear through the practice of viveka.

work, the saying from the *Īśā Upaniṣad* also appears as "soha", a derived form of sohamasmi, pronounced by Discrimination in verse 4.31.

As demonstrated in this chapter, the PcN and the RcM share numerous commonalities in their Vedantic and bhakti perspectives, those found in the latter serving as a source of inspiration in the former through adaptive reuse. However, it is important to mention some doctrinal differences briefly. First, the role of Rāma's/God's name as a mediator between his nirguṇa and saguṇa form is not as pronounced in the PcN compared to the RcM. Additionally, there is no explicit mention of Śiva in terms of subordination to Rāma or Kṛṣṇa. Instead, as demonstrated earlier, the PcN focuses on a godhead who embodies both a nirguṇa form referred to as Viṣṇu/Hari and saguṇa forms such as Kṛṣṇa and Rāma. Consequently, the harmonisation in the PcN occurs between the worship of these latter two gods. In this sense, it is telling that in the concluding portion of the PcN story, when Individual Self awakens to his true nature and encounters Bond-with-Viṣṇu, there is no explicit reference to Kṛṣṇa or Rāma. The mention of the servants of the brajabāsī – that is, Kṛṣṇa – occurs only after the PC story, strictly speaking, is concluded in verse 6.161, as it becomes clear that Kṛṣṇa is the chosen divinity for the author (6.168).

Brajvāsīdās's address to his fellow bhaktas underscores that a classical Vedantic drama like the PcN circulated and held value within communities such as Vallabha's. As discussed in Chapter 1, the PcN's opening praises the community of truth-knowers, the satsaṅga, and extols the value of bhakti. By doing so, it frames the PC story and situates it within a bhakti discourse, appealing to potential readers and listeners. The significance of the PC story is reinterpreted in the PcN as a means to consolidate bhakti while showing the way to peace and detachment through discrimination. For authors like Brajvāsīdās, recreating the PC story as a bhakti narrative that resonated with his audience also involved adaptively reusing the RcM, which to some ears could mean listening to both Kṛṣṇa's and Rāma's tales simultaneously.

Chapter Summary

This chapter has discussed the metaphysical framework and soteriological path of the PcN. The first section focused on analysing the philosophical concepts and the characters that personify them that elucidate the PcN's Vedānta, namely Supreme Self, Material Nature/Illusion, the phenomenal world, Individual Self and Mind. Through the examination of these principles and their portrayal in Brajvāsīdās's drama, it has become evident that the text draws inspiration from the

Upaniṣads, the BhG, and the Purāṇas but that the RcM emerges as its most significant source for Vedānta.

The PcN draws on the RcM across various concepts and to different degrees, and one effective way to identify these instances of adaptive reuse and simple re-use is by scrutinising the prosody. For instance, regarding Supreme Self, while the PcN aligns with the BhP's multiple ways of addressing this concept, it reuses the RcM's terminology to convey the paradoxical nature of the Supreme Self as both nirguṇa and saguṇa. Yet the PcN does not blindly repeat the RcM in every instance but instead maintains its own unique points of interest. An example of this is the emphasis it places on the essence of paramātama puruṣa being both bliss and equanimity.

Based on this exploration, it is evident that the PcN simultaneously recognises an unqualified or absolute and qualified brahman, and it acknowledges that brahman's qualified form can be identified with either Kṛṣṇa or Rāma. However, as this chapter has shown, Brajvāsī provides precise clues that his preferred form is Kṛṣṇa. This is achieved, for instance, through the adaptive reuse of a head-to-toe description of Rāma, where specific physical details pointing to Rāma are deliberately omitted, as well as the explicit address of the extradiegetic narration in the closing verses of the drama, after the completion of the diegetic and intradiegetic levels.

The PcN demonstrates sensitivity not only towards the treatment of metaphysical concepts found in the RcM but also towards other authors who wrote in other regional languages. This sensitivity is particularly evident in its portrayal of Māyā as a deceitful woman in competition with another woman, Bond-with-Viṣṇu. In this aspect, Brajvāsī appears to draw inspiration from the works of nirguṇa poets like Kabīr of the fifteenth century and, notably, Sundardās of the seventeenth. By contrast, Brajvāsī's affiliation with the Vallabhite tradition is not prominently highlighted in the analysis of the aforementioned ontological categories. For example, the PcN does not employ the term prapañca (extension, manifestation) to positively indicate the phenomenal world, as its orientation towards the world is generally negative. However, this should not be interpreted as conclusive evidence of a non-Vallabhite orientation because the overall aim of the PcN story must be considered. As we have observed, the PcN emphasises cultivating detachment from the world and fostering attachment to the personal brahman through love. Consequently, it is natural for the world to be characterised in a predominantly negative manner. What can be inferred more definitively is that the PcN presents conceptions shared by thinkers from various distinct schools, such as the idea of brahman as the locus cause of the phenomenal world, a theory that found support from scholars like Vijñānabhikṣu, an exponent of Bhedābheda philosophy in the sixteenth century, as well as from slightly later Advaita thinkers like Prakāśānanda.

From the sixteenth century onwards, there was a growing tendency towards harmonisation among the different schools of Vedānta. This trend is also discernible in the PcN's treatment of Individual Self. Despite considering jīva as a part of brahman, our author did not hesitate to employ images typically associated with opposing doctrines such as reflection and delimitation. In addition, the story of the PcN proposes the allegory of the Supreme Self as the father and the Individual Self as his son, an analogy that had been employed in Vedantic discourses by the Advaitin Bhāratītīrtha/Vidyāraṇya in the fourteenth century. Vijñānabhikṣu also used this imagery to express the notion that the jīva is a part of brahman, much like Brajvāsī. However, unlike Vijñānabhikṣu, Brajvāsī was able to incorporate this classical Vedantic analogy into his bhakti perspective.

The chapter's second section demonstrated how the son is reunited with the father through a final state called yoga or "union with Viṣṇu's being". This path combines elements of yoga, Vedānta, and bhakti. First, the PcN makes room for a bhakta who, like Bharata in the RcM, is both a yogi and an ascetic. Brajvāsīdās achieves this by portraying various figures, including paṇḍitas but also yogis and ascetics residing along the banks of the Ganges. These figures, depicted as practising bhakti or forging a bond with God, are referred to as both vivekī (discerning people) and bhaktas (devotees).

However, it is not limited to this aspect alone; the PcN also elaborates on its own bhaktified version of the eight-limbed yoga as Discrimination's ministers. This list is influenced by Pātañjala Yoga through the adaptive reuse of its definition of yoga, as well as by Haṭha Yoga and the yogas described in the Purāṇas. There are also parallels with the yogic practices found in other Brajbhāṣā works, such as Sundardās's *Sarvāṅgayogapradīpikā*. Notably, I have argued that the presence of sama (balance) as the designation of the third limb – in the YS, and traditionally, posture (āsana) – could be influenced by both the description of yogic practices in ŚvU and the RcM's blending of sama with yama and nema, the two preceding steps of the eightfold path.

I have associated sama with the mental state of samatā or equanimity, already mentioned as identified with the essence of brahman together with bliss. Equanimity in the sense of "seeing everything the same" implies no longer experiencing aversion or desire towards the world; that is, cultivating dispassion. This goal of the PcN connects it with the BhG's teachings and depicts the yogi as someone with unwavering insight. In fact, Mind must relinquish his inherent fickleness and attain awareness of his true nature. To achieve this, the PcN blends yoga with the knowledge obtained through listening, contemplation, and meditation on the words of Upaniṣad through Discrimination. However, Discrimination alone is not sufficient; he becomes empowered only through the assistance of Bond-with-Viṣṇu. When Wisdom manifests, Individual Self attains stability and is completely liberated from egoism and misconception. Finally, he is ready to encounter Bond-with-Viṣṇu and, consequently, to love only his father, Supreme Self.

Conclusion

Talking is easy; doing is difficult.
Talking without doing is worthless.
A cat eats an educated parrot,
And in the scholar's hands [only] a book remains.
Gorakhnāth, Sabad 119

The twelfth-century yoga master Gorakhnāth reminds his audience that practice is the only way to make sense of theory and that without practice, the very existence of theory may be at risk.[1] This book has made a point of demonstrating what and how the PcN has contributed to the history of Indian philosophy, specifically within Vedānta traditions. Through the study of Brajvāsīdās and his PcN, it becomes evident that Indian philosophical traditions are eclectic, and that being philosophically and literarily creative in eighteenth-century North India encompassed various aspects. The analysis presented in this book does not rely on a single predetermined method for reading and interpreting an early modern Brajbhāṣā text that combines Vedānta and bhakti. It takes a stance by not directly relating the intellectual intervention of the PcN to its most apparent comparison, the KPC, but finds a closer contextual connection with Tulsīdās's RcM.

Chapter 1 explored the figure of Brajvāsīdās as an author in the eighteenth century by examining his relationship to the Vallabhite tradition and his literary output, shedding new light on the edition history, circulation, and reception of his PcN. Both the PcN and the *Brajavilāsa* appear to confirm the shift of authority from Sanskrit to vernacular sources as a predominant literary practice in the century, the modalities of which scholars still do not fully understand.[2] The chapter then discussed key textual features upon which the critical analysis draws, foregrounding observations about the play's narrative structure, plot, performative and oral dimensions, and language.

In Chapter 2, the focus shifted to the intricate connections among texts within the PC tradition, particularly earlier works associated with the PcN. It highlighted that the transmission of the PC story, especially in the case of the GH, was not a linear progression from Sanskrit to Persian or vernacular languages but often involved bhāṣā sources, such as Nanddās's PC. However, it was demonstrated that Nanddās's PC was likely unknown to Brajvāsīdās or that he purposefully did not select it as a source in order to place less emphasis on Vallabhite theology. Moreover, the chapter stressed that the KPC is remembered both as a text and as an

1 The sabad comes from the corpus of the *Gorakhbāṇī*, translated in Djurdjevic/Singh 2019: 73.
2 See section 1.1.

https://doi.org/10.1515/9783111063836-006

oral tradition, blurring the boundaries between "historical truth" and "poetic truth". The examination did not attempt to disentangle them but rather to illustrate how Kṛṣṇamiśra and his text – or story – were (re)imagined within the context of retelling the PC story.[3]

Regarding the focal point of this book, the Vedānta and bhakti perspectives of the PcN and their sources, the introductory chapter posited that four key points could determine the position of the PcN on the spectrum between adaptive reuse and simple re-use. Modifying the order in which they were presented earlier, our criteria for adaptive reuse were that it involved (1) resuming the usage of (2) a clearly identifiable object after an interruption in its use (3) by at least one consciously acting agent (4) aiming to achieve a specific purpose.[4] Let us see how this criteria was analysed in Chapters 3 and 4 to test my earlier hypothesis that in the PcN, simple re-use fits into a broader framework of adaptive reuse.

Regarding point (1), the object whose use is resumed in the PcN *is* the RcM, a text Tulsī began composing in 1574, nearly two centuries before Brajvāsī wrote his play in 1760. Point (2) may be more challenging in terms of establishing how "clearly identifiable" that object is. Brajvāsī does not explicitly mention Tulsī, and there are few instances of "proper" quotations – in the sense of whole or isolated sentences reproduced verbatim – that would immediately allow us to identify the RcM as the source.[5] Chapter 4 demonstrated that the PcN's author frequently adaptively reuses verses from the RcM by, for example, reusing the first pāda of a dohā but combining it with a second pāda of his own invention, sometimes modifying the rhyme.[6] In other instances, his reuse involves less than an entire line. It was shown that the extent of reuse could often be identified through prosody, which governs word order and syntax and helps determine cases of adaptive reuse when the metres of the texts differ.[7]

The adaptive reuse of the RcM in the PcN involves not only complete and partial pādas but also single words, such as the comparison of the teacher to the doctor.[8] These latter cases, entailing individual terms rather than syntactic units and whose respective contexts are often dissimilar, may be seen as simple re-use. However, the relationship between the PcN and the RcM as a whole may be

3 See sections 2.1 and 2.2.

4 Freschi/Maas 2017: 13. See "Adaptive Reuse: Discussing a Method" in the introductory chapter above.

5 See section 3.3.

6 See subsection 4.1.1.

7 Refer to "Paramātama's Pervasiveness" in subsection 4.1.1 for a verse where the caesura allows for the isolated segment to be adaptively reused.

8 See section 4.2.

termed adaptive reuse. It is evident that the RcM serves not only as a source of words but also as a literary paradigm for the PcN from the way Brajvāsī adaptively reuses a physical description of Rāma in his own description of Viṣṇu/Kṛṣṇa, sometimes expanding patterns provided by Tulsī to enhance the ear-pleasing alliterations.[9] These elements demonstrate that it is Brajvāsīdās who is the "consciously acting agent" in point (3).

The purpose (point 4) behind resuming certain usages (point 1) was initially addressed in Chapter 3 through circumstantial evidence and further explored in Chapter 4 by investigating the mechanism of adaptive reuse and the conceptual parallels between the two texts. According to the definition of adaptive reuse provided by Freschi and Maas, it involves repurposing the reused item with the intention that the audience recognises both the item and its new function. In the PcN, the reuse of the RcM is detectable to such a degree that it suggests that Brajvāsī not only knew the RcM and consciously reused it but also intended his prospective audience to both recognise the reuse and recognise it *as repurposed*. This recognition is comparable to the change of genre observed in Maas's examination of the reuse of Patañjali's *Yogaśāstra* in Māgha's *Śiśupālavadha*, where the transition from shastra to kāvya indicates a clear case of repurposing. Maas argues that this adaptive reuse serves a twofold purpose: associating the *Yogaśāstra* with a Vaiṣṇava narrative and viewpoint, and enhancing the yogic mood of some passages within the poetic context, thereby conferring the prestige of a shastra to the kāvya.[10]

Although the PcN and the RcM differ in genre, with the former being a drama and the latter a long epic poem, both texts operate through dialogue and in many cases the PcN reuses elements from the RcM without radically repurposing. For example, Brajvāsī adaptively reuses content about the brahman from the RcM to discuss brahman in the PcN, and he defines personified bhakti (Bond-with-Viṣṇu) as "beloved" (pyārī) to the Supreme Self, mirroring the RcM.[11] While conceptual differences exist between the two texts, the adaptive reuse is primarily driven by the same intended purposes. This suggests that Brajvāsī adaptively reused the RcM to remind his audience of the earlier text. The prestige of the RcM in Brajvāsī's day is confirmed by sources contemporary with the PcN,[12] and it is possible that Brajvāsī wanted to leverage its appeal for his own composition. Chapter 3 demonstrated that the RcM contains the PC story and assigns a significant place to viveka. Brajvāsī's audience may well have drawn on their own memories

9 See subsection 4.1.1.
10 Maas 2017.
11 See subsection 4.1.2.
12 See Introduction and section 3.3.

of the RcM in receiving the PcN and interpreting its key aspects such as the emphasis on viveka. On the broader level, it is possible to say that the PcN repurposes the RcM text and story by placing it into its own narrative to underscore its distinct Vedānta and bhakti perspectives.

We will now explore how this book contributes to the "Greater Vedānta(s)" approach mentioned in the introduction and what insights it offers regarding early modern Vedāntas in bhāṣā. In Chapter 4, the examination revealed the philosophical and religious affinities between the PcN and RcM. Still, it also emphasised that the PcN presents its own philosophical and religious reflections independently of the RcM. The didactic aim of the PcN is to make philosophical exposition affirmative and teleological. However, this does not imply an unawareness of the debates that had (and still have) animated Vedānta thinkers for centuries, such as the distinction between māyā and avidyā.[13] At the same time, Brajvāsī is selective in the philosophical themes addressed in his drama. For instance, in passages discussing the embodiments of Viṣṇu/Hari as Kṛṣṇa or Rāma, he does not delve into questions regarding the nature of the body assumed by the deity or whether they are subject to karman. Instead, the focus is placed on the very assumption of a form or disguise (rūpa dhar-).

Moreover, the chapter demonstrated that Brajvāsī draws from the Sanskritic tradition to articulate his ideas but modifies it according to his bhakti sensibility. For instance, when describing Supreme Self paradoxically as having no feet but still walking and no hands but still grasping, the ŚvU follows this by reflecting on the cosmic puruṣa and his all-encompassing body. In contrast, Brajvāsī, along with Tulsī, concludes the image with a reflection on the nirguṇa and saguṇa brahman, which are of central concern for bhaktas of all classifications. This demonstrates that while the Upaniṣads were regarded as authoritative texts, their teachings and metaphors were continuously reinterpreted, and their texts were adapted to suit different contextual needs rather than being transmitted in unaltered form.[14]

The intellectual intervention that Brajvāsī sought to make in terms of Vedānta gains in depth by adding the RcM to the picture, for it is clear that Brajvāsī not only "read" the RcM but also read Vedānta through the RcM and other bhāṣā sources. Similar to what A. K. Ramanujan observes in folk versions of epic and Puranic narratives, the PcN appears to "localise" and "contemporise" the PC story, Vedānta,

13 See subsection 4.1.2.

14 See Freschi 2015: 92. For Veṅkaṭanātha's approach to employing quotations from sacred texts as support for independent arguments rather than mere proof in themselves, see Freschi 2019: 72–77.

and bhakti through the adaptive reuse of the RcM.[15] When studying early modern Vedānta, it is insufficient to approach it solely from a diachronic perspective that focuses on evolution since the teachings presented resonate and intersect with other pertinent concerns and affinities.

In fact, the PcN contributes to our understanding of how Kṛṣṇa and Rāma emerged as significant deities in modern-day North Indian Hinduism.[16] Similar to the RcM, Brajvāsī presents Kṛṣṇa and Rāma as two alternatives for worship without explicitly asserting the superiority of one over the other. Instead, he embeds references to both of them and their deeds within the PC story.[17] Therefore, in the case of the PcN, bhakti and bhāṣā literature become indispensable factors to consider when analysing the text's Vedantic dimension. To fully grasp the intellectual import of the PcN it does not appear possible to separate bhakti from Vedānta as two distinct entities, that is, "religion" and "philosophy". Through the adaptive reuse of the RcM, Brajvāsī's work redefines the boundaries of what scholars traditionally consider Vedānta philosophical texts and challenges the notions of what constituted a text or story suitable for teaching philosophy during his time. In a word, the PcN sheds light on what texts and ideas were critical in eighteenth-century North India.[18]

This analysis of the PcN has also emphasised the need for caution when evaluating early modern bhāṣā texts dealing with Vedānta. While for some such texts, it may appear straightforward to determine their support for Advaita Vedānta positions, we should refrain from making hasty evaluations. Michael Allen's criteria that an Advaita Vedānta perspective can be inferred if one finds a straightforward doctrine of unreality of the world and identity between brahman and jīva can be applied to the PcN to some extent. However, it is not possible to categorise the PcN as having a clear-cut Advaita orientation. Some passages assert the world's unreality, but others suggest the world is a transformation of brahman, akin to gold and ornaments.[19] Similarly, the understanding of brahman and jīva is presented in various ways that may seem contradictory. In this regard, Angelika Malinar's suggestion proves valuable. It cautions against evaluating doctrines

15 To make the PC story a bhakti story and thus relevant and interesting for his intended audience, Brajvāsī deemed it necessary to include vinaya verses at the beginning of the drama, aligning with the approaches taken by Tulsī and Sūrdās in their own works. See subsection 1.2.2. See also Ramanujan 1993: 103.

16 See Clooney/Stewart 2004: 165–166.

17 See subsection 4.1.1. Importantly, considering Rāma and Kṛṣṇa as the primary deities for veneration in the PcN does not imply that their martial aspects, such as Rāma's battles with rākṣasas or the BhG narrative associated with Kṛṣṇa, are entirely disregarded. See also Brockington 2005.

18 See Kaul 2020: xxv.

19 See subsection 4.1.2.

solely in absolute terms and highlights the importance of considering a text's respective contexts and overall purpose.[20] In the case of the PcN, the play's plot provides insights into Brajvāsī's position regarding brahman and jīva. Additionally, it is possible that the metaphors and similes used in the text, which can be perplexing when interpreted solely from a doctrinal standpoint, were employed for their imaginative power and to serve the encyclopedic aspirations of the text as well.

How does my study – any study – of the PcN improve our understanding of the modern perception of Vedānta? Recent scholarship has rekindled the discussion surrounding the prominence Advaita Vedānta enjoys today compared to other Vedantic traditions. In fact, when Vedānta is mentioned, it is often understood to refer exclusively to Advaita Vedānta. The focus of this renewed interest seems to lie in demonstrating that the popularity of Advaita Vedānta was not solely due to the work of Orientalist intellectuals but that some premodern thinkers had already begun to conceive of Vedantic traditions in a unified manner and that Advaita Vedānta was already somewhat popular not only among intellectuals but also among ordinary people.[21] These reflections are pivotal for better comprehending the relationship between the past and present. However, drawing definitive conclusions is challenging.

We encounter ambiguous elements when we attempt to assess Brajvāsī's views on Vedantic traditions. In Act V, Brajvāsī mentions (through Faith) both Śaṅkara (the eighth-century Advaita Vedāntin) and Bhāskara (the eighth/ninth-century Bhedābheda philosopher) as part of Discrimination's army in the doctrinal war with Bewilderment's faction. Brajvāsī depicts Śaṅkara as connected to the teachings of the Vedas (veda acāra), upholding the puruṣārthas, and defeating the nāstikas.[22] These elements can be interpreted in themselves as granting Śaṅkara a position of honour. In contrast, Bond-with-Viṣṇu's praise of Śaṅkara (5.51) after he gave the final blow to the Veda-Deniers might indicate the subordination of the teacher of the Vedas to bhakti. Additionally, Bhāskara's inclusion among an array of characters, including personified Mahābhārata and the six darśanas, might suggest that this is not a discourse about competing Vedantic traditions trying to establish the superiority of Advaita. Nevertheless, it remains challenging to

20 See Malinar 2017: 600–601.

21 See Nicholson 2010; Allen 2016, 2017. See also Maharaj 2020: "Introduction". Among those who had defended Advaita Vedānta and dismissed other Vedantic schools as its distortion, see, most notably, Deussen 1907: 46.

22 See PcN 5.47–5.51.

determine whether the mention of Śaṅkara and Bhāskara pertains to their historical personas or not.[23]

The analysis presented in this study of the PcN has considered just one aspect among many others. Undoubtedly, a deeper understanding of eighteenth-century North Indian history and literary culture could have provided more insights into Brajvāsī's text and its connection to broader questions of intellectual history. Nevertheless, this study challenges "the old ways of questioning" Vedānta philosophies[24] and emphasises the necessity of translating and considering early modern vernacular sources and authors like Brajvāsī who aimed to position themselves within philosophical discourse. In this regard, it is essential to highlight that refraining from extracting a "philosophy of the PcN" also means refraining from extracting the religious and ethical dimensions of the drama. The ontological and metaphysical issues addressed in the PcN are intricately intertwined with soteriological concerns, and neglecting them would render the doctrines abstract and detached from their historical context. It is their pairing with a bhakti viewpoint that offers us a glimpse into the vibrancy of Vedānta in early modern times.[25]

As we have observed, we cannot classify the PcN as a translation in the canonical sense. However, it can be seen as a crystallisation of the PC story, resonating with numerous thinkers and poets across the centuries. The Brajbhāṣā drama itself contains imagery that was also employed by nineteenth-century Vedāntins and continues to be remembered today, albeit with different significance. For instance, there is a concise tale in the first act of the play, where Discrimination explains Mind's forgetfulness to his wife. Just as a lion cub forgets its lion nature and lives among sheep, bleating and not eating them, similarly, Mind fails to remember his pauruṣa, his "belonging to puruṣa", his connection to Supreme Self.[26] I have been unable to trace a written source for this story. Still, it is primarily remembered in India because Swami Vivekananda (1863–1902) employed it frequently, such as in his "Paper on Hinduism", delivered at the Parliament of Reli-

---

23 The concise portrayal of Śaṅkara's actions resembles the digvijaya genre (biographical narration involving Śaṅkara's conquest of the quarters by defeating philosophical adversaries). For more information, refer to Bader 2000. In this context, the battle against the heretics can be seen as the development of a literary topos, considering that Buddhist and Jain adversaries were no longer posing actual threats by the eighteenth century. See Okita 2016 for insights into how the Gauḍīya philosopher Baladeva Vidyābhūṣaṇa critiqued Buddhist doctrines.

24 See Kaul 2020: xv.

25 Some scholars argue that highlighting the religious and ethical aspects of Indian philosophical thought may hinder attempts to produce an account that goes beyond the perception of a "mystical" India. See Torella 2011.

26 See subsection 4.1.4.

gions in Chicago in 1893.[27] While I refrain from drawing direct connections between these instances due to the need for more profound knowledge of each case, I believe this book has shed light on the central role that stories, tales, and narratives continue to play in Indic imagery.[28] While Brajvāsī's work can be seen as a Vedantic text, we must not forget that through the PcN, we also hear the voices of Tulsī, Sūrdās, Sundardās, and Kabīr echoing as they did for Brajvāsī. It is through this polyphony of voices that Brajvāsī, as Jonardon Ganeri would put it, found "a way to mediate a conversation between the past and the present".[29]

27 "Come up, O lions, and shake off the delusion that you are sheep; you are souls immortal, spirits free, blest and eternal; ye are not matter, ye are not bodies; matter is your servant, not you the servant of matter." See Vivekananda 1957: 11. It is noteworthy how this allegory, originally meant to symbolise the awakening of self-awareness and rejection of an inflated ego, has found resonance in the repertoire of "muscular nationalism". It has been utilised by figures like Narendra Modi in his "Man kī bāt" as well as other public religious personalities who align themselves with Hindutva politics, drawing inspiration from Vivekananda's usage. See Modi 2014.

28 See Ramanujan 1989.

29 Ganeri 2014.

Part II: **Annotated Translation of the *Prabodhacandrodaya Nāṭaka***

Note to the Translation

Details about the sources used for this translation and observations about the language employed in the PcN have been extensively discussed in Chapter 1.[1] This note addresses some of the challenges I encountered in approaching the PcN and the solutions adopted in translating it into English for the first time. The process of translating Brajvāsīdās's composition involved grappling with the scarcity of information about him and the literary landscape of eighteenth-century North India. Moreover, the text itself presents difficulties due in part to its being a nāṭaka in Brajbhāṣā, a genre that lacks sufficient translations and studies, as noted earlier.[2]

Walter Benjamin argued that translations carry forward the life of a text and, to my mind, this translation – indeed, any translation of works from the PC tradition – serves to preserve the paramparā from fading away into historical obscurity and revives its presence within important intellectual and cultural contexts.[3] As the translator of the PcN, it was necessary for me to constantly navigate between aspects I deemed of primary and secondary importance in my interpretation of Brajvāsīdās's drama.[4] First, the text is composed entirely in verse. My translation does not attempt to reproduce the rhymes and other rhetorical devices of sound and meaning present in the PcN. This is a major yet inevitable sacrifice when rendering text into English.[5] In the case of poetry, meaning is also conveyed through elements that go beyond the content, such as alliteration. Far from disregarding these aspects, I believe that Brajvāsīdās's skill as a poet can be better appreciated through the chapters of my critical study. To highlight Brajvāsīdās's creativity, my translation remains largely literal, prioritising semantic equivalence. While using the 1875 Naval Kishore Press lithographic copy as my primary source text, I also took into consideration the published translations of the KPC and a few other texts from the PC tradition, primarily in Sanskrit.[6] This English translation complements these texts, offering a new interpretation of the PC story.[7] In

1 See subsection 1.2.1.

2 See subsection 1.2.3.

3 Benjamin 2012: 77.

4 These considerations are prompted mainly by Umberto Eco's discussion of translation in Eco 2020.

5 See Eco 2020: 274.

6 See subsection 1.2.1 and section 2.1.

7 See Eco 2020: 247.

https://doi.org/10.1515/9783111063836-007

terms of translation style, it aims to faithfully convey the philosophical and theological passages while also honouring the literary qualities of the PcN.[8]

I was reluctant to compromise on the dramatic format of the play. However, treating the PcN as a Sanskrit drama proved challenging due to certain structural issues that surfaced. While the text I had at my disposal provides indications about the end of each act, there is no consistent method for denoting character entries, exits, and movements. Furthermore, specific stage locations and the individuals present onstage are often not marked at the beginning of each act. Initially, the absence of these expected features can be disorienting, and as a translator, it is tempting to supply them. However, I ultimately chose to live with these omissions and refrain from adding indications. This decision was motivated by two factors. First, I believe that these missing features may shed light on a performance context that is currently beyond our grasp.[9] Second, from a translation perspective, my aim was to render the text in accessible English while allowing these details to bring readers closer to the historical period and cultural milieu in which the PcN was composed and subsequently published.[10] By forgoing the rhyme scheme and preserving the structural omissions and occasional idiosyncrasies in the English translation, I sought to navigate "between making the strange familiar and the familiar strange".[11] In my view, this approach encourages reflection and deepens one's understanding of the text.

To navigate through the PcN, it is important to recall that the drama, as presented by Brajvāsīdās, is narrated by Kṛṣṇadās Bhaṭṭ and performed at the court of King Kīratibrahma.[12] Therefore, references to the sabhā or "(assembly) hall" of the king appear throughout the story.[13] For instance, at the beginning of Act II, the First Actor instructs his team to set the scene in Kāśī within the sabhā of King Kīratibrahma's court:

First Actor *(at* Deceit*'s words, wisely instructs his troupe)*: Set up Kāśī somewhere in the hall! (2.6)

8 See on this matter Kapstein 2009: xlviii.

9 See subsection 1.2.3.

10 See Eco 2020: 170.

11 See Polizzotti 2018: "Translation navigates between making the strange familiar and the familiar strange. It couches another's cultural expressions and assumptions in a form that we can understand, identify with, enter into. At the same time, it maintains its distance from too great an assimilation of those expressions and assumptions, so as not to homogenise the experience of the foreign."

12 See subsection 1.2.2.

13 Alternatively "darabāra" (1.26) and "majalisa" (2.6).

The specific locations within the world of the play are not explicitly mentioned for Act I, while Act II unfolds along the riverbanks of the Ganges and within Bewilderment's encampment. Act III's location is also unspecified, while Act IV takes place at Discrimination's court, which is not geographically specified, through the king later departs for Kāśī. Act V includes a description in the past tense of the war that took place in the holy city of Vārāṇasī, while the location of Act VI is unknown, though it is most probably also Vārāṇasī.[14]

References to the backstage area, separated from the stage by curtains, are common, as seen in Act II. Many actors first speak from this offstage location before entering the scene:

In that moment I-MAKER appeared from behind the curtains in the guise of an old priest of an exceptionally violent nature. (2.14)

In this way CONFUSION and FALSE VIEWPOINT conversed behind the curtains. Then they arrived in the hall and danced gracefully. (2.139)

The verses quoted above are stage directions presented in a narrative tone, using the past tense. As such, I have not placed them after the name of a character preceding their direct speech. Instead, I have left them as independent verses, as they appear in the Naval Kishore edition, which I follow for verse numbering as well. However, when a stage direction precedes direct speech within the same verse, I have retained its position and translated it into the present tense. In these cases, I have indicated the stage direction by enclosing it in parentheses and setting it in italics:

> BEWILDERMENT *(as* FALSE VIEWPOINT *is brought before him, he rejoices)*: O dear, love of my life, without you I was very sad. Now your arrival is a happy event. *(He motions for her to sit respectfully on his lap.)* (2.140)

Many of the "independent" stage directions are dohās, the functions of which in the PcN were discussed in Chapter 1. From the English translation, it is often not possible to discern that these narrative stage directions are rhyming couplets, yet they can prove useful for a synthetic reading of the text.[15]

The PcN exhibits a notable presence of long narrative and descriptive passages, and in this sense, the intradiegetic layer of the drama's structure, where Kṛṣṇadās narrates the story of the performance of the play to his disciple, is constantly present. However, this aspect should not be regarded as an idiosyncrasy

---

14 For the plot of the PcN, see subsection 1.2.2. A list of characters in order of appearance as well as a genealogy of the central characters of the PcN are provided after this Note to the Translation.

15 See subsection 1.2.3.

unique to the PcN. In an article on Kavikarṇapūra's *Caitanyacandrodaya* (1572), Gary Tubb notes that stage directions in some instances are written in the past tense, emphasising their descriptive nature rather than a purely utilitarian function. Additionally, the poet often develops the stage directions by providing further details. While these elements may present challenges for readers, they would not have hindered performances of the drama in the Gauḍīya environment following its composition.[16] As such, the fact that such elements in a play dedicated to the founder of the Gauḍīya tradition are also present in the PcN may serve as evidence supporting the possibility that Brajvāsīdās's drama was also performed.[17] Further research could whether this this "narrative turn" in dramas was specific to the PC tradition, to which Kavikarṇapūra's drama can be said to belong, or if it became a common practice from the sixteenth century onwards. Given the significance of acting and dramatic performance in Vaiṣṇava contexts, which intertwine aesthetic theory with theology, I have retained the indications of actors "taking the appearance" or assuming the disguises of characters within the story.[18]

A final aspect to address here is the editorial interventions made to the text. While the primary source of the translation, the Naval Kishore 1875 edition, did not present significant textual issues, certain emendations were made and are duly documented in the footnotes. To illustrate this, let's consider the segmentation of verse 36 in Act III. In the Naval Kishore edition, the beginning of the verse reads as "karunā sunike bisa maiṁ gahī". However, the usage of "maiṁ", a postposition indicating locative sense, with the verbal root gah- (Skt. grah-), meaning literally "to grasp", is not attested. I could only translate this sentence in the sense that Compassion (karunā/karuṇā) obtained poison (bisa). I have, therefore, conjectured that "bisa maiṁ" is a single word, "bisama" (Skt. viṣama), meaning "uneven" or "difficult") in the oblique case of "bisamaiṁ". This interpretation allows for the translation of the stage indication as: "Compassion *(listening, and finding it difficult to understand* [. . .])" (3.36). Thus, the translation aligns with the context, where Compassion is bewildered by the words of the Buddhist Monk and seeks assistance from Peace.

Besides addressing issues related to textual editing, the footnotes offer alternative translations of verses and shed light on the decisions made when encountering grammatically or semantically challenging passages. They also provide insights into the translation of certain characters' names (See Figure 1 as well)

16 See Tubb 2014: 699.
17 See also subsection 1.2.3.
18 See Wulff 1984: 7–24.

and offer a deeper understanding of the religious, philosophical, and literary world of the PcN. When discussing specific passages in the critical study, I had identified any additions made to the text (both to correct the English and to give contextual clarifications) using square brackets and, in some instances, discussed alternative possible translations. My hope is that this approach served to highlight the linguistic richness of the text and facilitate discussions regarding its content and meaning. In this annotated translation, alternative translations are again discussed, but I have limited the use of brackets to essential clarifications only in order to minimise distractions and enhance the reader's enjoyment of this wonderful play.

Selection of Dramatis Personae

(in order of appearance)

ACT I

Kṛṣṇadās Bhaṭṭ	Kṛṣṇamiśra as imagined by Brajvāsīdās
First Actor	Naṭa, also the stage-manager (sūtradhāra)
Actress	Naṭī, First Actor's wife
Desire	Kāma, Bewilderment's first minister
Passion	Rati, Desire's wife
(King) Discrimination	Viveka, the protagonist of the play
(Queen) Good Reason	Sumati or Mati, Discrimination's first wife

ACT II

Deceit	Dambha, one of Bewilderment's ministers, I-maker's grandson
I-maker	Ahaṃkāra, Mind and Engagement-in-Action's son
Deceit's Disciple	Dambha Śiṣya
Golden Thought	Betadhāra, Bewilderment's staff-holder
(King) Bewilderment	Moha/Mahāmoha, Engagement-with-Action's son, antagonist of Discrimination and Mind
Materialist	Cārvāka, favoured by Bewilderment
Materialist's Disciple	Śiṣya, Materialist's student
False Company	Asatsaṅgata, doorkeeper
Ignorance	Ajñāna, a messenger
Anger	Krodha, Bewilderment's minister
Greed	Lobha, Bewilderment's minister
Thirst	Tṛṣṇā, Greed's wife
Violence	Hiṃsā, Anger's wife
Confusion	Bharamāvatī, False Viewpoint's attendant
False Viewpoint	Mithyādṛṣṭi, Bewilderment's lover

ACT III

Peace	Śānti, Faith's daughter
Compassion	Karuṇā/Karunā, Peace's friend, one among the four sisters (the others being Friendship, Sympathy, and Gladness)
Jaina Ascetic	Jaina Jatī, an exponent of the Digambara belief
Jaina's Faith	Jaina Śraddhā

https://doi.org/10.1515/9783111063836-008

Buddhist Monk	Sevarā
Buddhist's Faith	Bauddha Śraddhā
Skullman	Kāpālī or Kapālī, a worshipper of Bhairava
Skullman's Faith	Kāpālī Saradhā or Kapālīnī

ACT IV

Friendship	Mayatrī, one among the four sisters (the others being Compassion, Sympathy, and Gladness)
Faith	Śraddhā or Saradhā, the Sāttvika faith belonging to Discrimination, Peace's mother
Knowledge-Of-The-Vedas	Vedavidyā, Discrimination's staff-bearer
Investigation-Into-Reality	Bastubicāra, Discrimination's minister
Patience	Chamā, Discrimination's minister
Contentment	Santoṣa, Discrimination's minister
Dispassion	Bairāga, Mind's long-forgotten son and Discrimination's minister, sent by Bond-with-Viṣṇu
Discrimination's Charioteer	Sārathī

ACT V

Kṛṣṇadās's disciple	The student for whom Kṛṣṇadās wrote the PC, according to Brajvāsīdās
Bond-with-Viṣṇu/Hari	Viṣṇubhakti/Haribhakti, Discrimination's strongest ally
(King) mind	Mana, son of Individual Self, husband of Engagement-with-Action (first wife) and Disengagement-with-Action (second wife), father of Bewilderment, Discrimination, and others
Intention	Saṅkalpa/Saṅkalapa, Mind's servant, as fickle as his master
Sarasvatī	The goddess of learning, also called Śārdā

ACT VI

Individual Self	Jīvātama Puruṣa, Mind's father and Discrimination and Bewilderment's grandfather, son of Supreme Self (Paramātama Puruṣa)
Upaniṣad	Discrimination's long-forgotten second wife
Deep Concentration	Nididhyāsana, sent by Bond-with-Viṣṇu
Knowledge	Vidyā, daughter of Discrimination and Upaniṣad, twin sister of Wisdom Moon
Wisdom Moon	Prabodhacandra, son of Discrimination and Upaniṣad

Selection of Additional Characters

(in order of appearance/mention)

(King) Kīratibrahma	The king in front of whom the PC was first staged
Gopāla/Karṇa	Kīratibrahma's minister and enemy
Moral Principle, Self-Restraint, Observance, etc.	Yama, Dama, Nema, etc., Discrimination's ministers
Intoxication, Envy, Falsehood	Māda, Matsara, Jūṭha, Bewilderment's ministers
Material Nature/Illusion	Prakṛti/Māyā, Supreme Self's wife and śakti, Individual Self's mother
Engagement-with-Action	Pravṛitti, Mind's first wife and Bewilderment's mother
Disengagement-with-Action	Nivṛtti, Mind's second wife and Discrimination's mother
Misconception	Avidyā, I-maker's twin sister
Kaliyuga	Bewilderment's ally
Firmness and Love	Dṛṛhatā and Prema, two among Bond-with-Viṣṇu's allies
Non-Violence	Ahiṃsā, Discrimination's ally
Duty	(Niṣkāma) Dharma, Discrimination's ally
Bhairava's-Science	Bhairavī Vidyā, the Skullman's daughter
Gladness	Muditā, one among the four sisters (the others being Compassion, Friendship, and Sympathy)
Relinquishment	Tyāga, Discrimination's ally
Sophistries	Kutarka Śāstra, Bewilderment's ally
Mahābhārata	*Mahābhārata* (the epic poem traditionally attributed to Vyāsa), Discrimination's ally
Bhāskara	Bhāskara, the famous Bhedābheda Vedānta philosopher, Discrimination's ally
Śaṅkarācāra	Śaṅkarācārya, the celebrated eighth-century Advaita philosopher, also called Teacher of the Vedas (Vedācāra)
Veda-Deniers	Nāstika, collective appellative for the traditions refusing the authority of the Vedas, also called Pākhaṇḍa
Supernatural Powers/Honeyed Knowledge	Siddhi or Madhumatī Vidyā, the powers that are obtained with yoga, a temptation along the path to salvation
Doubt	Vikalpa, Mind's servant
Sacrificial Science	Personification of Yajñavidyā
Hermeneutics	Personification of Mīmāṃsā.
Kumārala	Also called Kumāra Svāmī, the Mīmāṃsaka philosopher Kumārila (end of seventh century?)

https://doi.org/10.1515/9783111063836-009

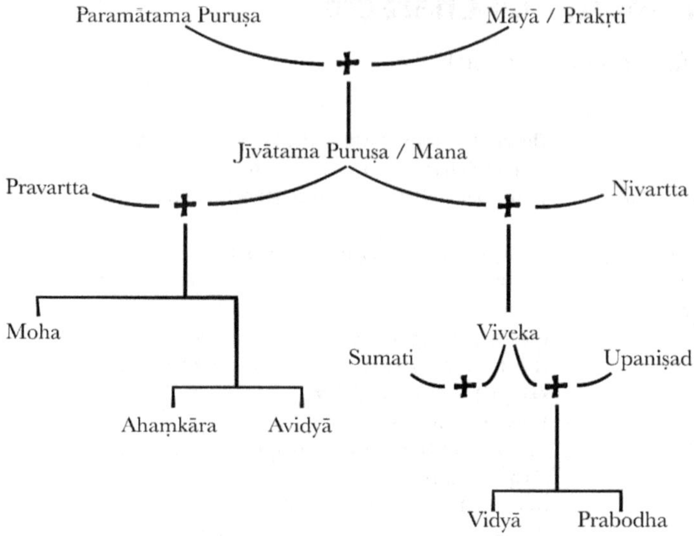

Figure 1: Genealogy of the PcN's family.

ACT I

I bow to the beautiful lotus-feet of the single-minded devotees of Hari,
by a merciful sidelong glance of theirs all the gods are pleased. [1]

In whose hearts, pure abodes, there is always peace and a sentiment of joy,[1]
there, Hari, who is existence, consciousness and bliss, dwells perpetually. [2]

The forms of Hari, one and many, unqualified and qualified,
the several doctrines, the various revealed and transmitted texts, sing about him in many ways. [3]

They all describe his manifold forms, but they cannot find a limit to him.
Therefore, they ascertained that he is "Neither this, nor that" and without a foundation. [4]

Bearing such a Lord in their hearts, they [the sants] think about the world.
Their conduct towards the whole universe is similar to water and lotus leaves.[2] [5]

For them another person's self-interest is their own self-interest, they bestow the Highest Goal,
with their company [the satsaṅga] the individual self obtains the four aims of human life. [6]

Compassionate to the wretched, merciful, pure, wise truth-knowers!
Please, listen to the devotee[3] Brajvāsīdās's prayer! [7]

O compassionate ones, considering me a wretched person,
please bestow me the refuge of the lotus-feet and the words of the company of the truth-knowers. [8]

In the end, how could I – impure, wretched, vile, lacking the merit connected to the aims of human life, full of sin – speak? Dependent on the objects of the senses, how far can I still suffer the blows of the so-called skilful Desire, Anger, Greed, Bewilder-

1 dhavala dhāma jinake hṛdai sadā śānti sukha rāsa | tahāṁ saccidānanda hari karata nirantara vāsa ||. I interpreted rāsa as rasa (sentiment) with an elongated *a* due to metrical reasons as the metre of this verse is a dohā. The phrase śānti sukha rāsa could also be translated as "there are always the sentiments of peace and joy" or "there is always the joy of peace".
2 That is, the fact that even when they are placed in water, the leaves of the lotus do not become wet. In the same way the sant is not attached to the world even while living in it. See BhG 5.10.
3 The term translated as "devotee" is jana.

https://doi.org/10.1515/9783111063836-010

ment, and Pride?[4] It is your power to save those wicked like me; you are the saviour of the vilest and the river I desire.[5] Please bestow me the offering of agreeable, simple, wise truth-knowers. Having come for refuge, I will stay in the company of the truth-knowers.[6] [9]

For a living being there is no greater attainment than the recognition of one's own true form and nothing more crucial than the disappearance of error. There is no greater joy than a bond with God,[7] no other suffering – Rādhikā's beloved must be known.[8] The four aims of human life will not be recognised without the company of virtuous men – tell it adequately! The Vedas prescribe and the world sees that the company of the truth-knowers is worth honouring, it is right for all indeed. [10]

Suffering vanishes, those bound to God experience an ocean of joy, this is the fourfold fruit of the bond with God.
Being[9] in the company of virtuous men, a man crosses this ocean of existence. [11]

In the company of the truth-knowers I bowed, keeping my head on the lotus-feet of my teacher.
First of all, he described[10] the extremely enjoyable origin of the book, which I report in the way I listened. [12]

4 From the Braj text, the adjective "skilful" could in fact be applied to either the narrative voice of Brajvāsī or the personifications of Bewilderment's ministers. An alternative translation could thus read: "Dependent on the objects of the senses, how far can I still suffer the blows of Desire, Anger, Greed, Bewilderment, and Pride, and be called skilful?".

5 The part translated as "you are the river I desire" is completely tentative. See Brajvāsīdās 1875: [. . .] aura te nadā cahoṁ | [. . .]. The *Śivsiṃh Saroj* has a slightly different text of the last part of this verse: [. . .] aura te na jāṁcahauṁ | [. . .]. In neither case can I offer a translation. See Semgar 1926: 217.

6 [. . .] | sarala sujāna santa pyāre kī nichāvara mohiṁ dījai saranāgata santasaṅga moṁ paro rahoṁ ||. The verbal compound paro rahoṁ is not clear. It can be translated literally as "[. . .] fell in the company of the truth-knowers, I will stay".

7 I translate the word bhakti as "bond with god" or "bond" and bhakta as "bound to God". The term "bond" seems specific enough to convey a connection, something that binds one person to another, and general enough to leave open the nature of the connection. Brajvāsīdās's bhakti entails shifting from a bond with the world to a bond with a god. It thus seems apt to translate the term as such.

8 [. . .] | nāhiṁ baṛo sukha bhakti te dūsaro duḥkha na jānibo rādhikā pī ko ||. We could read the first nāhiṁ together with the dūsaro duḥkha and translate as: "There is no greater joy than a bond with god, no other suffering than not knowing Rādhikā['s] beloved."

9 [. . .] | sādha saṅga ho nara tarai yaha bhava pārāvāra ||. I translated ho as an absolutive of ho-.

10 [. . .] | prathama su kaha atiramya granthotpati jihi bidhi sunī ||. I translated kaha as perfective form of kahnā, attested in TS 79.

In the south there was a scholar, adorned by a bond with God, with insight, knowledge, and virtue. Extremely compassionate with the wretched, a benefactor who wanted the fulfilment of their lives. With teachings based on his own knowledge he educated disciples, and their impurities linked to birth and death were erased. Like a sunbeam dispelling the obscurity of bewilderment was Kṛṣṇadās Bhaṭṭ, such was his name. His disciple was a twice-born youth, an observer of his duties, come for refuge in the favour of the teacher. Extremely full of sympathy for him, the teacher desired to save him from the ocean of existence. The teacher taught him about Vedānta[11] and reinforced the bond with God, knowledge, and dispassion. He was of a young age, with ever-new passions, his heart was moist by sentiments like love, etc. The teachings of the teacher did not enter his heart; like someone taken by fever, he did not enjoy a healthy meal. When he begged, he obtained unhealthy food, sour; as soon as he ate it, his temperature increased. To him, a physician, experienced and wise, gave a powder mixed with sour food. When he ate it, the fever was eradicated, and there was no way for him to distinguish between impure and pure. [13]

In that way the wise teacher made then a new book,
 an amusing treasury of art, perfected with regard to meaning, consisting of
 Vedānta. [14]

He called the book the *Rise of the Wisdom Moon*; listening to it is extremely sweet, its understanding brings happiness. Churning the ocean of the Vedas, Kṛṣṇadās extracted this river of nectar. He made his disciple a vessel, a reader[12] of this book, a drama of literary refinement. [15]

I heard in this way about the origin of the book in the company of the truth-knowers. The work is a variegated composition, with descriptions of beautiful things that are true. There will be the destruction of ego, the sense of possession, bewilderment, and the obtainment of the prosperity of knowledge. For one who listens, understands, and reads this work with delight, the adversity of the world will vanish. [16]

---

11 I think here the meaning of Vedānta is literal, to be understood as the Upaniṣads, which form a central part of the didactic content of the PC tale and of the PcN.

12 [. . .] | rīta nāṭaka tāsu pāṭhaka śiṣya kīnhoṁ pota ||. The term pāṭhaka also means "teacher" (HSS 2934), but I chose to translate it as "reader" and referred to the disciple for contextual reasons. Verse 13, in fact, states that the student's mind has imbibed the rasas, hence I imagine the source of such "juices" to be reading works of kāvya literature.

At the time, indeed, I could not reflect in the Sanskrit and Prakrit languages,
since in order to understand them, the intellect needs an incomparable learning. [17]

Balīrāma made a book of it in the Persian language,
but that knowledge as well is extremely difficult, not easily understandable. [18]

One friend said that if it were bhāṣā, it would be easy for everyone,
and then, listening, people would find bliss. [19]

Therefore, I rendered it in bhāṣā, according to my understanding,
with the glory of the company of the truth-knowers and a wide range of metres. [20]

Since I am not an expert, nor the best connoisseur, nor a gifted and skilful poet,
at the end I will say to the devotees of Hari: "O virtuous ones, please improve my composition!" [21]

Hari's devotees observe compassionately the mark on the body of those
whose forehead has been drawn by the creator, be it an auspicious or inauspicious sign. [22]

In the year 1817, full of sentiment and with pleasure,
the devotee Brajvāsīdās composed this book in bhāṣā. [23]

Kṛṣṇadās Bhaṭṭ narrated to his disciple a story, awakening him,
under the pretext of a play of actors, consisting of the Highest Reality, purifying. [24]

KṚṢṆADĀS BHAṬṬ: Listen, disciple, to a beautiful story, variegated and pleasant. There was a king called Kīratibrahma of incomparable aspect. To what extent can I describe the greatness of this king? The abundance of his prosperity in every respect: his strength, his glory, his reputation, and his majesty[13] pervaded the whole universe – similar to brahman. How can I describe his lordship of the realm? The universe was the body, the king was the self. The king had a minister called Gopāla. He was an excellent expert in the ways of kingly conduct. Concilia-

13 jāsu teja jasa nāma pratāpā | brahma samāna sakala jaga vyāpā ||. As jasa can stand also for jaisā (HSS 1724) we could read the first pāda as: jiskā teja jaisā [uskā] nāma pratāpa, that is: "The majesty of his name was equal to his strength, it pervaded the whole universe, similar to the brahman."

tion and gifts, dissension and military force[14] – he knew all the rules, a skilled hero. He thought that kingship belongs to the king,[15] as extreme egoism is shelter to an enemy of bliss. Looking at the people the king swelled with joy and staying focused; he forgot even himself. [25]

The excellent First Actor, extremely wise, was a receptacle of all the qualities and arts. He arrived in the courtly hall, in order to meet the wise men. [26]

With much expertise, and as beautiful as a song of passion, someone brings the tāla and a mṛdaṅga, someone else brings a small ḍhola and the muhacaṅga;[16] someone a flute, a vīṇā, and an upāṅga, someone else a large sārangī. Someone brings a sitāra, someone else a khanjarī and a kartāla. A big company of beautiful young women; looking at them even the bodiless Desire, thirsty, blushes. Extremely lovely, delicate, youthful, long-haired, all pretty-faced and exceptionally expert, with lively eyes like a wagtail or a fish. Well-dressed and decorated with ornaments, the whole assembly is incredibly charming. All versed in dancing and singing, similar to the women of the celestial singers. Looking at them, even Urvaśī would hesitate. All competent in the ways of producing amusement, they describe various stories. They assume different kinds of disguise, according to the stories they narrate. Clever, they bring into the performance many speculations; looking at them, the heart rejoices. Such a unique assembly, they can take on distinct appearances. They all arrived in front of the king and pronounced auspicious words. [27]

All the assembly started singing in chorus, first the benediction, the salutary praise of the venerable Lord. [28]

Then, setting up a stage, the actors kept there the acting equipment. The First Actor and the Actress started dancing, wishing for Supreme Love. [29]

First Actor *(after dancing for a moment, raising his arm)*: Friends, stop the ḍholas a little and all be quiet! [30]

When they stopped singing, only the sound of the instruments could be heard. [31]

First Actor *(to Actress)*: O deer-eyed, cuckoo-voiced, beguiling beloved. Today a voice was heard in the sky, really extraordinary and pleasant. My head was

---

14 The four upāyas, strategies of foreign politics described in *Kauṭiliya Arthaśāstra* 2.10.47 and in *Mānava Dharmaśāstra* 7.108–7.109, 7.200.

15 tāke mate rājya nṛpa kerā | [. . .] could also mean: "The king ruled following his advice."

16 It is not clear to me what musical instrument or instruments the muhacaṅga might refer to, perhaps some kind of woodwind instrument?

heavy, full with concern for my own self and my possessions, causing great pain. The voice was gone after I listened to it, then, feeling relaxed, I slept. [32]

ACTRESS *(smiling)*: O my life, husband, lord!
Tell me, who spoke? What did he or she say through it? [33]

FIRST ACTOR: O woman, after having experienced the world I heard this voice, root of joy.
As a female cuckoo is thirsty for the raindrops of svāti, I was thirsty for it, as well. [34]

Whose gander, praised by the Vedas, the celebrated Supreme Self, consists of light,
inscrutable, indestructible, who manifests the world, in his every pore there are multiple universes
Bestowing all happiness, abode of every joy, pervading, ultimate bliss,
limitless, infinite, unborn Venerable One, without parts, existence, consciousness, and bliss. [35]

The revealed texts praise him, yet, not being able to define him, they say, defeated: "He is neither this, nor that."
Beyond the senses and speech, his pure deeds are said to be infinite, they do not have an end.
Through his Material Nature[17] he manifested the world, present in all, yet separated from all.
Who is without qualities, and yet a unique form with many qualities, a wonderful beauty, infinite. [36]

No one could describe his incredible actions! Without feet and hands, he moves and grasps.
Without skin, he touches; without eyes, he sees; without a nose, he smells.
Without ears, he listens; without mind, he ponders; without a tongue, he tastes.
Without a body, he is still beautiful, an abode of wonder; without a voice, he utters many words. [37]

17 The word employed here is Māyā. I have translated this term sometimes as "Material Nature", other times as "Illusion", depending on the context: "Material Nature" in the case of cosmogony, "Illusion" if the concealing and confusing aspect of the brahman's power is intended. We also sometimes find Prakṛti, which I have translated throughout as "Material Nature".

For the sake of the devotees he took a form with qualities, the One living in Braj, fond of worldly enjoyment, a root of joy.
Nanda's son Kṛṣṇa, delighting in the rāsa dance in a matchless couple with the venerable, fair Rādhā.
Loving fresh butter, flute-holder, his look catches the mind like a bird, Lord of the three worlds, inner controller, his capital is Vṛndāvana. [38]

Such Supreme Self, imperishable, a receptacle of all wonders,
although voiceless, he spoke, with the desire of benefiting the devotees. [39]

The Merciful One spoke to me in such a way and I shall tell you about it. Listen, O beauty, he said: "King Kīratibrahma is very fortunate, attached to kingly duty. His heart has made such a superior wish: to set foot on the path to the Highest Goal. He has recognised that worldly objects are false, like a mirage of water caused by the sun. Man wanders thirsty for the enjoyment of worldly objects in this way, like prey in the power of the I-maker. Therefore, he is like the I-maker's female servant: breaking any relation with it, the Indestructible is to be worshipped. The king thought in this way; such a condition existed within him. [40]

However, the minister Gopāla gave such a seed-formula to him,
that the king was caught again in the web of kingly duty. [41]

He, who desired to run away to avoid being caught in the web, stayed in the middle of it and exulted. If a suffering knowledgeable man falls in the web of women, he is like a gander rejecting milk after drinking water. A big army being prepared and the war signal sounding energetically, he took the company of how many powerful heroes! He won all the battles on earth; the kings he met on his way, they accepted [him] or he expanded using force. How many, who came in confrontation with him, he banished! How many he let stay, were called 'his own'! How many, forgotten, he imprisoned! King Kīratibrahma became the sovereign of the world. He enjoyed the many objects of senses, defeating Indra in that; of the many who opposed him, no king survived. Then sorrow spread in the heart of the king, he caught a severe illness, there was no contentment. The sleep of bewilderment fell, he considered dreams as truth; on a false way, he forgot the true path. [42]

Such a desire of contentment arose in the heart of the king,
therefore, sing something connected to the sentiment of peace in front of him. [43]

Sing the *Drama of the Rise of the Wisdom Moon* with a pure disposition. The venerable Kṛṣṇa Bhaṭṭ, who composed it, is the best among the scholars of belles-lettres. The three sufferings, the concern for his self and his possessions, burning the heart of the king, will soon go away. He will be calm, obtain peace, and enter an ocean of joy. [44]

What happens thanks to a good company is known in the history of the whole world:
without the flower the perfumed sesame oil cannot be produced by the seed." [45]

Such a voice manifested in the sky; by listening to it I became immersed in an ocean of joy. Therefore, make yourself beautiful on one side and on the other side prepare the other actors who will take part in the drama.

As soon as the First Actor stopped talking, the pretty-eyed, intelligent Actress answered. [46]

Actress: Lord, what you said is right, but both my heart and intellect cannot fathom it: the king's thought is ensnared by the sentiments caused by the perception of the sense objects. Tell me, how can peace dwell in his heart? He has a taste for love and heroism, whereas peace dwells in the mind of the yogis. When a king enjoys himself there is no peace; how could the sun shine together with the night? This "king" stands for "earth-protector, lord", and a kingdom with good fortune is filled with good inhabitants. It is impossible to describe the preparations kings make for pleasure; they rival[18] in wealth to Indra. They are bound to sons, wives, and friends; a king's thoughts are tied by love for them. [47]

How can a king's thoughts escape these attachments just once?
They are obstacles to joy and peace; how, indeed, can there be the true path? [48]

The king shall obtain equanimity because he knows the truth: pleasure and pain are the same, interposed between the true path and himself. With Great Bewilderment being destroyed, Pride, Desire, and Anger are buried, too. Leaving bad com-

18 sukha ko sāja kahata nahiṁ bane | sampati meṁ surapati te pane ||. The meaning of pane is somewhat unclear. For my translation, I took as a locative of pana, a tadbhava coming from Skt. paṇa m. "(ifc. f. ā) play, gaming, playing for a stake, a bet or a wager (paṇe ni-as, to stake at play)", MW 580.

pany, he shall accept peace[19] as company, the mind immersed in concentration in the divine sound. Thus he shall approach a truth-knower, keeping his head at their feet: is not the body delicate and weak next to such a person?[20] [49]

He shall win the organs of sense thanks to his own strength, shall leave behind any desire linked to this world. He shall combine love for God and observance, he shall live with the teacher, thirsty for the sentiment of beholding God. Destroying his own self, he shall reduce it to dust; laying down his life, he shall watch the spectacle of the world. When he acts in such a way, he will obtain peace – the world will be but a sugar-cake in milk! [50]

Subjugated by Desire, Anger, Pride, and Greed, and dependent on the enjoyment
of the objects of the senses,
how can he attain the sentiment of peace that the silent sages bestow? [51]

FIRST ACTOR: O woman, such a doubt is not to be entertained, for what can be done without being entitled? This human body is difficult to obtain, yet it is fit for all kinds of accomplishment. It is acquired through many good actions; thus the Vedas say and repeat. The individual self belongs eternally to Hari but, meeting with Illusion, became agitated. How is he in the power of Illusion? It is like the parrot's error about the lotuses.[21] When, in this way, a lot of mental effort accumulates, then Hari's compassion appears. When Murāri gives a human body,[22] then one is entitled to pure knowledge. Going on pilgrimage, making vows, food – without fervent love for the Lord's feet, there is no progress. [52]

Even if the night of ego pervades the abode of the heart for a long time,
all the same, when the light of knowledge shines, obscurity drifts away. [53]

Now, King Kīratibrahma took pleasure in many ways; his heart was fulfilled, having accomplished all the desires of this world.[23] And from the moment the army of the

---

19 Since samatā, which I have translated as "equanimity", is already mentioned in the first line of verse 49, I translated sama as synonymous with śānti in this case, following HSS 4959.

20 Maybe the feeling of one's body weakening in the presence of a sant may allude to a decline of physical impulses, connected to worldly desires, owing to the influence of a spiritually accomplished person.

21 This alludes to a popular simile in Indian literature, used for example by Kabīr. See poem 241 in Kabīr 1968: 669.

22 taba nara tana deta murāri | [. . .]. Grammatically, there are no case markers so the half-sentence could also be translated as: "When a man offers [his] body to Murāri."

23 [. . .] | bhau pūraṇa mana kāma sakala laukika ke je sidhi ||. I have tried to translate this complicated sentence interpreting bhau as the perfect tense of the verb ho-. However, it could also be a noun meaning saṃsāra, jagat, or duniyā (HSS 3706). The concordance of laukika ke is

king had killed all in Karṇa's army, those who were his enemies had retired and he had made his family happy. Now he wishes that, obtaining the company of the truth-knowers, fervent love for the Lord's feet may grow: in the same way in which King Discrimination, after defeating Bewilderment's faction, generated Wisdom. [54]

Just as the First Actor was saying these words to his wife,
the voice of Desire, who stirs the heart,[24] rang out from backstage. [55]

DESIRE *(angrily)*: Vile actor! How can you say these things?! Talking about King Bewilderment's enmity, you, wicked! How could Discrimination beat him?[25] I caused his [Bewilderment's] ascent, with a bow and flower-arrows in my hands. How many women and men are in the world, ignorant, praise all of my qualities? As long as I am powerful in the world, I will make the king happy. While I am alive in the world, tell me, who can triumph over Bewilderment? [56]

FIRST ACTOR *(afraid, emphatically to* ACTRESS*)*: This is the warrior Desire, mighty, extremely victorious, King Bewilderment's hero, chief of the army. He is experienced in achieving Bewilderment's desires. He was incinerated and survived, but the vow about making Śiva fall in love was not burned.[26] He has conquered the world, reducing it to subjection. Even silent sages are extremely and continually scared of him. I was telling you about the defeat of Bewilderment, and Passion's husband [Desire], listening, is angry. Everyone in the world is afraid of him; that is why we must leave immediately. [57]

After talking in this way the FIRST ACTOR ran away with his wife, out of fear of DESIRE. PASSION's husband entered the assembly of the royal hall with her. [58]

Covered by a thin cloth of flowers, a majestic beauty appears, with the delight of ornaments of flowers. A bunch of flowers on the head was resplendent and with the left hand he brought a bouquet of flowers. In his right hand the bow and flower-arrows, his arm around Passion's neck, taking delight in the sentiment of

---

also unclear, and I have paired it with the masculine kāma since sidhi (Skt. siddhi) is a feminine noun.

24 There are several occurrences of the different appellatives for Kāma (Desire). I chose to write "Desire" every time followed by the translation of the appellative, as in this case of manamatha, meaning "who stirs the heart".

25 ati kopa kahyo paṭa bhītara taiṁ | naṭa nīca kahī so kahā yaha taiṁ || nṛpa moha ko droha kahe saṭha taiṁ | kimi jīta bibeka sakai tina taiṁ ||. The word taiṁ can have the meanings of the postposition se and that of the personal pronoun tū (HSS 2136), both of which I have employed in the translation of these verses.

26 This may refer to the fact that Śiva eventually fell in love with Pārvatī, so Kāma's task was accomplished.

love with his wife. In such a disguise, Desire, the heart-born, arrived as if he had come from the abode of Spring.[27] [59]

Arriving in the middle of the royal hall, the two began to dance. All the instruments started up, and the company of the FIRST ACTOR started to sing. [60]

DESIRE *(smiling)*: Look, that lowly First Actor, a fool, where did he go, leaving the hall? Declaring in front of me that the good Discrimination has defeated Bewilderment, he has run away for today, with his face completely bent downwards. My authority is over the whole world, as far as the species possessing a body exists. Conscious and unconscious beings of the three worlds are under the power of the sword that are women. [61]

Superior or inferior, charming, young, jewel-studded with many precious gems; the nose-ring and the eyes are the edge of the sword, the lampblack is its whetstone, a woman is double-edged. Then youth is polished by rubbing ubṭan paste and powder; after wiping with a beautiful perfume on a cloth, she looks lovely in the sheath-veil.[28] When my woman-sword is shining formidably, who can be so composed? Accept it, only Passion is known as true in the world! [62]

In Spring passion thrives, extreme beauty flourishes in the garden. The bee wanders, makes many thoughts wander. The mind hums like a peacock singing; a perfumed, slow, fresh wind and the composure of discerning people is torn. The

---

27 phūlanahīṁ ke dukūla mahā chavi bhūṣaṇa phūlana ke abhirāma te | phūlana ko śira gukṣa lasai kanduka phūlana ke kara bāma te || phūla sarāsana sāyaka pāni bhujā rati grīva rame rasa bāma te | aisī sarūpa manobhava ko uṭhi āyo hai māno vasanta ke dhāma te ||. This whole passage is not completely clear, but it is undoubtedly a physical description of the character of Desire, who enters the scene with his wife Passion. Here it looks like Brajvāsīdās plays on the double meaning of bāma as the adjective "left" and noun "woman" (Skt. vāma/vāmā). See BHK 1026. The meaning of kanduka is not clear either. Skt. kanduka means "a boiler, saucepan; a ball of wood or pith for playing with; a pillow" (MW 250). However, kandula would be the Kanchan tree flower in Marathi. See Kulkarni 1993: 147. I used "Spring" with a capital letter since it is a character in the play, Desire's ally.

28 paramāparama bicitra mūṛha maṇi jaṭita ratana gana | nathanatha locana ghāṭa sāna añjana dudhāra jana || yobana kīnhoṁ sikila bahuri ubṭana raja mañjana | śubhaga gandha paṭa poṁchi myāna ghūṅghaṭa madhi rañjana ||. The translation of the first pāda of this verse is tentative. Mūṛha can be a synonym for mugdha according to BHK 611, an adjective which in Sanskrit can mean "inexperienced, simple, innocent, artless, attractive or charming (from youthfulness), lovely, beautiful, tender, young" (MW 825). I interpreted it in this sense, together with bicitra as "charming" (MW 959), since the subject of this chanda is a comparison between a woman and a sword. Coincidentally, mugdhā is also a typology of nāyikā in the literary genre of nāyikā bheda, developed especially by court poets. It is also relevant that Brajvāsī reused the popular comparison between a woman and a sword, also common in rīti poetry. See e.g. Busch 2011: 70.

pollen spreads and delights the flowers; upon seeing it, Discrimination's resoluteness vanishes. The cuckoo cries, saying "kuhu kuhu", and upon hearing it, Discrimination's army longs for pleasure. Then various temples and rooftops, the myriads of moonbeams illuminates them in the night. The clouds thunder resoundingly with a loud rumble, the sparkle of lightning, a light rain. The cuckoos and the peacocks wail, the crickets chirp; the wind blows and the trees shake. Ornaments and clothes shine in auspicious places, the deer intoxicated by the perfume of the musk. Dishes endowed with the six flavours, good water, mouth-perfumes;[29] a shining, bright mirror for reflecting beauty. Dance, songs, music, musical overtures, a combination of melodies, wails, and lamentations. The soft "dhū dhū" of the mṛdaṅga, thought variations, differences of musical tone and thoughts. Delicate, immaculate wife, lovely-faced, my heart's delight, lake of sentiment! [63][30]

What is this great power of mine? When the human body is desired primarily. If this condition is obtained just for an instant, then what is the knowledge of Discrimination? What the authority of the Purāṇas? Who is a such a resolute truth-knower? My love, I swear by you![31] [64]

Passion: O husband, of all the heroes you are the mightiest, for your extensive supremacy reigns in the world. But, dear, Bewilderment is the chief leader, yet his mind is delicate, naturally pure. King Discrimination is severe, his body solid, people call him a brave chief. Now the enmity within him has increased exponentially: when will something good – not bad – happen, when indeed? [65]

---

29 The exact word used is sukhabāsana, which corresponds to mukhavāsana, according to BHK 1263. See also BHK 902.

30 This verse, uttered by Desire, describes the universe of kāma. Brajvāsīdās represents desire not as a feeling pertaining merely to an individual's inner dimension but as a dynamic and palpable presence in one's life within the phenomenal world. The embodied and affective material culture of kāma is highlighted through its various "constituents": specific weather conditions, perfumes, food, clothing, music, etc. The list in itself recalls the encyclopedic style of several medieval Kāmaśāstras, such as Someśvara's *Mānasollāsa* (twelfth century), which build a whole world of pleasure and its pursuit. Even from the perspective of offering a general critique of kāma, which is one of the main factors binding men and women to this world and causing them to forsake the pursuit of metaphysical insight into their own real self, Brajvāsīdās's interpretation does not trivialise it but draws attention to its all-encompassing dimension. Furthermore, this passage shows that he was familiar with the literary and courtly discourses surrounding desire and pleasure. For an analysis of the material culture of the Kāmaśāstras and a reflection on the evolving nature of the genre, see McHugh 2011. See also note 32 in this chapter.

31 The expression is tuva śīsa pāna, literally, "the hand [on] your head".

DESIRE *(smiling)*: I know that women's nature is fearful; that is why your heart, trembling, almost stopped. Tell, O beloved, who is equal to me in the world? Who is such in the three worlds that has not been pierced by my arrow? Who has not been wounded by my woman-sword? The edge of womanly eyes cuts for good. [66]

The chief witness is the king of the gods [Indra] who, intoxicated, blind, did not
understand anything.
Assuming the appearance of Gautama, he cheated with the sage's wife. [67]

The second witness is the lord of the night [Moon] who seduced his teacher's
[Bṛhaspati] wife,
to whom Discrimination pays homage with his hands, clutching
the greatness of his feet. [68]

Then Brahmā, father of the world, acknowledging my terrible power,
he assaulted his own daughter. Who is as mighty as I am? [69]

And how could I describe inferior beings? Gods, men, anti-gods, I know them as my people. Koka did not disrepute me in his treatise.[32] Whom did not Bewilderment overpower? The king trusts me, for I am Bewilderment's servant in words and actions. Who in the world is not won by me? King Bewilderment's shadow oppresses everyone. Passion, you are fearful – that's womanly nature – but what living being can defeat Bewilderment? [70]

PASSION *(sweetly, joining her hands)*: O lord of my life, I know about the true might you describe. Although all are in your power and no one can defeat you in the world, nevertheless, please listen to such political science: never evaluate the enemy while boasting. [71]

32 Koka, also known as Kokkoka, is recognised as the author of the *Ratirahasya*, a Sanskrit text belonging to the Kāmaśāstra tradition and possibly dating to the thirteenth century. This work, along with Kalyāṇamalla's *Anaṅgaraṅga* (fifteenth century) and Padmaśrī's *Nāgarasarvasva*, was widely known in premodern India. By mentioning Koka, Brajvāsīdās acknowledges the existence of other authoritative texts on matters of kāma during the second millennium, distinct from Vātsyāyana's *Kāmasūtra*. For a discussion on re-evaluating the literary landscape of kāma and recognising the plurality of Kāmaśāstras, see Ali 2011a. Dating the composition of the *Nāgarasarvasva* is challenging, but it likely falls somewhere between 800 and 1300 CE. On this, see Ali 2011b. See also footnote 30 in this chapter for another verse pertaining to kāma.

A war, a disease, an enemy, a fire, a debt, a king, a great ascetic, a serpent –
consider them rightly, do not diminish them, stay alert at all times.[33] [72]

They say that in Discrimination's company there are eight strong ministers.
They are firm, brave, of profound intelligence, expert in the tasks assigned by their lord. [73]

I know that Moral Principle is the first minister. I consider him as endowed with ten characteristics: non-violence in thoughts, words, and deeds, and the highest respect for true words. He avoids theft by body and mind, attached to a firm abstinence. Moreover, he is forbearing with the poor, enduring all kinds of people. He is also extremely compassionate with everyone and with a gentle heart, looks, and words. He stays always pure and destroys falsity. Now, please, listen to me about the second minister, who has become known in the world as Observance. It is told that he has such characteristics: he does not enjoy the taste of food with the tongue, etc. Being hostile to worldly pleasure, he is content with what he obtains. He believes in the Vedas and the teacher and is a donor without desiring the fruit of his action. He listens to preferably essential matters and ponders over many discussions and disputations. He considers the respect of holy people as authority, not the regard of public opinion. He performs worship in the mind and meditates on Hari with the sixteen ways. After that, he stays silent in the repetition of Hari's names; at the same time, he meditates in silent repetition. Moreover, there is always the offering of the senses in the fire of knowledge. [74]

The third minister is called Balance, his features are known as such: yoga is described as an incomparable practice that restrains the mind. Mastering the postures such as the lotus and so on, he fixes the view on the three points at the base of the nose. Afterwards, meditating on the five elements, he removes diseases from his body for good. [75]

Now I shall describe the fourth minister, whose name is Breath Control. He has three types of activity, is extremely strong, and is a receptacle of good qualities. The channels called piṅgalā, suṣumnā, iḍā, and so on – by performing inhalation, retention, and exhalation, the breath is made to mingle in them. [76]

I know the fifth minister as Withdrawal; I believe this is his known story: the so-called five senses of knowledge, their qualities are experienced separately. Sound, touch, form, taste, smell – they are bound together in reverse order. The sixth min-

33 This appears to be a subhāṣita, employed also in RcM 3.21ka. See also verse 3.99 of the *Pañcatantra*: "Remnant of a debt, remnant of a fire, remnant of a sickness, remnant of foes – | When he leaves them all without a remnant, a wise man will not founder." Olivelle 2009: 140.

ister is called Concentration, these characteristics of him are obtained: he knows the five elements after distinguishing them. The vital breath comes among them. The ten elements move along the backbone and he holds his breath at each of them. [77]

Meditation is the seventh minister. Extremely resolute, he is firm in his strength. He holds in his heart the lotus-feet of the guru as Hari.[34] He turns, gazing, and crosses over to a place where there is no differentiation. The eighth minister is Absorption. He lives after having destroyed the deceit of the world. Where he goes, his ego being destroyed, indeed, there is no difference between the servant and the Lord.[35] [78]

These ministers of Discrimination always stay with their beloved.

Listening to the fearful PASSION's words, DESIRE, the bodiless, spoke. [79]

DESIRE *(smiling)*: Until now I had never heard such qualities as those of Discrimination's ministers. Conquering the world, they subdued it. Listen now, lovely beloved, to these powerful warriors' names [of Bewilderment's ministers], for I am going to describe them to you. Mighty, brave, heroes, soldiers, civil – no one is similar to them. [80]

I am the first and I bind the world through the rope of young women. Gods, men, demons, serpents, animals, birds, insects are blind for my intoxication. Many ascetics, excellent sages, wise men, truth-knowers, and learned people always dwell in the forest, respectful of King Discrimination. [81]

If they see the face of a woman in a dream only once, they will sing just my glory, forsaking any respect for Discrimination. If a young man and a beautiful young woman live together in a house,[36] they become my slaves without a price, whom I will sell at the market![37] [82]

---

34 [. . .] | hariguru pada paṅkaja ura dharai || can alternatively be translated as "He holds [in] his heart the lotus-feet of Hari and of the guru" but also "of the guru, Hari".

35 āpāhū miṭi jāta jahāṁ | svāmī sevaka bhedana tahāṁ ||. Brajvāsīdās 1875 presents bhedana as a single word, and I have previously taken bhedana as such, interpreting it as "particular feature, specificity", derived from a literal "breaking, cleaving, splitting" (MW 766). See Pastore (forthcoming a). However, I am more inclined to think now that bheda has to be separated from na. This does not nullify my argument in Pastore (forthcoming a), but it implies a readjustment. See subsection 4.2.1.

36 A reference to the causes (hetu) of erotic sentiment (śṛṅgāra rasa) being a man and a woman of young age. See NS 6.44–6.45.

37 te mere bina dāmana cere beñcoṁ jahāṁ bikāhīṁ; literally, "They [become] my slaves without a price, [whom] I will sell where they are sold." It is an idiomatic expression, employed also by Sūrdās; for example, see SK 5:840.

The second minister, after Desire himself, Anger, is known to the whole world. His sword, indeed, oppresses Discrimination, his force conquers all men and women who commit and say perverse things. [83]

Know the third minister as Violence and consider him extremely fearsome. Then the fourth minister is Greed, beautiful next to King Bewilderment. A man who walks in his company is extremely wretched and helpless. Know that the fifth minister is Intoxication; there is no one equal to him. [84]

Know that King Envy, Intoxication's brother, is the sixth minister, who has the habit of feeling happy upon seeing another's prosperity. However, extreme dishonesty befits him, and he is intent on blaming the other in reality. All the embodied beings are in his utter subservience. [85]

King Deceit is the seventh minister, expert in the counsel of Bewilderment. Deceivingly, he is pristine on the outside but extremely dirty on the inside. He utters sweet words while inside he is tainted. Like a crane, yet stealing the appearance of the gander, he looks at the fishes. Men and women, looking at him, are astounded and burned by his radiance from the beginning. Who can defeat Deceit, he who subdues the whole world? [86]

Know that the eight minister is Falsehood, by whom the world is tied. Falsehood, indeed, determines all thought. To what extent can one describe his behaviour? Meaning becomes certainly meaningless, there is no world without words.[38] [87]

To what extent could Discrimination's group win the entire world one by one? The king, emperor, Bewilderment persists, obscuring the world. Who can defeat the king, emperor, Bewilderment? No one. Because of his power, sages, like Brahmā, etc., live in the forest, subject to fear. Discrimination hopes that they will all gather but how many will be capable, starting with Observance? Who will confront? No one. [88]

DESIRE thus described BEWILDERMENT's principal ministers. Then his wife enquired. [89]

PASSION: People say, dear, that Discrimination and Bewilderment were born in the same family,
please tell me how the things stand with both kings. [90]

DESIRE: I tell you, beloved, that what you have heard is true.
But how can one describe a family with one father and two brothers? [91]

38 artha nirarthahu hoī | jaga binu bola na koī ||. Though the meaning of the phrase stays somewhat obscure, alternative literal translations include "No one [is] without words [in] the world" or "Without the world [there are] no words."

The two, the wise Bewilderment and Discrimination, are blood brothers.
Now, listen, I will describe to you their lineage. [92]

That Hari, indestructible, dwelling in every body, enjoying himself, is the beginning. Whose lordliness cannot be known, he stays always as the ground of everything. Who is described as without attributes, without external support, without form, infinite, boundless. Brahmā, Śiva, Śārada, Śeṣa, and Nārada sing his qualities, yet they do not reach the end. [93]

That brahman, pervading, pure, unborn, without parts, desireless, beginningless,
eternal receptacle of being, consciousness, and bliss, Illusion's husband,
beginning of all. [94]

Such Lord of three worlds, light of supreme lordliness,
touching Material Nature's body, there was a brief movement of
his eyebrows.[39] [95]

From that movement, Material Nature became pregnant. She gave birth to a son, visible, named Mind. Mind, indeed, is also called Self; Mind is the field of Self. But Self and Mind are not distinct; they are just one – like body and shadow. Mind became the king of the world; hell and heaven were his decorations. They say Mind has two queens; one called Engagement-with-Action, the other Disengagement-with-Action. [96]

Bewilderment and the other heroes were born from Engagement's womb.
In the same way, Disengagement's children include the firm Discrimination,
whose reason is steady. [97]

PASSION: O lord, please now explain further. From one father two kings were born. For what reason are they filled with enmity that they want to fight each other? [98]

DESIRE: Listen, beloved, this course of action perpetually repeats in every age of the world. The realm and the treasury are a king's duties and enmity suits brothers. Listen, dear, to the reason why they are all hostile towards each other. [99]

The great King, Emperor Mind, is the father of both kings.
His power and majesty spread in the three worlds in every way. [100]

39 Movement of the eyebrows is a symptom (anubhāva) of erotic sentiment. See NS 6.44–6.45.

King Bewilderment is our master and he follows Mind very much. With folded hands, together with his ministers, he keeps looking for Mind's instruction. Whatever wish King Mind makes, he accomplishes it immediately, out of affection for his father. Therefore, Mind shows him particular favour and has bestowed upon Bewilderment the sovereignty on the whole world. Discrimination does not follow Mind's orders; that is why Mind does not respect his son. He gave him just a trifling sovereignty and he lacks wealth. [101]

The wealth, prosperity, and sport of the eminent king Great Bewilderment,
Discrimination considers them as vice and displeasure in his heart, hence he
wants Bewilderment's destruction. [102]

PASSION *(saddened by the hostility between relatives, sweetly)*: There is no profound cause; the enmity between the brothers has increased vainly. Now they want to defeat each other, this is contrary to Mind's family. Hostility between relatives is not a positive development: the sprout which will destroy the family has germinated. Now please, dear, explain to me for I cannot understand Discrimination's conduct. You said that Discrimination is powerless, deprived of realm and treasure; the respect of the people for him is rather feeble. Tell me, why is Discrimination fiercely hostile? [103]

DESIRE: Listen, Passion, as I think about the matter you are enquiring about my body shakes. It is said that there is a woman called Upaniṣad, the charming wife of King Discrimination. They say that if she gets pregnant with twins, they will be the murderers of my family. [104]

Her first daughter has the aspect of a large demoness,
known as Knowledge, she will devour Bewilderment's family
in particular. [105]

Discrimination's powerful ministers will gain strength from her;
they say that in this way they will win, if time assists them. [106]

PASSION became tearful, her face pale. She fell to the ground as if faint, unconscious. DESIRE embraced her and held her to his chest.

DESIRE: O beloved, how much are you afraid? This story may be true or false. How much do you worry in your heart? How can it be known if it will be true or not? As long as I live in this world, who can defeat Bewilderment's fiery group? [107]

PASSION: Listening to Knowledge's evil qualities, dear, I tremble.
Tell me, who will be born as second child from Upaniṣad? [108]

DESIRE: Listen, woman, this fact is well known in the world: if someone digs a well of deceit, that well will be in front of him in the future. They want our destruction; they will get their reward. After Knowledge, an evil son will be born from Upaniṣad's womb. Mind's family is like a beautiful bouquet of lotuses; he is like the Himālaya. His good name is Wisdom Moon, an abode of the qualities of bad conduct. [109]

Know that as soon as this son is born, it will mean the end of the family.
That is why may Discrimination and his ministers be annihilated promptly. [110]

PASSION *(lamenting)*: A thousand curses to the family,

where will such a progeny be generated?! [111]

Knowledge and Wisdom – you told me that they are wicked.
Why are they called truth-knowers, virtuous and righteous? Please explain, dear. [112]

DESIRE: O Passion, evil-minded men keep causing trouble. Inside they are tainted, while their outward appearance is luminous. The flaws they possess in the end will be harmful to them. Rain and lightning may destroy the harvest, but then they destroy themselves. Listen, O beauty, just as the fire produces smoke, which is toxic, with time it will become water and will extinguish the fire. [113]

When King DESIRE uttered such words, the actor playing DISCRIMINATION arrived.

DISCRIMINATION *(from backstage, angrily)*: Ah, wicked DESIRE, stirring the heart, extremely vulgar, why do you talk nonsense, liar? Ah, you traverse a despicable path, impure, by making the whole world blind and wretched! Listen, fool, to how "we" are! Externally brilliant and internally obscure! This may be the case sometimes, but I do not accept such words about us, my dear. On the contrary, these words are true – truth-knowers all say this precept: "Be it father, mother, brother, teacher, master, or some relative, who does not distinguish between virtue and vice, does not understand one's own gain and loss, who – leaving the customs of his or her family – wanders with a bad company and is attached to a bad path. In their company one dwells in hell, if one does not fear them. Abandon them for they are similar to enemies." Knowledgeable people say only this is true. [114]

King Mind is the son of the eminent father of the universe.
Forgetting his own self, he brought forth a bad company. [115]

A handful of earth assumed as a servant body, Mind said: "My body is delightful." By means of it he knows pleasure and pain through a continuous mental process, but its evolution are ashes or lowly worms. Sensual enjoyments are like scratching a ringworm. At first there is relief then, in the end, a great pain, but – still – scratching is considered a respite from pain. He [Mind] took the company of the organs of sense, forgot his own nature, and became fond of bad company; what is useful in it for him? [116]

Just as a lion cub lives among the sheep, oblivious to his own nature, it bleats together with them, in this way Mind has forgotten that he belongs to the Individual Self.[40] Having forgotten this, he assumes their qualities. It [the lion cub] does not understand well that all the sheep are its nourishment.[41] In this way Mind roams with a cowherd, the fetters of error at his feet; he has become a sheep among sheep, his own self lost. [117]

My father commits another big offence:
his own father is anguished by him; he does not comprehend anything. [118]

Individual Self is my grandfather, that is, Mind's father. He loves his son dearly, caught up in the relationship with him. Bound by the filial relation with his son, he has found that Mind behaves as his master. He bears manifold sufferings and torments caused by Mind. Mind does not think he is at fault, and has stubbornly fallen into Bewilderment's obscurity. He does not realise that Individual Self is in pain because of him. [119]

Therefore, I have abandoned my father and stay away from him.
How am I corrupt? Tell me, O cruel heart-born DESIRE! [120]

Upon hearing DISCRIMINATION's speech, DESIRE, alarmed, stopped his dance with PASSION.

DESIRE *(afraid, to* PASSION*)*: This is King Discrimination. With a large but weak body he practises austerities; he mutters prayers and suffers terrible strain. [121]

He brings with him the queen, a beautiful woman called Good Reason. They have come together after listening to the facts I was narrating to you. He is the eldest

40 It could also be "his being Puruṣa", as pauruṣa may mean both puruṣatva, "the state of being Puruṣa", and puruṣa sambandhī, "connected with Puruṣa", as an adjective (HSS 3125). I exclude the translation of pauruṣa as "manhood, virility" (see also MW 651) or "humanity" since the context deals with ontology.

41 The Braj text here has direct speech: [. . .] | e nahiṁ samujhai neka sakala e mere bhakṣana ||.

in the family, a king with a position higher than mine. It is not appropriate for me to face him, let's leave at once. [122]

DESIRE grabbed PASSION's hand and both left. DISCRIMINATION and GOOD REASON entered the hall. [123]

Extremely pure, excellent, with the aspect of a great ascetic,
the king talked to Good Reason, like a couple of royal geese. [124]

DISCRIMINATION: Beloved, look at this show! Desire has left speaking such things which cause distress and derision. He speaks about me, saying that I follow a corrupted path and do not pay heed to my father's command. He describes himself as devoted to our father and at his righteous service. [125]

GOOD REASON: Please listen, O king, you are wise and modest. You are well acquainted with the Vedas and the texts related to them, the Purāṇas, the scientific literature. Many men in the world are stupid, ignorant, full of their own evil-doings; those who are virtuous and blissful do not look for nor see others. The theft of the jewel of Lord Kṛṣṇa – O dear, how can a stain be brilliant?[42] Those not mindful of duty, who do not ponder, are always called wicked. Looking for someone wretched[43] and to irritate them is their nature, they never accept a teaching, witnesses to this fact are Hari and Duryodhana.[44] The dust falls right on the face of those who throw it over the moon: your actions are on yourself, your good glory and majesty on earth. [126]

DISCRIMINATION: O Good Reason, see, I myself have seen that King Mind, my handsome father, was manifested by what is called brahman, a brilliant compact mass, illuminating the world. Who is changeless, impenetrable, unique. He has no beginning, no auspicious end. Indivisible, shining, receptacle of consciousness and bliss.

---

42 Maybe a reference to the Śyamantaka jewel story. It was a magic gem given by the Sun to his devotee, Satrājit. Kṛṣṇa was accused by Satrājit of having stolen the jewel, which is why he ventured to look for it – to prove his innocence. Kṛṣṇa returned the jewel to Satrājit who gave his daughter Satyabhāmā in marriage to Kṛṣṇa in order to be forgiven for his false accusation. The story is narrated, for example, in *Viṣṇu Purāṇa* 4.16 and BhP 10.56.

43 apanī būjharījha hai jinake khījata dīne sodhana | [. . .]. I am unsure about how to interpret sodhana, which I have omitted from the translation.

44 It is difficult to say to what epic story this sentence refers to precisely. It may be to an episode in the *Udyoga Parva* of the *Mahābhārata*, where, before the war between the Pāṇdavas and Kauravas ensues, Kṛṣṇa goes to Duryodhana's court as a mediator in order to avoid war and preserve peace. However, Duryodhana refuses the proposal and does not believe even when Kṛṣṇa manifests his universal form, considering it a mere magic trick. The implicit comparison is between Kṛṣṇa/Discrimination vs. Duryodhana/Desire.

Speculation cannot fathom him; it cannot be expressed with any voice. The Vedas sing, "He is neither this, nor that", he does not have any particular feature. [127]

Unfathomable, unconquerable, the best, complete, superior light.
By a movement of his eyebrows the universe was manifested, without effort. [128]

A drop in the ocean, a particle of such mass,
Mind, miserable in bad company, is dependent on Bewilderment, like a juggler's toy.[45] [129]

Mind has fallen in with the company of Bewilderment, and abandoned such greatness. Forgetting, he wanders in the pleasure of sensual objects, just as if Indra milked the celestial cow and then abandoned it. He does not reflect about himself even for a bit; by his own mistake, he does not remember himself. [130]

And Bewilderment's ministers, starting with Desire, their preparations are for deceit.
They are major wrongdoers; they do not have any shame. [131]

Such unsullied, pure Individual Self, whose auspicious song[46] is extremely purifying. They made him commit millions of disgraceful actions, some so extremely perverse that they cannot be described. Subjugated, he bears multiple troubles, the threefold afflictions, many kinds of anguish. His eyes cannot see what is in front of him clearly. Seeing his condition, I said: "Well, I want to free him from suffering." That is why they paint me as a cheat. [132]

Listening to DISCRIMINATION's honourable speech, the wise GOOD REASON spoke, with a humble voice, clear as if mixed with the juice of nectar. [133]

GOOD REASON: Listen, king, you have your glory, while he is not renowned nor a king. How could Desire make something up against you? How can Passion be the

45 [. . .] so mana saṅgati manda moha vivasa naṭa ko baṭā ||. The expression naṭa ko baṭā is not easy to translate. It also appears in the *Satsaī* by Bihārī, dohā 285: jhaṭaki caṛhati utarati aṭā naiku na thākatī deha | bhaī rahati naṭa kau baṭā aṭakī nāgari neha ||, which is translated as "Awaiting his arrival, she quickly goes up to the attic and comes down in an instant but is not weary at all. Stuck in love with him, the wise lady had become like a juggler's play-pulley going up and down" in Kapur 2011: 124. Aarti Gandhi gives cakaī as a synonym for baṭā. According to HSS 1439, a cakaī is a pulley-like or a toy of circular form, a sort of yo-yo (OHD 296). See Gandhi 2017. Since I am not able to understand precisely what sort of object it is (a pulley, a ball, a sort of yo-yo), I chose to keep the generic "toy" in the translation.

46 śubha gītā may be a reference to the BhG.

highest good of one's self-interest? He [Desire] does not realise his own vice; the wicked think faults are in virtuous people. Such a corruptor cannot prove anything nor convince anyone of anything. The blind does not listen, nor does he believe his own shortcomings. [134]

Mind's own nature is described as incomparable, extraordinarily radiant, with boundless greatness. His light is the brilliance within the universe. King Bewilderment's people are not fit to be with him. They cheated him, but this matter is kept secret. [135]

DISCRIMINATION: O well-behaved queen Good Reason, listen: the association of men with women is unfortunate. The many doubts, grief, suffering, and anguish are caused by the company kept between men and women. A woman is a part of the unborn Illusion in the world; she produces error in the heart of a man for sure.[47] When a man is in the company of such a woman, is there something improper he does not like? His own nature, unsullied, root of joy; Supreme Self forgot it because of the unborn Illusion's company. The harmful woman burns him day and night, he tolerates millions of troubles in many ways. [136]

The pristine water of the Ganges is pure and purifies the world.
    But in the company of a distiller of spirits, it becomes an extremely impure liquor. [137]

GOOD REASON: What the great king says is true. Even if he [Mind] has good eyes, he is not able to see Bewilderment. An ocean of form, uniquely brilliant, a thousand suns are not even the tiniest fragment of the Supreme Self. What is Illusion there? Tell me, O king, for she is somewhat obscure, she defames him. [138]

DISCRIMINATION: O well-behaved Reason, about what you have said:
    listen to her other characteristics, I am going to explain them to you. [139]

Actually, she was manifested from Supreme Self, she is ignorance with the nature of a servant.[48] Whose infinite power is, however, wonderful. No one finds a limit to her infinite power: she creates ten million worlds, then she causes their destruction in an instant. A collection full of infinite qualities, of multiple, wondrous actions. Why would she not mislead everyone? She is Supreme Self's Illusion! [140]

---

47 On women as the manifest appearance of the unmanifest Prakṛiti/Māyā, see subsection 4.1.2.
48 It could also be "with the nature of an ignorant servant", since ajāna can be both noun and adjective. See HSS 124.

She is from beginning to end very beautiful, generated by Supreme Self. He has a pure body like a sparkling, unbroken, jewel of one colour which is known only by keeping a colour beneath it: through the colours red, yellow, green, and blue, it is well recognised. [141]

If one sees this with the eyes of reflection, pondering,
then one understands that there is no modification without that shining light.[49] [142]

Just like everything is reflected in a mirror, yet the mirror is separated from everything,
in this way Supreme Self was not tinged by the union with Illusion. [143]

GOOD REASON: O lord, omniscient master, please explain me this, teacher. If Supreme Self is a receptacle of bliss, from his body Individual Self was born, indestructible. For what reason is Illusion separated from such an ocean of bliss? He [Individual Self] sank in an ocean of doubt and bears several sufferings that cannot be described. He is not tranquil even for a moment, wanders agitated without finding a resting place. [144]

Please, lord, tell me all these secrets, for my head is full of doubts.
After listening, DISCRIMINATION smiled and spoke lovingly. [145]

DISCRIMINATION: O well-behaved Good Reason, this Illusion cannot be described; I know just that she is the root, a collection of extraordinary qualities. The unborn Illusion saw Supreme Self, thinking in her heart of him: "Supreme Self is full of praises. He is pure in every way, the most powerful, extremely wise. If he showed me his favour, I would have a son by him, then he would make of him the emperor of the three worlds." Such Illusion, extremely fond of him, desired him. She obtained a part of him and became pregnant – such is the unspoken tale. Her wish was fulfilled, her heart increased its bliss. A lovely son was born, called Individual Self, by her. He built multiple kinds of lovely temples in the five elements. He gave them a variegated beauty by decorating them with nine new doors. He taught everything to his son Mind, including the many characteristics of Illusion. He [Individual Self] desired to be king of the three worlds, to be beautiful with the enjoyment of happiness and sorrow. [146]

GOOD REASON: What you say is true, O king:

49 Without the sparkling light of Puruṣa (that is like the jewel), no change would be possible. Puruṣa is the animating conscious principle.

the qualities of a mother, auspicious or inauspicious, are always transmitted to the son. [147]

DISCRIMINATION: Oh farsighted Good Reason, Supreme Self is a receptacle of form.
By him, the Individual Self was manifested through the woman, he is the light in the universe.[50] [148]

From Individual Self, Mind was manifested, who has one queen.
Her name is Engagement, I am going to tell you about his lineage, listen. [149]

In the way the light of the sun is reflected in a mirror, in the way the light of the mirror is the reflection in the water. Then one must consider the light coming from the light of the water, indeed, as the light in the temple. Think that, in that way, from the Supreme Self comes the Individual Self, from the Individual Self comes Mind, from Mind comes the brilliance in the universe. The observer, the Supreme Self, undivided, is present in all, yet separated from all, always a compact mass of bliss. [150]

That Individual Self, forgetting the Supreme Self, was caught by Mind.
By the suffering of Mind, he was saddened; by the joy of Mind, he obtained joy. [151]

Mind's first-born from Engagement-with-Action was a son, I-maker. A daughter was delivered with him, Misconception, his twin. The affection of Mind towards them was excessive; he was not away from them for an instant. Taken by love only for them, he forgot even his own father Individual Self. Now, together with them, won by affection, he sleeps on the bed of unawareness. Fallen into a profound sleep, he forgot his supreme form. [152]

Now he is like someone who sleeps and experiences many things in dreams: for example, he dies fighting with someone, which he would not believe if awake. [153]

In this way, indeed, Mind sleeps a heavy sleep, together with I-maker.
Fallen on the bed of unawareness, he sees terrible dreams. [154]

---

50 aho dūri daraśī sumati puruṣa rūpa kī rāśa | tāte hai yaha prakaṭa jiya tiya te jagata prakāśa ||. The referent of the tāte is not clear. It could just have an adverbial sense, so that the second line of the dohā could be translated as: "That is why Individual Self was manifested from the woman, he [Individual Self] is the light in the universe." The sentence could also mean: "From him Individual Self was manifested, from the woman the light in the universe." However, this latter interpretation does not seem to be consistent with the fact that light, i.e. consciousness, is a property originating in the conscious principle, i.e. Paramātama or Jīvātama Puruṣa, not Māyā/Prakṛti.

In his dreams he is the best among the seven seers and silent sages, or a beggar; in all his dreams he wanders the world in the four stages of life. In his dreams he talks in this way: "I have a father, a mother, a brother, a wife, a son, this is my house, my village, my name." In his dreams he is carried away, whirling, in the middle of the ocean of existence; then again he swims, tires, then drowns and swims. Without waking, he does not know that he, indeed, has become everything by himself, he observes his own self, he is absorbed in his own self. [155]

Forgetting his real self, he fell in difficulty just by himself.
He made of himself both the groom and the wedding procession.[51] [156]

GOOD REASON: O lord, your father Mind has become crazy, won by his love for I-maker [ego]. I am aware now of all these secrets, that Mind is happy in this slumber. That is why Mind believes stubbornly in the unreal world. Now do some effort, O king, by which Mind may be awakened. By what means will he consider dreams as such and recognise his own form and become aware? [157]

The queen pronounced such words, at which the king was embarrassed. He lowered his head without replying. [158]

GOOD REASON: O lord, please tell me all the secrets concerning Individual Self. Again, you do not answer – what is the cause now? Turning your eyes away, you have become silent – what is this agitation and embarrassment? What's the difficult matter which covers your heart like a shadow? [159]

DISCRIMINATION *(he spoke, knowing* REASON *well)*: O enlightening, auspicious woman, there is something concealed, and I am going to name this guilt of mine. I respect you a lot, so I listen to you. The nature of women is that, listening to a second woman's name, they become sad. [160]

Upon hearing about a co-wife in the house of her own husband, a woman's heart feels a pang.
That is why I worry in saying its reason, young woman. [161]

GOOD REASON: Listen, King Discrimination, there are two kinds of woman in the world. One is superior, has the body consisting of light, the other is inferior, con-

51 āpa bisāryo āpu ko āpuhi paryo kavāta | āpuhi tau dūlaha banyo āpuhi banyo barāta ||. The term I translated as "difficulty" is kavāta. The noun is not attested as such in dictionaries, where I have found kavāṭa, meaning "door, leaf of door" (HSS 861), but I could not make sense of it in the sentence. I have found, however, the word qabāhata, which comes from Arabic (HSS 789) and means "difficulty". Since this sense suits the sentence, and qabāhata is built with verbs such as par- (MSH paḍnā), I interpret kavāta as a short form of qabāhata.

sisting of darkness What makes the beloved happy makes happy a superior heart as well. A woman is called inferior when she looks for happiness opposed to the husband. [162]

They are happy, indeed, for the reason the beloved is happy, just by seeing it. They love those whom the beloved loves as well. The desire of the beloved stays in their mind; without the beloved desiring they do not desire. The beloved's behaviour is excellent when his woman in the world is superior. [163]

I think that the wives acting against the will of their husbands
reach hell by themselves; they are roots of big evil. [164]

A man has a superior woman when he is not overpowered by her. It is not appropriate to use my power on you, a king. I am, O lord, at your orders. In whatever way the kingly tasks will be accomplished, that way shall be told and it shall be carried out without any obstacles, O beloved. [165]

DISCRIMINATION *(happy with* GOOD REASON*'s words)*: O righteous queen, if this is fine for you, I will not keep any secret from you, but I am going to explain everything as it is. There's a woman called Upaniṣad, she is my second wife. I left her a long time ago because she behaved with some arrogance towards me. Now she lives separated from me. How, indeed, could I meet the lovely woman now? If Faith assists me and approaches her on my behalf, she will talk sweetly with her and let her know my desire; she will make her stop being arrogant by praising her and will bring her to me. [166]

When she arrives, dear, do not mind it.
For there should be no obstacle in my union with her. [167]

In your house, O beauty, there are three conditions, namely wakefulness, dream, and deep sleep – they are there with your king [Mind]. I will leave you upon entering the fourth state of consciousness [turīya], where I will go with Upaniṣad for a few days. There, with the blessing of the Lord, I hope that the son named Wisdom, splendour of the beautiful moon, will be generated from her. He is self-possessed, knowledgeable, a receptacle of qualities and bliss, learned and wise. He is the dawn bringing light to the night of Mind's sleep. [168]

GOOD REASON: If this is the method by which a storehouse of virtue may abandon Misconception, for through her, Mind has power on you, then I am very fortunate. [169]

I, lord, am at your orders. Helping your father befits you, please send for Upaniṣad at once; Faith will go and make her understand. Please, lord, give her a constant love – kings are endowed with rightful conduct for the sake of duty. Please do not postpone an auspicious task; just as a small blanket becomes wet, its burden will increase.[52] [170]

Discrimination *(at her pleasing, humble words, mixed with virtue and love, the king rejoiced)*: Excellent, dear, you are extremely upright. [171]

If you will assist me in this task,
then all other tasks will be accomplished naturally, without effort. [172]

Now if King Mind sleeps, in dreams many things take place. He, son of Supreme Self, does not need such sleep. This is first of all the fruit of affection for I-maker and, secondly, Bewilderment keeps misleading him. Desire and the others are all their helpers, they do not want Mind to be awakened. Therefore, with the dear wife called Upaniṣad, I will dwell in the fourth condition, turiya. As soon as Knowledge is born from her, all my ministers will become powerful. Many wicked people exist, beginning with Desire and Anger, who deceive Mind. Fighting them, I will kill them first – I am going to tell you all. Then I will destroy Bewilderment; I will not let him live. At that moment my son Wisdom will be born, and he will awaken Mind. When Mind is woken from that slumber, he will cease to live in dreams. He will recognise his primary own form; he will know his own father. Since Individual Self was born from Material Nature, he is not aware of his father's nature. Illusion makes him play a destructive game, so that he does not remember his father. The venerable Hari, father of the universe, also knows that she [Illusion] is playing with Individual Self. She [Illusion] never wants Individual self to wake, this way staying between them, next to the father. Bond-with-Viṣṇu, beloved of Supreme Self, is extremely merciful and the benefactor of the world. She will tell Individual Self everything, then he will take her company. Bond-with-Hari[53] will make him meet with Hari, and only at that moment will he obtain his nature.[54] When Bond-with-

52 This means that even a trifling task should not be delayed because it may become a bigger issue. Even a light piece of cloth becomes heavy if one does not stop it getting wet. The same expression has been used in many different contexts, for example by Tulsīdās in *Śrī Kṛṣṇa Gītāvalī* 46.3. See Tulsīdās 2006: 53. It is also found in ramainī 15 in Kabīr's *Bījaka*. See Kabīr 2013: 14. Also Kabīr 1950: 9.

53 Another name for Bond-with-Viṣṇu. The two names are interchangeable throughout.

54 I have translated the term marayāda (Skt. maryādā) as "nature". It is not a canonical interpretation of the word, which is usually translated as "limit, boundary", or secondarily, "correct behaviour, decorum" (OHD 794). While I think that marayāda here does include all of these meanings, I have preferred "nature" because it can convey something "giving or containing clear marks or signs", as the literal meaning of the term indicates (MW 791). That Individual Self finds his marayāda certainly does not mean that he finds himself limited; rather, he finds the infinity of Supreme Self in himself. He realises his true nature.

Viṣṇu makes them meet, she will make the father embrace the son. The son will sit on his father's lap, they will be joyful and there will be supreme bliss. By the meeting of father and son, there will be goodness, there will be a surge of bliss in the ten directions. Father and son will stay together, they will experience supreme joy and eternal bliss. In this way Supreme and Individual Self meet, the father embraces the son. [173]

I have resolved upon doing what's best for my father in this way, by destroying all the family of the pernicious Bewilderment. I have made such a promise. When Mind finds his place, leaving the perishing body I will enter in that same brahman. [174]

GOOD REASON: Lord, what you suggested I will also keep in mind. When you will leave this body, then what shall I do alone, miserable? I also took a decision: I want to leave my body as well, together with my husband. Otherwise the fire of separation will arise because by parting from you I, attached to your feet, will burn quickly. [175]

DISCRIMINATION: You conduct is excellent, Good Reason, I trust you.
After taking leave from the hall, let's go and visit the temples. [176]

GOOD REASON and DISCRIMINATION left the hall, conferring in this way.
The FIRSTACTOR's disciples started playing the stringed instruments. [177]

Thus concludes Act I of the "Drama of the Rise of the Wisdom Moon" in Brajbhāṣā.

ACT II

First Actor *(addressing* King Kīratibrahma*)*: O king, Bewilderment got to know about Discrimination's plans. [1]

Therefore, Bewilderment ordered his ministers
to do everything in their power to prevent Discrimination's plan. [2]

The actor playing Deceit arrived and, after preparing himself, spoke loudly from backstage. [3]

Deceit: The supreme king Bewilderment has given this order: "King Discrimination made such a vow, he said that he wants to make Upaniṣad give birth to Wisdom. He wants to destroy me and all the family with his help. He sent his army as support to every place of pilgrimage. This information provokes all bad presentiments. Therefore, how many soldiers I consider mine among the fighters, they should all be alert, leaving the idleness of sleep. Wear an armour, constrict the body by keeping firm the cloth around the hips! Take all weapons after whetting them, be watchful day and night; do not let the enemy's strength increase! First of all, go to the places of pilgrimage and make them your outposts." And to me, Deceit, servant of the king, he commanded: "Encamp in Kāśī. It is called the city of liberation and it has always been Śiva's abode. A large number of discerning people and accomplishers of perfection in many ways dwell there. Interfere secretly in their prayers, austerities, and observances – make them opposed to the Vedas with your power. Settle in that main place so that it becomes my outpost. In this way your authority will increase." [4]

Having listened to the king's order, I've come to Kāśī, and I dwell here now.
Be watchful, people eating the king's salt![1] [5]

First Actor *(at* Deceit*'s words, he wisely instructed his troupe)*: Set up Kāśī somewhere in the hall! [6]

Somewhere there is someone disguised as a scholar. His forehead is adorned with the auspicious mark, with folded dhotī, shawl, lotus leaves. Holding the pure sacred thread, the *Bhagavadgītā*, in his hand, he says authoritative and love-provoking words: "Intelligent people, listen carefully!" [7]

Somewhere else, there is a renunciant, devoted to Hari, with the auspicious mark on his forehead – his appearance is lovely. Wearing a necklace, an instrument for

1 An idiom meaning those who depend on him, who are his servants. See HSS 4325.

https://doi.org/10.1515/9783111063836-011

remembering in the hand, he is exquisite and beautiful.[2] With smiling eyes, perturbed words, overcome by divine love and frenzied, he sings the praises of Hari's qualities, claps, and gives various instructions. [8]

Many actors have taken the appearance of silent sages. They glow with the radiance of austerity. On their heads, gleaming matted hair; on their bodies, they rub ashes and fasten a loincloth. Someone prays silently with the help of prayer beads; another practises the fivefold heat austerity. Someone keeps their arms raised; someone else looks at the sun in meditation. [9]

Such actors playing discerning people sat, having disguised themselves in variegated costumes. In the highest, most elevated and purified place there is a lion-throne where DECEIT, arriving, takes his seat. [10]

His waist is beautiful with a long dhotī, and so is the shawl with the monogram of his name. A brilliant red hat, the shining auspicious mark on his forehead. The beautiful sacred thread and the prayer beads near his heart, eyes full of sentiment and lips full of melodious words. Many books scattered here and there, on pillows as supports. [11]

The best discerning people, with the guise of devotees, walk on a very fortunate path,
where DECEIT stays, sitting, hiding like a thief. [12]

DECEIT's beauty cannot be described; he is like a pair of scissors covered in honey. His disciples are two capable tricksters, standing in front of him with folded hands. He knows many magic tricks, by which desires are fulfilled at once. He is endowed of knowledge and learning, omniscience and accomplishment. Like Vālmīki and Vyāsa, he illuminates the disciples' faces.[3] Men and women came to behold him, they prostrated with their full body from afar. The favour gained from his auspicious beholding is impure, for he is dependent on the objects of senses in his mind. His external appearance is that of a gander, but his mind is fixed on the crow's deeds. Just as the beautiful, lovely, sweet-voiced peacock is a snake-devourer,[4] in

2 The sentence reads kaṇṭhī mālā sumirana ālā kara meṁ lalita birājai. Sumirana ālā kara meṁ, rendered here as "an instrument for remembering in the hand", is unclear. Ālā is a term of Arabic origin, which I interpreted as equivalent to Skt. upakaraṇa (tool, instrument). See OHD 94. It seems difficult to see the expression as an attribute to kaṇṭhī mālā (lit. "neck-beads") since, as mentioned by the term itself, Vaiṣṇavas usually wear it around their necks. There is reason to believe that the ambiguity is due to the fact that Brajvāsī reworked the GH, which at this point depicts people with prayer beads in their hands (a common Islamic practice), absorbed in recitation. See Cappello 2019: 86.

3 Maybe in the sense that they are astounded by his prowess. The Brajbhāṣā expression is: śiṣyana ke mukha kiyo prakāsā.

4 See RcM 1.161b.

the same way DECEIT's attractiveness is superficial, inside him the desire of worldly enjoyment spreads. His disguise as someone bound to God is highly charming, but inside he is separate from any feeling of bond with Hari. Just as sages stay separated from the world, similar to lotus leaves which stay separated from water. In the same way, in DECEIT's aspect as being bound to God, there is no trace of the nature of someone who really is bound to God. Sitting on the lion-throne, he gazed at the waves of the Ganges. [13]

In that moment I-MAKER appeared from behind the curtains, in the guise of an old priest of an exceptionally violent nature. [14]

With frowning, irascible eyes he looks at someone as if he would kill him. If he calls someone dear, it is as if he is abusing them. Taken by restlessness, he has arrived in Śiva's city. He had gone to the banks of the Ganges but did not find any calm. There he had enjoyed himself with young women and jumped into the water with his clothes under his armpit. He had sank, splashing, and reached the bank on the other side with toil. There, how many things did he say to the Ganges!

I-MAKER: How miserable are you, for I am able to cross the vast ocean! [15]

DECEIT *(still gazing at the river)*: Listen, everyone! Look at that man, he has no measure! He has but one brain, yet he is burned by the fire of foolish conduct. [16]

I think he comes from the town of Rādhā,
where many evil tricksters, loose women, and fools live. [17]

My grandfather resides in that town,
so some information about him may be obtained from this fellow. [18]

I-MAKER *(he proceeds into the hall and sees many learned people immersed in love for God's story)*: Ah, they do not know their own nature! Tell me, reading the Vedas, what is its point of view? Just as the crane spends its life in vain looking at the frog. Tell me what, indeed, is churned of a cloud? Those whose enjoyment and leisure are visible, staying away from them, they restrain body and mind. Those who listen to stories, consider them stupid – closing the eyes, they see the path of liberation. [19]

Advancing further he saw a group of renunciants, like a big flock of ganders: their aspect pristine, their words clear. [20]

Someone described Hari's qualities, someone else – listening – treasured them in their heart. Someone, sitting, sang devotional songs and filled his heart with an ocean of love and bliss. Someone was repeating Kṛṣṇa's names, another was paying homage, someone else was meditating on the lotus-feet – the worship was of various kinds. One was inclined to the attitude of service and obeyed the Lord's will.[5] Someone else had turned to the attitude of friendship, with Hari in his heart. Someone else had dedicated his body, mind, and wealth as a sacrifice to Rādhā's groom. In these ways, I-maker saw the renunciants sharing a bond with God[6] and became purple with rage. [21]

I-MAKER *(to the renunciants)*: O wicked, stupid people!
These people live perturbed and restless, bear burden and strain. [22]

That visible body – that should be considered eternal – they call it impermanent and perishing. On the contrary, they will praise what is not seen by their eyes perceiving it! These abodes of intelligence say they will establish the measure of the sky! [23]

They think the words of the Vedas dispel error,
but in my view they [the Vedas] have created a web of error for those fools. [24]

After this statement, he moved further, jumping. Upon seeing the actors playing the group of ascetics he burst into laughter. [25]

I-MAKER *(laughing)*: All are to be seen as thieves, yet they are considered without punishment. In order to gain some food and clothes, they mortify the body with austerity. Enjoyments such as women, etc., are made useless by their actions. O fools, this is yoga! It is like a bodily disease. Just by looking at these people one suffers, because they are burning themselves. [26]

In this way I-MAKER criticised the actors playing the devotees of Hari. Advancing, he saw the dwelling place of DECEIT and rejoiced. [27]

How many deceivers were sitting there, assuming an auspicious aspect. Someone, having built a pit for a fire sacrifice, was completing a sacrifice in front of it. Someone else had closed their eyes and was recalling and hearing Hari, mentally absorbed. Someone else was distracted, making the prayer beads lay idle and concentrating their mind on a girl. [28]

Many people passed by, like embodiments of sensual enjoyment. Somewhere the smoke of the homa sacrifice appeared, and its scent spread all around. [29]

A polished place, spotless, here and there a clean sky. Everyone described the condition of perfected people: everything they say happens. The one who re-

5 Prabhu kī icchā soī śirai, literally "[to put] the Lord's will on [his/her] head".
6 I have translated bhakti karai as "sharing a bond with god", which retains the etymological nuance given by the root bhaj-.

ceived the master's favour – his food and wealth increased. How many barren women got a father – the master knew their secret wish! [30]

I-MAKER *(standing still and seeing* DECEIT *sitting on the throne, he says to himself)*: I must know if in this place lives some amiable important man who knows the secret of the Highest Reality and nurtures the body in a good way. In this case, I am going to stay here for some days; then I will get to know about some other place. [31]

(He enters the house and approaches DECEIT.*)* Take my blessing. [32]

DECEIT'S DISCIPLE *(seeing the priest* [I-MAKER] *approaching* DECEIT, *he stands up, shouting)*: Hey, do not go close to the mahant! Take your blessing from a distance! [33]

I-MAKER *(displeased, looking at* DECEIT*)*: What country is this? Foreigners are not respected where cruel-minded men live. [34]

DECEIT *(reassures him with a wave of the hand, and to his servant)*: It is not his fault. He has come from some distant place; hence, he does not know the differences in custom. Ask him what the custom of his family is; did he become a priest or were his father or grandfather also priests? Make him wash his feet and hands with pure water, and when he comes near ask him about his predicament, please. [35]

I-MAKER: You do not know about my family, of which I am the best! I come from a town called Rādhā, among the biggest places on earth.

He kicked the dust of the path, and sat on DECEIT's seat without even paying respect to him. [36]

DECEIT *(angrily)*: Why do you not understand, idiot? How dare you come to me without washing your feet? If a drop of sweat inadvertently falls from your body on to my clothes, then I will have to bathe again – together with my clothes. [37]

I-MAKER *(laughing)*: I have been travelling through many places, such things there are not well-grounded. I have never heard nor seen it! [38]

DECEIT'S DISCIPLE *(standing)*: I say, priest, you know nothing! A lot of kings come here but they cannot touch the mahant's feet! Miserable beggars and poor people like you, no one ever dared to ask something of the mahant! Nonetheless, the master showed you regard and talked to you with affection. Now do not talk nonsense, give your blessing from afar! [39]

I-MAKER *(listening to these words, he thinks to himself)*: Well done! I must get to know if it is Deceit living here. [40]

He then sat next to DECEIT. DECEIT'S DISCIPLE, protesting, scolded him again. [41]

DECEIT'S DISCIPLE: I say, brahmin, you never listen! I have explained to you already! Everyone bows to this seat, for here the mahant prays! [42]

I-MAKER *(in a fit of rage, answers as if they were alone)*: Fool, your mahant is no more important than I! [43]

You do not know about me. Listen, I will tell you about my own reputation! My mother is very virtuous, but she was born in a low-caste family. I raised my family's status through marriage, as high as the Himālaya is high. So, my rank in society rose and it is far greater than my father's. Many people of high or low caste exist in the world – my control extends to them all. I'm going to say more, listen, as I am very modest. [44]

My brother-in-law, a relative, had a friend whose female servant had a daughter, whose name I cannot recall. She gave him a false bad reputation, about which I heard somewhere from someone. [45]

Know that as soon as I heard about it, I felt ashamed and greatly disgusted. Leaving realm and array, wealth and house, I set off. [46]

DECEIT: One day I went in the assembly of the excellent Brahmā. Upon seeing me, he rose and quickly stood standing. And all the gods and perfected beings present in the hall came quickly to touch my feet. Bringing there a graceful seat of gold, they begged me to sit. I told the god of fire: "Please purify this, then wash it in the water of the Ganges, and only then please touch it." [47]

Then Brahmā washed his back with the water of the Ganges,
    and begged me to sit on it. All the gods were astounded! [48]

I-MAKER *(to himself)*: Excellent, Deceit, liar! On the one hand, you assumed a man's body;
    on the other hand, you seized Brahmā's place! [49]

(To DECEIT*)* You fill your heart with the deference of one Brahmā,
    but innumerable Brahmās keep falling at my feet! [50]

They are always afraid of a frown in my look!
    But I never look at them, not even in my dreams! [51]

DECEIT: I see you are a very generous man!
    I must know if you are my grandfather, I-maker. [52]

He fell at I-MAKER's feet who took him on his lap.

I-MAKER *(cleaning* DECEIT*'s face, he embraces him)*: How are you? A long time has passed since I saw you, son, it was the Dvāpara age. [53]

Back then, son, you were a child; in this age, you are a man.
I've grown old so I could not recognise you now. [54]

Tell me, son, is your own son, Nonsense, fine? In what village does he dwell? Tell, is your father, Greed, happy? Where does he stay together with Thirst? Recount, O Deceit, to your grandfather, are all my relatives happy? [55]

Arriving in this town, son, one can see much.
After some days, you will tell me in what ways I have corrupted it. [56]

DECEIT: You know everything about the plan of King Discrimination. He wants to stay in this town and cause the birth of Wisdom Moon. If he is born, the family will perish – you know that this is perverse. The men you saw while approaching are sent by the good Discrimination and are very powerful. I want to mix among them so that, in this way, Wisdom's birth will not take place. [57]

I-MAKER: Now the enmity has increased greatly among them,
both kings, Discrimination and Bewilderment, want to do battle. [58]

DECEIT: O dear, such matters are surely not easily forgotten.
This was the proper time for you to come to this town. [59]

King Great Bewilderment as well wants to come here, today or tomorrow.
This city will be made the capital of his reign by our own king. [60]

AS I-MAKER and DECEIT talked, BEWILDERMENT emerged from the curtains. [61]

GOLDEN THOUGHT (*the staff-bearer enters, loudly*): O men and women, get ready, listen carefully! Today the supreme king Bewilderment arrives. Wipe the dust from the road, make every threshold beautiful. Spread good perfume in the middle of the roads. [62]

Wearing good clothes with precious stones in the crown, along with parasol holders with great royal pomp, arriving in the middle of the town. Where the FIRST ACTOR, DECEIT, and the others were singing, there BEWILDERMENT, the king, is an adornment to the lion-throne. I-MAKER and DECEIT bow to him and King BEWILDERMENT, joining the assembly, sits and a staff-bearer shades the king with a parasol and a flywhisk.

BEWILDERMENT *(to himself)*: It is necessary that discerning people leave this place. *(Seeing* DECEIT*)* Tell me, what is the state of this town? [63]

DECEIT *(stands up with folded hands)*: Your majesty is strong and stable. Since this servant of yours holds sway here, the great king's rule has come back. Following each and every of our master's orders, the evil Discrimination's party is oppressed. [64]

Some people, leaving Discrimination, have become ours.
Those wretched stay disgraced, as if they do not receive any respect. [65]

BEWILDERMENT *(rejoices upon hearing* DECEIT's *words, but then continues to worry and lowers his head.) (Aside)*: It is necessary to get rid of the people of the enemy. The more I am able to subjugate, the more I will make of myself the undisputed king. [66]

(To DECEIT, *smiling)* I will start giving my teaching towards the discerning people. [67]

Please, look at these foolish men! Even if they possess eyes, consider them blind! They do not consider the body and the self to be one; rather they accept that the self is distinct from the body. The thing that is nothing, indeed, they say that it certainly exists. What is proven visible through sight they maintain is unreal![7] [68]

They also say that this body is generated by the five elements –
the individual self is distinct from it and very luminous. [69]

I have ascertained that other things are generated from the five elements. In the five elements the whole world exists, which, being manifold, thrives: a body is manifested from a body, a man from a man, an animal from an animal. A bird is born from a bird, a tree originates from a tree. For this reason, the self is not different but one with the body – this is true. [70]

And the opinion that the body is unconscious, that the self is conscious –
all lies! Listen, for I shall tell how what is seen as separated is actually undivided. [71]

When all the elements are joined, then the conscious appears;
and when they are separated, the conscious is contained in them. [72]

Just like lime, kattha extract, betel nut, and betel leaf – these four things, mixed, become pāna. When these four elements become one, then the colour red appears. Until the four are separated, the redness stays concealed. Therefore, the opinion that the self is something different from the body, it is but a lie – everyone worships the body.[8] [73]

---

7 The teaching given by Bewilderment belongs to what was recognised as the Lokāyata/Cārvāka doctrine. It follows the few principles outlined in doxographical texts such as the *Ṣaḍdarśanasamuccaya* by Haribhadra Sūri (eighth century) or the *Sarvadarśanasaṅgraha* by Mādhavācārya/Vidyāraṇya. See Chattopadhyaya/Gangopadhyaya 1990: 246–257, 257–297.

8 [. . .] | so saba jhūṭha deha saba pūjī ||. I could not find pūjī as such but translated it as the verbal root pūj- followed by the affix -ī in order to form a noun expressing an agent. For the varied origins and functions of this affix, see Strnad 2013: 235–236.

And they say that in the Vedas there is the proof of the existence of the four castes,
and of their distinct duties – this is the cause of all lies! [74]

However men exist, all are equal: same hands, same feet, same eyes, same face, same nose, same ears; they hear and look in the same way, their limbs all bend in the same way. Is there a mark on their bodies, by which all living beings are differentiated? [75]

And finding a fault in lust for other men's women and possessions,
the whole world is afraid, but what is really destroyed? [76]

What is the difference between one's own and another's women and wealth?
Colours, limbs, qualities are the same – do the Vedas explain this? [77]

What is the fault in violence? What is the fault if one eats meat? Meat is nourished by meat,
there is no guilt in that. [78]

Those who have forgotten their own world, listening to the Purāṇas and the Vedas, with many discerning people, frauds, who make a fool of them, driving them crazy. Following the authority of their words, they stay indifferent to worldly passions, forsaking everything: pleasure, wealth, and moon-faced women. They stubbornly differentiate among types of salvation, but do not know anything about it. I think that salvation is death in this very world! [79]

When one dies in this world, it's his end.
Why, then, there's no other visible sign of him left? [80]

And those who maintain that the self is eternal, there is no one more subjugated than them. The five elements which are visible, what is seen by the eye is eternal. There is a fruit of such life: it is constant enjoyment, day and night. Nourish this doctrine through all the ways in which the body is satisfied – by eating and drinking, by music and merriment, by the company of beautiful deer-eyed women! [81]

As Bewilderment spoke, one actor, taking the aspect of Materialist, arrived. [82]

Materialist *(teaching to the disciple accompanying him)*: Son, never listen to the authority of the Vedas. There is no essence in them, they are completely false ideas. They say, in fact, that the one who performs a sacrifice goes to heaven. These are completely empty words – understand it with your heart, dear. If some

Bel tree is burning in the mouth of sacrificial fire, they want from it flowers and fruits – this is the root of utter folly! [83]

And I am very amused by the men who carry out funeral rites,
how can an extinguished lamp be filled with oil and light?[9] [84]

MATERIALIST'S DISCIPLE *(having listened, enquires)*: Master, if the Vedas are false, then I am surprised: why do the excellent sages stay distressed? They abstain from enjoyments like women, etc., the six tasty flavours of food and drinks. Beautiful ornaments, clothes, perfumes, music, and merriment do not appease them. They live alone in the forest; as food they eat roots of different sorts and fruits. With ornaments and clothes made out of tree bark they sing the name of Rāma untarnished. In this way they take extremely strenuous pains and carry out prayers, austerities, and fire offerings in conformity with the Vedas. All this for the sake of knowledge of the unchanging brahman, for which these followers are called best. [85]

MATERIALIST *(laughing)*: The author of the Vedas –
he was a chief prankster, who deceived the world.[10] [86]

What such greedy individual said, making it up, I will explain to you in this way: as if there is someone thirsty, sitting next to sweet water. Someone tells him, "Go away from here, this water smells very bad." I reply: "The water is good; you are seeing a mirage." When he gets tired in it, then he looks at it, rejoicing.[11] Some thinking is needed: why is there no sportive gander? Where is the beautiful couch for rest with the young body of a woman? How can you stay in the forest alone, sitting on deerskin on the surface of the earth? If there are many varieties of cooked food endowed with the six flavours, then why eat roots, fruits, and

9 The sense should be that if the light of a lamp is out, one can fill it with as much oil as one wishes, but the lamp will not ignite again. Referring to funeral rites, this means that they are of no use, since they cannot bring back to life the dead.

10 Materialist, a nāstika, rejects the authority of the Vedas by attributing them to a flawed human author. The Vedas as śruti were conceived by Mīmāṃsā as "without human author" (apauruṣeya) and were accepted as an authoritative means of knowledge (pramāṇa) by the Vedāntins. See Bilimoria 1989. Materialist's view does not support testimony (śabda) as pramāṇa but only perception (pratyakṣa) and, to an extent, inference (anumāna). This has been stated through examples by Bewilderment, in verses such as 2.68–2.73. See Gokhale 1993. As conventional as the representation of the adversaries' point of view by Brajvāsīdās may be, it nevertheless points to an awareness of the claims advanced by opposing traditions to what Federico Squarcini calls the "monopoly on veridiction". See Squarcini 2011.

11 jaise jo kou pyāsī toya | miṣṭa nīra baiṭhī hoya || koū kahe tāhi soṁ jāya | yaha jala ati duragandha basāya || mai jala ācho deu batāya | rabi marīca kahaṁ dei dikhāya || jaba vaha vāmeṁ ati śrama pāya | taba vaha vāko lakhi harakhāya || [. . .] The meaning of this first part is not clear to me.

flowers as food? If soft garments giving pleasure to the body exist, then, say, why wear the clothes of tree bark? [87]

Who says that from forsaking the sense objects forever there is heaven,
it appears clearly to them that life in this way is spent in a place similar to hell. [88]

BEWILDERMENT *(hearing* MATERIALIST*'s words, rejoicing, to himself)*: This man seems wise. *(Summoning a staff-bearer)* Call that man and bring him to me. *(After* MATERIALIST *salutes.)* [89]

I'm very delighted by your words.
Where do you come from? Who are you, what is your name, virtuous one? [90]

MATERIALIST *(joining his hands, he says his name and adds)*: A well-wisher and always a servant of this reign. [91]

Kaliyuga sent me here to give you a message. He said he prostrates in many praises in front of you. After my master told me his message, I left, focused on that. With my head lowered, your majesty, I listened to your words. There is some task left which I will accomplish at once, with your blessing, as I came some days ago to see your lotus-feet. [91a]

BEWILDERMENT: Report in detail how much work has been achieved and tell me how much is left of the tasks Kaliyuga gave you. [91b]

MATERIALIST *(joining his hands)*: The so-called path of the Vedas is not widespread now in any way; there is no discussion about it. All men, ceasing mutual regard, deliberately choose fraud. Disciple, teacher, father, brother, and son, all act in their own interest. They nourish their body; they know what they like. No one asks about righteousness, possessed rather by a liking for unrighteousness. The four social classes relinquish their distinct duty; they do actions contrary to it. The excellent brahmin undertakes service, agriculture, and trade, while the outcast recites the Vedas. And the places of pilgrimage like Kurukṣetra, full with many discerning people: after sending them away, now you are praised there with the beat of a drum. Now there should be no worry in your heart, O king, about Discrimination and his wife. Where this servant of yours stays, Wisdom will not be born. How many, such as Balance, Self-Restraint, Observance, Moral Principle, etc., run away from here and there? There are no opportunities to act for Lord Discrimination or his servants. [92]

BEWILDERMENT *(listening to Kaliyuga's deeds, rejoices greatly)*: I know well the valour of Kaliyuga. If Kaliyuga brought our rule to the places of pilgrimage, then it is clear that he will accomplish these other tasks easily. [93]

MATERIALIST: Great king, Kaliyuga told me something else,
to be said at the appropriate time. [94]

BEWILDERMENT: Tell me, what did the auspicious Kaliyuga say more?

MATERIALIST then reported some other details. [95]

MATERIALIST: The great general Kaliyuga further said: "The tasks you gave me are all completed. However, in the houses where Bond-with-Viṣṇu resides, I can describe how our force does not thrive there. She always stays united with her allies, she joins Firmness together with Love. The Supreme Brahman, the primaeval cause of the universe, assumes a form in order to remove the weight of evil people/ wickedness from the earth. The venerable king Rāma killed Rāvaṇa of Laṅka; then, Kṛṣṇa severed Kaṃsa's head. All the deeds he carried out according to the sentiment evoked, indeed, his name was acquired in such a way, according to their cause. Ocean of Mercy, Lord of the Cows, Holder of the Mountain, Enemy of Mura, Yaśodā and Nanda's son, playful in the courtyard. The great king Daśaratha's son, who broke the bow, decoration of the right conduct of the Vedas, who freed Ahalyā from the curse of the stone.[12] Bond-with-Viṣṇu creates a mental image of him [brahman] and concentrates on it with emotion – in how many ways she honours and pays homage to brahman. She sings and does not hold distinction among brahman's names; she relishes the sweet form of her friend with her eyes. She makes the fluctuations of the mind absorbed in the mental image, just as swift lightning is hidden by dark clouds. She ties the anklets to her feet and dances in front of the image. She forgets the condition of the body through the threads of Love. Day and night, always, she carries on in this way; she never desires the enjoyment of sense objects, even in dreams. Even if I create many obstacles against her, nonetheless, the mighty warriors do not come back, and they are won by her." [96]

Upon hearing this, fear rose in the king's heart. Even so, being firm, he pronounced fearless words. [97]

BEWILDERMENT: Listening to the behaviour of Bond-with-Viṣṇu, I do not consider her a friend –
I believe she has always been my enemy. [98]

---

12 Ahalyā, the wife of Gautama, was cursed by her husband after she betrayed him with Indra. Following attributes that remind the reader or listener of Rāma the prince, Brajvāsīdās hints at an episode involving Rāma as a merciful god. While the motif of the stone is not found in Vālmīki's *Rāmāyana*, it appears in the Purāṇas and in the Tamil work *Irāmāvatāram* by Kampan (twelfth century), on which the RcM also draws. The differences among the Sanskrit and Tamil interpretations of the Ahalyā have been analysed by Ramanujan 2004: 134–144. For the various versions of the story in the Purāṇas, see e.g. Söhnen-Thieme 1996.

Someone else from my side is to be found to help Kaliyuga
to capture Bond-with-Viṣṇu and bring her here or kill her on the spot. [99]

(He calls a doorkeeper named FALSE COMPANY *and issues orders)* Inform Anger and Greed that they have to attack Bond-with-Viṣṇu. They must bring her to me, captive, or kill her there. [100]

While the king made such a plan, IGNORANT, a messenger, arrived. He saluted the king and while holding his feet handed him a letter. BEWILDERMENT enquired of his provenance.

The messenger replied: "I have come from the east, the very pure country of Utkala, where the ocean with beautiful shore is. There everyone worships the venerable Supreme Self, from whose majesty there is the best of men (Narottama). The mighty Great Pride sent this account from there, o king. I am a young footman named Ignorant, a former messenger of this house." [101]

The king picked up the letter and gave it to I-maker.
He ordered him to read it aloud, to make its content clear to everyone. [102]

I-MAKER: INTOXICATION and PRIDE have written such account for BEWILDERMENT: "Faith fled from here with his daughter Peace. Discrimination has sent them to persuade Upaniṣad about giving birth to Wisdom – you know about it. Sitting next to her day and night, they praise her. They want to take her in the presence of king Discrimination. Moreover, Discrimination's virtuous friend, Disinterested Duty, it seems that he went to help them." [103]

BEWILDERMENT *(at* PRIDE *and* INTOXICATION*'s message, he grows distressed and looks at everyone with fear)*: Listen, all generals whose arms are the best, do not be afraid of Faith! Wherever in the world there is created nature, my power fearfully confounds everything existing. Where would Faith find a way to join Discrimination and Upaniṣad? [104]

Aye, Ignorant, listen!
Quick, approach Pride and Intoxication, tell them about my condition. [105]

After that, Disinterested Duty, who lives to help Discrimination,
capture him with a trick and send him here. [106]

And regarding Faith, close to Discrimination's Upaniṣad,
an army is going after her, to capture her. [107]

IGNORANT left and FALSE COMPANY returned and saluted the king. He announced that GREED was coming with ANGER as aid. [108]

They arrived and bowed to the king. At first ANGER boasted about his strength. [109]

ANGER: They say that the great king somewhat fears Bond-with-Viṣṇu,
and Faith and Peace are like a painful sigh. [110]

Do you have a servant braver or more powerful than I? How many are the capable servants[13] who can oppress the poor Faith and Peace? In whose heart I reside, whoever he may be, I make him instantly blind, dumb, and deaf. If some time I enter in the heart of a scholar who knows the Vedas and Purāṇas, then that knowledge, intelligence, and skill he possesses I scatter as smoke. [111]

GREED: The world is kept in my power. People always return to me, not frightened of death even a little. Day and night, they keep thinking: "How could I obtain much money?" They say: "I will make great effort to accrue many possessions. [112]

Now that I've gathered so much and more I will gather in future; taking that riches I will go away and increase my wealth further. After going in another place, I will gather twice as much wealth."

They keep thinking in this way, day and night, without Faith and Peace. Someone says: "Who serves the king, stays always present and does not neglect an order beneficial to the Highness. If he says to take the house of a priest, looting it, and to kill the cows, obeying the order instantly, one does not delay it. If some time someone comes into confrontation, a fight, either one kills or one dies. In this way one makes the king happy and then obtains some land." If one keeps thinking like that, Faith and Peace will not spring in one's heart. Someone says that he will make a hole in a big house, stealing the wealth. Someone says they will steal money from travellers. Think of many other efforts – I possess all: dancing, singing, studying, teaching – I have achieved the fourteen knowledges. I have subjugated how many who possess a body in the world? They always stay dependent on me, they all focus on me. In this way they waste life, worrying about the accumulation of wealth. If they do not embrace Faith even in their dreams, how can they obtain Peace? [113]

ANGER *(coming again before the king)*: I did not say much, how much do you know my strength and courage, O king? Viśvamitra, even if of royal family and extremely pure, killed the hundred sons of the sage Vaśiṣṭa without thinking. As if he was not afraid of killing brahmins, a grave sin. This is all the result of my courage, nothing is hidden. After that Indra, so-called king of the gods, lost his honour:

---

13 [. . .] | kitaka samartha sīva jo dābai saradhā sānti bicārī ||. The word sīva either means "limit, boundary" or is a variation for Śiva (BHK 1260); however, I cannot make sense of it in this sentence. In the previous half we have sevaka; therefore, it seems coherent to translate sīva as if it were sevi (Skt. sevin).

he killed with his own hands two brahmins, Viśvarūpa and the demon Vṛtra. Śaṅkara separated with the trident the fifth head from the body of Brahmā. Even if the father is to be worshipped by everyone, indeed he [Śaṅkara] did not think well of it. This greatness is all due to the strength of my arms, should I tell you more? If the gods, starting with Brahmā, are under my sway, how can I describe what happens to men? Even if discerning people, scholars, are called "extremely virtuous", they are nevertheless my subjects as they lose all awareness and understanding out of love for wickedness. [114]

GREED called his wife, THIRST. A smiling woman entered the hall and began to dance. [115]

GREED: O love of my life, do not stay away from me. They say something about Discrimination's consort; because of it Bewilderment has become worried. Therefore, I am telling you about me, O beauty, listen carefully. If such wealth appeared in the world and spread in other worlds, all the same my stomach would not be filled even with fourteen worlds. [116]

My majesty is like a thick cloud, shadowing the whole world.
The sky, the stars, and Wisdom Moon are nowhere to be seen. [117]

THIRST: What you said about yourself is true, O lord. Your glory is like a large net where the world is entangled, like fishes. All the same, I also am at your service, O beloved, and no one finds my end. If someone reduced the universe to ten million morsels and threw them in my stomach, even then they would not find any limit to it.[14] Even if countless universes burned in my stomach's fire, all the same it would not be satisfied. It increases more and more, Faith and Peace cannot even come near. [118]

At the moment ANGER called VIOLENCE. VIOLENCE entered, looking hideous and fearsome. [119]

VIOLENCE danced in an extraordinary way, frightening, like death with a body.

ANGER (to VIOLENCE): O beloved, you know well that the world is afraid of me. King Bewilderment summoned us today because of a task. If you help me, then all the tasks given by the king will be accomplished. If you ensnare someone, he will

14 Brajvāsīdās 1875 presents the verse as jo kadāpi kou koṭi kora | brahmāṇḍa karai jū eka ṭhaura || but I have interpreted the sentence reading kaura in place of kora, which rhymes properly with ṭhaura and also makes sense in the context of the translation. Kora can be considered a synonym for koṭi. Therefore, it is not completely incorrect; however kaura gives a more precise meaning, that of reducing the universe to food, to "morsels".

leave every thought and forget sorrow, doubts, fear, dishonour, the world, and the Vedas at once. [120]

VIOLENCE: Listen, my lord, consider me your servant. Mother, father, brahmin – whoever it is – I will kill him or her at once, at your order. Birds, deer, fishes, insects – their bodies I destroy easily. If I dwell in the hearts of wise people, they will have no other interest but me. [121]

When in this way the four actors had exposed their thoughts, BEWILDERMENT gave them an order. [122]

BEWILDERMENT: You four, leave quickly and get Peace and Faith killed. No matter how, by force or trick, carry out your task properly. *(He gives them an advance and bids them farewell. They bow to the king and leave.)* [123]

(Worry increases in his heart) Faith is extremely powerful, in intellect and strength. My wife, False Viewpoint, is the authority in stratagems. I will send her there so that, being close to Upaniṣad, she will capture her [Faith]. With Faith imprisoned, Peace will naturally suffer. Peace is very delicate, she will die separated from Faith, without giving me the need to kill her. *(He calls the attendant* CONFUSION. CONFUSION, *in an exceptional disguise, enters. The king addresses her)* Quickly approach the woman False Viewpoint and inform her of my desire, summon her respectfully, make her come here. [124]

Following the king's order, the attendant approached the woman. An actor taking the guise of FALSE VIEWPOINT was sitting, concealed by the curtains, dressed in brocade. [125]

The attendant CONFUSION went next to her and sat. After touching FALSE VIEWPOINT's feet she started reporting the king's order. [126]

CONFUSION: The king has invited you, dear. He sent me to you with great desire. The god of love has arrived. Get up and hurry, O woman. [127]

FALSE VIEWPOINT: O friend, much time has passed since I separated from the king. Now I am full of shyness. In what way should I approach the king? If asked, what pretext should I mention? *(She arches her body, stretching her limbs, and yawns. Her eyes are wet with the moisture of sleep, her limbs languid.)* [128]

CONFUSION: O dear, tell me this secret:
what is the cause of your weak limbs and sleepy eyes? [129]

FALSE VIEWPOINT *(smiling)*: When there is desire with just one partner, a woman cannot sleep or be lazy. Many partners feel passion for me, hence I do not take any rest at all, I am constantly sleepless. [130]

CONFUSION: Dear, say his name –
who wants to take rest with you always? [131]

FALSE VIEWPOINT *(smiling)*: Listen, wise attendant. The king Bewilderment is the first, then Desire, Anger, Intoxication, Pride, Deceit, Greed, Envy, and so on. The men of Engagement's family all love me and desire only my face. [132]

CONFUSION: Dear, I have listened to your words and I am very surprised because in King Bewilderment's house there is a lovely woman called Envy. In addition, Desire's wife Passion is well known as being very beautiful. In Greed's house there is a woman, Thirst, a charming beloved. Moreover, in Anger's abode there is another lovely woman, Violence; while the woman in Kaliyuga's house is called Misfortune. [133]

All of them consider precious the company of their own wife,
how, leaving them, they keep the habit of meeting you? [134]

FALSE VIEWPOINT: O friend, even if in everyone's house there is a woman,
still, they are restless without meeting me. [135]

The infatuation of my beauty charms the world,
all are thoughtlessly attracted to me; they lose all sanity. [136]

CONFUSION *(blessing her)*: May God make you his favourite,
may the love with all the partners increase, may it double day after day. [137]

Now go, do not be late. Your lover yearns for you,
like the cakora longs for the rise of the moon at the end of the day. [138]

In this way CONFUSION and FALSE VIEWPOINT conversed behind the curtains. Then they arrived in the hall and danced gracefully. [139]

BEWILDERMENT *(as* FALSE VIEWPOINT *is brought before him, he rejoices)*: O dear, love of my life, without you I was very sad. Now your arrival is a happy event. *(He motions for her to sit respectfully on his lap.)* [140]

FALSE VIEWPOINT sat with both pride and modesty on the king's lap. Embracing and kissing her body, the king was delighted. [141]

FALSE VIEWPOINT *(smiling)*: For what task has the king summoned me? [142]

BEWILDERMENT: O woman, you know it well: all my tasks without you are just dull. You always dwell in my heart. Remember, I do not forget you even for an instant! If today my tasks were all complete, then tell, beloved, for what cause would I have summoned you? [143]

Listen, Faith, together with Peace, has moved closer to Upaniṣad.
Making her meet Discrimination, she wants the appearance of Wisdom. [144]

His rise will cause the destruction of the family. Upon hearing this, I've become extremely worried. Therefore, dear, go there and, through deceit, grasp her forcefully. Grasping her hair, bring her here; if she is jailed, my distress will vanish. [145]

FALSE VIEWPOINT: Listen, O king, for your majesty I would carry out many other tasks! After tying her with the aid of the illustrious false treatises, I will bring her here at once. I will keep her among heretics, then Discrimination will not find any support. [146]

Upaniṣad has risen, similar to the moon, but I will make her sink.
If only my beloved commands, then how many dwellings could Faith find? [147]

BEWILDERMENT *(rejoicing and embracing her again)*: You, wise and brave, have knowledge and might. I am firmly convinced, O queen, that the task I could not accomplish, I am entrusting it to you for good. [148]

After these words the great king felt extreme joy. Talking, he kissed her face, touched her breast, and caressed her hair. [149]

FALSE VIEWPOINT: I say, king! Like this, in the middle of the hall, I feel embarrassed. If in your heart there is such a desire, please, let's go to the comfortable bedchamber. [150]

Hearing FALSE VIEWPOINT's words, BEWILDERMENT, through the pretext of the bedchamber, fulfilled his wish. [151]

Thus concludes Act II, entitled "Description of Bewilderment's Nature" of the "Drama of the Rise of the Wisdom Moon" in Brajbhāṣā.

ACT III

KṚṢṆADĀS BHAṬṬ: Having shown in this way Great Bewilderment's strength,
the First Actor arrived again on the scene and spoke to King Kīratibrahma. [1]

FIRST ACTOR: O king, False Viewpoint, sent by King Bewilderment, drew closer to Upaniṣad, taking along her heretical offspring. The heretical progeny surrounded Faith, who was terrified and ran away to hide somewhere. Looking at the fearsome people of False Viewpoint, Peace was truly miserable. Now she searches, agitated, for Faith. [2]

KṚṢṆADĀS BHAṬṬ: Thus did the excellent First Actor inform King Kīratibrahma.
In that moment the actor playing Peace appeared from backstage. [3]

A young woman, of extremely good and beautiful aspect. Faith's daughter, Peace, was delicate. She arrived in the hall, taking with her the pretty attendant Compassion. Peace was extremely shaken and visibly upset. Saying "Mother, mother!" She cried, unsteady. [4]

PEACE *(tearfully)*: Ah, mother, my receptacle of mercy. Where have you gone, abandoning me? Ah, mother, you that relieve one from suffering, wise, bestowing joy, how can you forget me? You could not stay without me even for a second, how could your heart become so cruel? False Viewpoint caused me great pain – ah, mother, what have you done, then? A holy place where the Vedas are sung, a pure temple, beautiful and attractive. Where Hari's story is told and there is praise of God, where people sing absorbed in love for God. You always reside there. So now, mother, where is your dwelling? Devoted to Hari and bestowing joy to truth-knowers; always affectionate towards Rāma and Kṛṣṇa. Wherever you stay, the heart blooms. You dwell there, mother, like a bee on a flower. [5]

Now she, my mother Faith, is among the heretics.
How will you live there and how will I? [6]

(Lovingly, to COMPASSION*)* Come here, I cannot bear the pain of separation from my mother. O friend, without Faith life is like a thousand curses – without a root a tree will not bear fruit. Without Faith, friend, if you want my good, then collect some wood and burn this body of mine. [7]

Watching the cloud of separation, PEACE, restless, sank, unstable. In that moment COMPASSION spoke, deeply fond of PEACE. [8]

COMPASSION: O Peace, Faith's daughter, extremely wise, please do not fear. These perturbed words of yours, how could you say them, abandoning courage? It is

https://doi.org/10.1515/9783111063836-012

known to the world that Faith, your mother, is considered the jewel among good women. Who could describe her glory apart from Discrimination?[1] False Viewpoint can never win her by force. Who can trap her? Why are you carried away in an ocean of affliction? Nowadays Bewilderment's army, starting with Kaliyuga, has become powerful. For this reason, the fearless Faith must dwell in a proper place, somewhere in order to hinder Kaliyuga. [9]

Make an effort to be calm, you will meet your mother, O young girl.
Wherever you go, search for your mother, be focused and patient. [10]

PEACE *(lamenting)*: Ah, Compassion, I call my mother's name in every place I go, crying. I cannot find any trace of Faith anywhere, I am exhausted after searching in every region. Now I cannot live anymore without Faith, how will I keep searching if I am like a fish without water? Displeased with that, I suffer greatly. O friend, how can I be brave? The four classes and four stages of life, those for whom religious austerities are fundamental and performers of sacrifice – I kept searching among them, but I could not see Faith anywhere. If in some way she is among the heretics, then there is no hope that she is alive. For, in such a circumstance, even a powerful person can become wretched, as Sītā became impure, living with the nightly demon. Therefore, let us search among them [the heretics], for without Faith there is no solace for me. [11]

In that moment a man arrived, having disguised himself as an ascetic of the Jaina teaching. [12]

Completely naked, a dirty body, hair dishevelled in four directions. Carrying in his hand a peacock feather, like a ghost under the guise of a man. [13]

COMPASSION *(afraid)*: Oh, young girl, who is that man? He seems extremely shameless. A bad aspect, dirty, full of terror. Or is it an evil spirit who has assumed a visible body? His way of thinking shall not be known; it is highly perverse and should not be described. [14]

PEACE: Looking at him, I think –
this hellish person arrives after having escaped from hell. [15]

Just a moment! See, his distinctive feature is being a sky-clad!
Of what is he the master? Of what place does he praise the Vedas? [16]

1 [. . .] | hita sena vesa viveka kevala jāsu yaśa saka ko bhanī ||. I could not find a possible translation for the terms hita sena vesa in the sentence.

The Jaina Ascetic entered the hall and sat, after having spread his rug under a tree. Some laymen went to him to listen to his words. Approaching him, they bent their heads at his feet. He blessed them and invited them to sit. He told them words of excellent advice[2] and began his teaching. First, he invoked the enlightened beings and said that one has to stop worshipping a chosen deity.

Jaina Ascetic: O brothers, listen! I will explain to you what knowledge is. [17]

The individual self is the brahman, whose form is indescribable.
It is always visible, with qualities, beautiful, scented, excellent. [18]

Do whatever gives happiness to the individual self, knowing that it is beneficial. Considering this a teaching, take all wealth and houses, with body, mind, and words. This world is impermanent but, still, donating something here is good. This is the occasion on which you can do it, so donate something in my hand and then go away. [19]

I will say one more thing about bathing and washing the body. The man who purifies the body by scrubbing it, he is a receptacle of extreme ignorance. To what extent can I describe him? All listen carefully: this body is made of the five elements, and inside is filled with dross and dust. It is manifested with nine doors; it has impurities like urine, mucus, etc. [20]

Even if it is washed on the outside, it does not become pure on the inside.
Therefore, consider all of this bathing and washing useless, be clever! [21]

If you wish to know the Lord, I am going to tell you how you can do it, listen! If there's an ascetic similar to me, thirsty, worship him with reverence. Bring him respectfully to your house, make him rest in every way. Prepare various meals with six flavours and, touching him deferentially, feed him. Accomplish every order he gives. Satisfy his orders, fulfil them, considering it your good fortune. [22]

If sometimes he takes delight with women or commits some sin,
all the same do not get angry with him! [23]

The essence of the universe – see it and ponder. There is no other merit similar to the good of other people, no similar burden to injuring another. Recognise your own good, do not know other people's good. The men of this world, know all as dogs! [24]

From whose house the ascetic leaves without having taken rest, that man is the lowest in the world; all his merit is destroyed at once. *(Upon hearing such words,*

2 bacana svadeśa. I interpreted svadeśa as the prefix su- joined to the noun ādeśa, since svadeśa as "one's own place, country" did not make sense in this passage.

the laymen touch his feet. He pats them on the back, gives blessings, and asks everyone to sit.) [25]

(Looking offstage and laughing) Faith! Go in their company and stay always with them, so that their mind will never stray elsewhere, even for an instant. [26]

PEACE stood there with COMPASSION looking at him. When she heard FAITH's name, she grew very sad.

PEACE *(dejected, reflects to herself)*: Is she Faith or some deceiver?

COMPASSION *(determined to make the worried* PEACE *understand)*: How can you stand in worry? [27]

This is not your mother, Faith. Why do you distress your heart?
Non-Violence told me that there are three kinds of Faith. [28]

One Faith, O young girl, is of sattva, and your mother. She belongs to King Discrimination and has many kinds of greatness. The second Faith is of rajas, to whom Bewilderment's authority applies. The third belongs to tamas, who loves Desire and Anger. [29]

The Faith of sattva never comes into their hands.
The Faiths of rajas and tamas always stay with the heretics. [30]

COMPASSION took PEACE's hand and comforted her. A BUDDHIST MONK entered.[3] [31]

BUDDHIST MONK *(holding a book in one hand, with a cheerful disposition, he dances briefly and then sits and addresses everyone in the hall)*: O men and living beings, you listen! Get to know about my beneficial teaching. [32]

The self, that is light, defined as eternal, is all a lie brought about by one's mind. Eternal is this body and worldly enjoyment – all the rest is just useless, like a noose of dry straw. There is nothing superior to these, see it for yourself everywhere! A beautiful house with the appearance of a painting, clean and beautiful bedding

3 sevarā is a tadbhava of the Prakrit sevaḍā (Skt. śvetapaṭa) and an appellative for Śvētāmbara Jainas (HSS 5278). It appears that it was so in critiques of Jaina believers in religious texts of the sixteenth and seventeenth centuries; see Horstmann 2017: 6–13. In this act, however, we have first the arrival of a Jaina ascetic called Bihamba in one instance (sky-clad, 3.16) and Jatī (Skt. yati, a Jaina ascetic) in all other occurrences; while Sevarā refers to a Buddhist monk, as the two quarrel about their doctrines. The Buddhist monk is, in fact, also once called Bhikṣuka (3.44) and once Faqīra (3.45), both translatable as "beggar" and used in a derogatory sense. The GH also presents this ambiguity, with the second character named Srevaraḥ. See Chand/Abidi 1961: 76. In the KPC, the former is called Digambara or Kṣapaṇaka and the latter Bhikṣu. See Kṛṣṇamiśra 2009: lxviii.

spread around. Various foods endowed with the six flavours and a moon-faced girl sitting in the house. Dancing, jumping, music, and merriment; rose-essence, alcohol, bhāṁg: all these preparations give pleasure, they make the body relaxed. [33]

What salvation? What Vaikunṭha? What other world? See!
If something exists, it is nothing but nourishing this body abundantly. [34]

And the individual self is an enjoyer in this body for a short while.
Every instant it is new, ever new, not a single entity. [35]

COMPASSION *(listening and finding it difficult to understand,*[4] *addresses* PEACE*)*: What is he saying? This man is really opposed to the Vedas. I am staying still for a moment, so that I may grasp his way of thinking. What is in his heart? What is the book he holds in his hand? *(Both keep watching. The monk remains sitting.)* [36]

Some other actors, disguised as intoxicated BUDDHISTS, sat around carrying out austerities. The monk started talking to them, showing them his book. [36a]

BUDDHIST MONK: This world is like the stream of a river. From where it comes or where it goes – this no one knows, indeed. What existed yesterday today is not visible, what exists today will not be found tomorrow. In a moment it becomes something else and no one can be said to be permanent. Therefore, this body has to be established as supreme. It is necessary to keep nourishing it by means of excellent eating and drinking. Keep jewels and take loans, please eat milk's cream, do whatever nourishes this body. The enjoyments of a man, like women and so on, are not bodiless. Therefore, do not leave them, rather make them multiply! [37]

They leave enjoyments, toil in vain practising austerities, they build a painful body.
Such men are like beasts without horns,[5] their lives ruined. [38]

Like a small flower-garden, the world receives the beautiful scent of flowers.
It does not last more than two days, after which the flowers become perfumeless. [39]

---

4 Brajvāsīdās 1875 presents the verse as [. . .] | karunā sunike bisa maiṁ gahī ||. However, the postposition maiṁ does not seem to make much sense paired with the verbal root gah- (Skt. grah-). I have chosen to interpret bisa maiṁ as a single word, bisamaiṁ, an oblique case of the masculine singular bisama (Skt. viṣama). Taking bisama in the sense of saṅkaṭa (HSS 4556), the sentence's meaning appears clearer. This reading is corroborated by contextual reasons, as Compassion's next words express confusion at what she has just heard.

5 This idiom is repeated in 6.58. It was also used by Tulsīdās in RcM 5.49.1, and 7.78a with a slightly different wording.

BUDDHISTS *(rejoicing at his words)*: Good, good! This discourse of yours, O master, it is true, we know it well.

With this praise, the BUDDHIST MONK became happy. He called offstage to his FAITH who, joyous, entered. [40]

BUDDHIST FAITH *(dances for a moment, then draws closer to him)*: What order do you have for me? O lord, tell me. [41]

BUDDHIST MONK *(laughing)*: These men, who are of my same belief, make them so attached to you that, separated from you, they would be in pain as if wounded. Do not stay away from them even for an instant, in order that they continue to be our servants. [42]

BUDDHIST FAITH: As you command, O lord. *(She exits, taking the* BUDDHISTS *with her.)* [43]

JAINA ASCETIC *(looking at the* BUDDHIST MONK *and in a disdainful tone*[6]*)*: Oh, beggar, explain to me your doctrine! [44]

What is proven by it? I cannot understand anything. You have described this point of view thoroughly: the self is enjoyer for a limited time. Further, you believe the enjoyment of this body to be salvation. If what you said is true, then why do you wander, after having left your house? Why do you make your body suffer with austerities, why did you become a beggar, leaving all enjoyment? [45]

The elders say that the men who stretch the body, practising austerities, obtain the best fruit. They destroy the suffering of birth and death. [46]

If in your view the self's birth and death take place at every second,
then the performer of austerities is someone and, at the time of death, he is
another person. [47]

Therefore, who gains the fruit of the austerities? Solve my question,
otherwise it will be clear that your doctrine is false. [48]

BUDDHIST MONK: You do not understand my doctrine, ignorant! In my doctrine, intellect is chief. [49]

6 I added this adjective to emphasise the sense of the re bhikṣuka present in the original text.

The individual self, temporary enjoyer, in proximity to intellect is liberated.
At the time of death, the individual self that attains the state of being enlightened is freed.[7] [50]

Jaina Ascetic: Your doctrine is not correct, who taught you it? Just consider it false, join my belief now. How could you fall, fool, for this bad doctrine? [51]

First of all, astrology is pre-eminent in our path.
Thanks to it, the individual self, sitting on the earth, obtains the way to heaven. [52]

And in my doctrine the individual self is proven to exist in each body,
that is why one has to nourish the body in a proper way. [53]

Buddhist Monk: Listen, you ignoramus, if in your doctrine it is maintained that the self is similar to the body, then how is the way to heaven to be known? If it is proven that the self is corporeal, then what is the difference between it and the body? Do not tell such lies, O fool. My doctrine is more profound than yours. [54]

They continued to argue and exaggerate. Meanwhile the actor playing Skullman entered. [55]

He wore massive prayer beads made of human bones around his neck. His body was smeared with ashes from a crematorium. His eyes were bloody, as if inflamed. He drank liquor and ate meat daily using the skull of a man. [56]

In such a frightening attire, the Skullman arrived in the hall and danced, joyful, blowing a horn. [57]

Buddhist Monk *(contemptuously)*: Who are you, ugly fool? What is your path, what is your goal? What you have obtained is extreme stupidity! I've never felt

---

7 nikaṭa buddhi ke ātamā chinaka bilāsī mukta | anta samai jo ātamā buddha lahai so mukta ||. The verse is not easy to translate. In the KPC the position defended by the monk is one of Yogācāra Buddhism. See Kapstein's commentary on verses 3.41, 3.57–3.58 in Kapstein 2009: 302–303. If Kṛṣṇamiśra employed the term vijñāna in 3.58, I think that here the doctrinal position is less clearly ascertained. We could take buddhi as meaning "consciousness", and the verse to mean that an individual self is enlightened (and freed) when it realises that the sole reality is consciousness. The second pāda could also be translated as: "At the time of death, the enlightened individual self attains [the state of] being freed."

more disgusted – what kind of begging is this?! You, ignorant, tell: what is the meaning of salvation? I want to know how you are grateful.[8] [58]

SKULLMAN *(laughing)*: O resentful Monk,

you do not know anything about the beauty of my way, it is very far from you! [59]

My path is eternal and all the other paths are envious of it.

While they are busy with useless, wicked arguments, my way reveals the attainment of salvation. [60]

Now I will describe my conduct to you: everyone is afraid of me. The sacrificial fire is my abode night and day, they are indistinguishable. There the sacrifice of meat of deer, bird, and fish takes place, together with alcohol. Then, finding the skull of a brahmin, filling it with liquor, I drink from it every day. On the bed of ashes left from the sacrifice, after roasting it near that fire, food will be placed. No one knows about these details. [61]

And every day after killing many living beings with my hands,

I worship Bhairava, head of the gods, with blood and liquor. [62]

I stay engrossed in this way night and day, avoiding bad thoughts. I have other nice enjoyments, such as bliss and happiness in seeing and touching. In my heart I think, "Is there a similar bestower of pleasure?" It is not possible to compare my doctrine with others, there will never be another like it. [63]

BUDDHIST MONK: Hey, listen, O ignorant Skullman! Even if bodily pleasure for you is fine, yet, when I reason about it, I find your doctrine dull. What is this way? What is this trickery? What is this thought, how was it produced? You, praising such a doctrine, are envious and fearsome! [64]

Just as the meat-eater group of dog, jackal, crow, and vulture,

your aspect is dirty and frightening, a demon's sport. [65]

SKULLMAN *(turning to leave)*: You do not understand! Listen, stupid! My path is the most powerful: even God, by whom the world originated, praises this path, pleased. What is your standing? What is your course?[9] Could you point out some-

8 [. . .] | jāni parai tū kasa kṛtajña ||. The term kṛtajña is problematic to translate in this context. It might mean the monk wants to know to whom the Śaiva shows his gratefulness; that is, which god does he believe in.

9 teri bisāta kahā tuhi kā gama | [. . .], "What is your chess-cloth and your game?" Although I kept the reading involving the primary meanings of the terms, this interpretation is equally sup-

thing similar? My path is proficient in the fulfilment of desires, I do not know any other like it. If its power is not clear to you, then sit a bit and I will describe to you its value. [66]

If I only say it, then Brahmā, Viṣṇu, Śiva, sun, moon, and stars – everything is at my order!
If I ask for them to be on earth, they will be taken away from the sky at once! [67]

BUDDHIST MONK: Hey, fool, I know everything about enjoyers! Your claims are deceptive like a ghost's sweets![10] Your doctrine roams in the world, deceiving, like an evil spirit jumping on someone! Your thought is nothing but Indra's net, the magic that exists in the world. Your actions are just as those of a sorceress who misleads the whole world. When you know your own self, then you will recognise the path of salvation. [68]

I must know: did an evil spirit give you this teaching?
It is at fault, there is no trace of salvation in it, only self-conceit. [69]

At this speech the SKULLMAN became furious, his eyes sparkling with anger. [70]

SKULLMAN *(starting to run after the* MONK, *drawing his sword)*: Fool, you made of my teacher an evil spirit. I will cut your head off and make it fly in the air. You criticise, O stupid, yet you do not possess any judgement.

The ASCETIC was nearby, teaching his doctrine. The MONK, frightened, ran to him and put his head in his lap. [71]

JAINA ASCETIC *(to the* SKULLMAN*)*: You really are a formidable man, aren't you? Now forgive him, do not kill this monk! [72]

SKULLMAN: Hey, the idiot does not understand the salvation of my path. Until one lives in the world the day passes happily. Love with a moon-faced, lotus-eyed girl who fills with water a beautiful skull, taking it to her master. Lord Śiva keeps the girl, Pārvatī, as his inseparable half, and remains isolated from everything else, eating poison and hemp. In the same way, drinking alcohol, the company of excellent beautiful women, day and night spent in love games – this is salvation, you fool! [73]

JAINA ASCETIC: Brother, listen to what I say: without relinquishment of worldly enjoyment,

ported by bisāta and gama. See HSS 3514 for the former term, HSS 1231 and MW 348 for the latter.

10 An idiom employed to refer to illusory objects, which are not real. See HSS 3679.

without practising austerities, desireless, there is not even a fraction of salvation! [74]

SKULLMAN *(aside)*: They are still not convinced by my path. [75]

Hearing his words, SKULLMAN'S FAITH entered the assembly from backstage. Her body had the radiance of the moon. Long, loose hair to her feet, shiny – with jewellery made of bones. Flirtatious eyes, as if she had drunk liquor. Such was the aspect of the SKULLMAN'S woman.[76]

She danced joyfully, then, bowing her head and smiling, addressed her master. [77]

SKULLMAN'S FAITH: Please say, lord, what order do you have for me? Please tell me.
Whatever pleases you, I will do it immediately. [78]

SKULLMAN: My life, these are the ignorant Buddhist Monk and Jaina Ascetic. They do not accept my doctrine; therefore, approach them. Mix with them as water and milk do. [79]

SKULLMAN'S FAITH ran at the MONK and patted him on the cheek with pleasure and laughter. [80]

At her touch the MONK jumped up delighted and took her in his arms, keeping her close to his heart. [81]

BUDDHIST MONK *(absorbed in the embrace)*: O Skullman, you are very fortunate! Beautiful, charming, agreeable young lady! A joy to the heart, she procures happiness and gives pleasure. For me, she has a thousand virtues, I have never heard of something more delightful! Your doctrine is quite full of love and my heart, seeing this, has become incredibly happy. Millions of curses to my doctrine, in which this happiness does not exist even in dreams! I am spending my life in vain by following it, as I will never obtain a taste of happiness such as this. [82]

Now, after initiating me, show me your favour, O compassionate.
Teach me the exceptionally loving way of your path. [83]

JAINA ASCETIC: This is a false doctrine, Monk! A little taste of woman was sufficient for you to refute your own doctrine and focus on a corrupt doctrine! [84]

BUDDHIST MONK: Hey, Ascetic, what are you babbling? How would you know the pleasure-loving way of this path, being yourself deprived of actions? Liberation in life is what I found in this Skullman's woman. Do not ask me about the way of this path, for I have obtained the fruit of life in the satisfaction of one moment! [85]

The SKULLMAN gave a signal and his FAITH approached the ASCETIC. She wrapped her arms around his neck and they embraced.

JAINA ASCETIC *(passionately)*: O pleasure-giver, sit for a moment next to me, for the sake of my virtue![11] *(Seeing her naked body, he is embarrassed and hides his own body with the peacock feather. Then, as if fallen in love at first sight)* I spent all my life in vain, unfortunate, on the Jaina path. In this way I could never find contentment, that the Skullman's woman made me know in an instant. [86]

Now, O master, true teacher, Skullman, do not be angry.
Make me your disciple, have mercy. [87]

SKULLMAN *(rejoicing, he fills the skull with liquor and drinks a little)*: Come, drink! Then you will be freed from this immovable snare.[12] Revel in the advantages of the human body!

BUDDHIST MONK: I am not going to drink; I do not touch bones with my hand. [88]

SKULLMAN *(to his* FAITH*)*: There is still arrogance in his heart. [89]

Get up and give them the skull. Grasp it with your hand and take them in your power. [90]

SKULLMAN'S FAITH got up, rejoicing. She grasped the skull and filled it with liquor. Approaching the MONK she drank first then handed it to him. He drank respectfully. [91]

JAINA ASCETIC: Hey, Monk, I say this is not correct!
You are drinking such sacred remnants all alone! [92]

The MONK called the ASCETIC to him and made him drink. [93]

Intoxicated, both laughing, they started describing the encounter with the SKULLMAN. [94]

Their drunkenness increasing, their limbs grew unsteady, their eyes drowsy. They fell to the ground, unconscious. The SKULLMAN rejoiced and, regarding them equally, accepted both as disciples. Then he started dancing with the woman.

SKULLMAN *(laughing)*: Today I have obtained two slaves, without paying.

He danced with great passion and joy. The two, hearing him, woke up. All four danced, then, tired, they sat. [95]

---

11 See RcM 5.47.4.

12 taba yahi nagata phānsa te chūṭahu | [. . .]. I cannot retrace what the nagata given by Brajvāsīdās 1875 could be. Therefore, I interpreted it as na gata.

JAINA ASCETIC and BUDDHIST MONK *(in unison, drunk, touching the* SKULLMAN*'s feet)*: O Lord, your doctrine is good! It is the chief of all doctrines. And this Faith of yours, bestower of joy – of all the doctrines she is the most charming and lovely. [96]

SKULLMAN: How could I explain this?
There are several other qualities in me that you do not know. [97]

Killing, infatuating, subjugating, driving away, attracting women even from distant regions. To attract or do whatever is needed. The illusion of Indra's net, incomparable, that no one else experiences. The eight supernatural powers and nine treasures are standing at my door. I'm accomplished in alchemy and can kill other people. [98]

JAINA ASCETIC and BUDDHIST MONK *(in unison, their hands joined in prayer)*: We are your devotees, O teacher, completely. By luck we obtained a beautiful vision, joy flourished, pain vanished. Now we are intoxicated, our heads heavy, we will forget your words. With some formula take away our drunkenness, then we will tell you in detail. [99]

The SKULLMAN rejoiced and ate a betel leaf. Taking their open hands, he spat it out on them. They ate it quickly and became sober. The two joined their hands again and started praising him.

JAINA ASCETIC and BUDDHIST MONK: We believe now, your doctrine is a receptacle of happiness. Now make us experts in it, please listen to our request. [100]

Give us the powers of attraction and subjugation,
thanks to them, those who want to obtain a desirable woman just by calling her. [101]

O master, showing your favour, please reveal such way,
by which the Urvaśī we desire will come immediately to our hands. [102]

Dear, divine teacher, one more thing, tell us about the qualities in the body. Meanwhile, the enmity between Discrimination and Bewilderment has increased. Tell, there must be a way to take revenge! Discrimination wants to carry out his duties, then he wants to conquer Bewilderment's kingdom. We are devoted to you now and we will describe all about Bewilderment's condition. [103]

These powers, us, and all tasks, indeed, have come together today.
What is to be done to prevent the destruction of Bewilderment's reign? [104]

We have heard about Duty's Faith,
    sent by Discrimination to find the woman Upaniṣad. [105]

Calling her, he wishes to make her generate Wisdom,
    therefore, King Bewilderment has ordered to look for Duty's Faith. [106]

If Upaniṣad meets Discrimination, then Bewilderment will not be pleased. The cause of this reunion would be Faith, as she does no other effort. Therefore, we should try thinking how we can go and kill Faith or, catching her, make her ours. We must drive her away from Upaniṣad.

SKULLMAN *(laughing)*: How many means I possess to kill her! If only, searching, I could get to know in what place Faith stays.

JAINA ASCETIC: This I know indeed. I am well versed in astrology. I must know what is going on in her mind, and therefore I am following her.[13] Faith is with Bond-with-Viṣṇu, as she fled, fearing False Viewpoint. She dwells in the heart of those who stay focused on Hari's feet, obtaining ultimate joy.

Hearing this, PEACE rejoiced like a peahen hearing the sound of the clouds.[14] She felt better. COMPASSION, rejoicing, congratulated her. [107]

SKULLMAN: Tell me, where has Duty fled for fear of Desire? [108]

JAINA ASCETIC *(after counting with his fingers)*: He went to Bond-with-Viṣṇu's place, taking refuge in her arms. [109]

SKULLMAN *(worried)*: If what you say is true, dear, and Faith has taken refuge with Bond-with-Viṣṇu, if she has the support of Bond-with-Viṣṇu, then she will accomplish Discrimination's plan with ease. We have no power over her; Desire and the others are scared as well. Even Kaliyuga has no power over Bond-with-Viṣṇu, notwithstanding his remarkable authority. The immense greatness of Bond-with-Viṣṇu is long current throughout the four ages. [110]

And until we live dependent on King Bewilderment,
    and the tasks he assigns are not called off, surely there will be a reward for her
        head. [111]

---

13 [. . .] | au pai mero pāṁvana paraī ||. The meaning of au is not clear: in previous occurrences it appears to stand as the conjunction aura (HSS 707). See verses 103–104 in this act. I have thus also interpreted it in this way here.

14 Adaptive reuse of a pāda from RcM 1.294.2. See subsection 3.3.1, where I have rendered the alternative translation of "Hearing the news, Peace rejoiced like a peahen hearing the sound of the clouds."

My daughter Bhairava's-Science is extremely mighty:
Śābara formulas, outward darkness – she has all means for carrying out the task. [112]

With the power of Bhairava's-Science I will take away
Faith and Duty from Bond-with-Viṣṇu, for the sake of the king. [113]

With the pretext of going to send for BHAIRAVA'S-SCIENCE, the four left the assembly. [114]

PEACE *(to* COMPASSION*)*: Friend, now we know about Faith! Come, quickly, we have to inform Bond-with-Viṣṇu that there is a risk to her, devised by the Skullman with the others.

COMPASSION: By your power, they will be destroyed. [115]

PEACE and COMPASSION exited the hall, talking with each other. The FIRST ACTOR stood to sing. [116]

Thus concludes Act III, entitled "Description of the Nature of the Heretics" of the "Drama of the Rise of the Wisdom Moon" in Brajbhāṣā.

ACT IV

Kṛṣṇadās Bhaṭṭ: Thus the best First Actor showed the king the nature of the heretics.

He mentioned as well that now King Discrimination has some support. [1]

Friendship *(her actor entered the hall, a lovely woman)*: I've come to hear such news from Gladness: [2]

After some time Bhairava's-Science reached out suddenly,
and, catching Faith, she flew in the sky at once. [3]

Had she wanted, she could have made her fall to the ground from up there; then no one could have ever found Faith. But Faith had Bond-with-Viṣṇu's help: immediately, having mercy, she saved her. When this fact came to my ears, it was as if a fire had engulfed me. I found no rest. My worried heart wants to see her safe and sound, and it is my good luck that I will see her on occasion arising. [4]

The actor playing Faith arrived on the scene. Friendship, happy, met her with love. [5]

Friendship (*enquiring after the other's well-being, affectionately*): Faith, my friend. [6]

The insolence of Bhairava's-Science towards you,
I got to know about it from Gladness. [7].

Even so I would like to hear it from you,
how was she, that disgraced one, violent to you? [8]

Faith: You are about to know a hidden secret, O my dearest, life of my heart! How could I recount what happened to me? Whenever I remember the way it happened, I shake. I believe that by the favour of the venerable Bond-with-Viṣṇu life came back to me, the world knows about it. That powerful calamity was unfortunate, since I was saved and I am meeting you, lucky one. [9]

Friendship: O wise woman, now do not harbour such fear. [10]

That Bhairava's-Science was similar to a mountain,
but now, burned by the fire of Bond-with-Viṣṇu, she has become ashes. [11]

Now do not worry at all,
describe in detail the crookedness of that cruel woman! [12]

https://doi.org/10.1515/9783111063836-013

FAITH: I am devoted to Bond-with-Viṣṇu's name.[1] She [Bhairava's-Science], that ruthless ignorance, manifested as a blinding light. A brilliant lightning, she descended on us in this way: as a white falcon attacks for the purpose of killing the partridge and the quail. Catching Duty with one hand, with the other she grasped my head. Seeing her I was afraid, no hope of living left. Then we cried, calling, invoking the venerable Bond-with-Viṣṇu, our voices distressed. She heard them, abode of mercy, the heart similar to an ocean of compassion, a treasury of waves. [13]

Seeing us confused and agitated, hearing the highly distressed voices,
Bond-with-Hari frowned at Bhairava's-Science, her eyebrows bent,[2] ready to punish her. [14]

Her power pervaded her [Bhairava's-Science], so that she trembled just at seeing her. She fell, unconscious, to the ground – no strength to get up was left to her. Seeing me and Duty anguished, Bond-with-Viṣṇu, the very abode of mercy, grew furious at Bewilderment and said angrily: "That evil was insolent to the point of sending Bhairava's-Science to my place! He knows that Faith is dependent on me, that is why the arrogant Bewilderment wants to kill her. Knowing that Discrimination is helpless, subjugating Mind, he makes a fool of him. Now this villain, no one considers him the same as before. He will reap what he sows.[3] [15]

I have placed my hand on Discrimination's head today.
I will give a lesson to that villain [Bewilderment], make preparations for victory! [16]

Faith, go to Discrimination!
Tell him to prepare the army and to leave any doubt – I will be of help in every way. [17]

Tell him this as well: the stupid Bewilderment has become quite impertinent, hence it is necessary to punish him. Now fight against him with strength! Do not be

1 saradhā kahi bali jāṁva | haribhakti kere nāṁva ||. The expression employed here is bali jā-, to sacrifice or dedicate one's life to someone, usually a god. See HSS 3410. In this case it refers to Hari's/Viṣṇu's bhakti herself, considered in the play as having the status of a goddess. With this sentence Brajvāsīdās builds an unbreakable bond between faith and bhakti, with the former depending on the latter for her life. Complete dedication was expressed, among others, in a similar way by Dādū Dayāl (1545?–1603) in the *Dādū vāṇī* (15.46): je jana rāte rama sau tinaki mai bali jāṁva | dādū una para vāraṇe je lāga rahe hari nāṁva ||, in Dādū Dayāl n.d.: 306.
2 bhṛkuṭi vikaṭa: in Brajvāsīdās 1875 the word vikaṭa is not clearly readable; it appears as vivaṭa, which, however, is not attested by dictionaries. Therefore, I amended it to vikaṭa, as bhṛkuṭi vikaṭa was a widely used expression by other authors as well. See HSS 4445.
3 [. . .] | so apane boye phala khāhū ||. Literally, "he will eat the fruit he planted".

afraid of Time, etc. My powerful heroes, starting with Dispassion, will help you for good. At the time of war, I will also join. I will burn that fool in an instant. Take good care of yourself, make the effort thanks to which Wisdom will manifest." [18]

I will approach King Discrimination bringing this message from Bond-with-Viṣṇu.
Tell, Friendship, since when do you stay here? [19]

FRIENDSHIP *(listening to* FAITH*'s beautiful voice and to the greatness of* BOND-WITH-VIṢṆU, *rejoices)*: If Bond-with-Viṣṇu has this in mind, then the arrogant Bewilderment – I know that this will be the end of his life. At the right moment, approach the king [Discrimination] and remove all his doubts. Recommending the right conduct to relatives and aids, he will defeat Bewilderment's party by the power of Bond-with-Viṣṇu. [20]

Until now the group of Sympathy, Gladness, Compassion, and myself,
the four sisters, watchful in each other's company. [21]

We reside in people's pure hearts,[4]
in order to arrange the preparations of the war of King Discrimination. [22]

FAITH: Good girl, go and please halt in a good spot: I heard from Bond-with-Viṣṇu that there is a place of pilgrimage[5] near the Ganges. It is called by everyone Fortunate Wheel. Discrimination will soon live there. There he will find the way to meet Upaniṣad. Without her, he is unhappy. [23]

His other wife, Good Reason, sits with him. She recites Hermeneutics' words over and over in many ways. She wants him to sacrifice for the enjoyment of Indra, etc. For this reason, the king cannot find the occasion to meet Upaniṣad. [24]

FRIENDSHIP *(joining hands)*: Please go back to what you were doing.
Whatever Bond-with-Viṣṇu ordered, do it immediately. [25]

If the king, attracted by the words of Good Reason,
forgets Upaniṣad, then all our objectives will be shattered. [26]

Talking in this way the two actors playing FAITH and FRIENDSHIP exited the hall, then the FIRST ACTOR came and spoke.[6] [27]

4 In the KPC, the four sisters dwell in Puruṣa's heart. See Kṛṣṇamiśra 2009: 147.

5 In Brajvāsīdās 1875 the word which I have translated as "place of pilgrimage" is tīragha, which is not attested by dictionaries. For contextual reasons, since Faith is suggesting to Friendship to stop at a temple, it seemed most obvious to amend tīragha to tīratha.

6 This verse has almost the same wording as 3.116.

FIRST ACTOR: Listening to Bond-with-Viṣṇu's message, Discrimination, king among kings, rejoiced.

At present he is camped in an appropriate spot with his army. [28]

DISCRIMINATION arrived, looking powerful. Ahead of him, a staff-bearer alerted all to the king's presence. A lion-throne was installed, and the king took his seat. He was furious and began to describe BEWILDERMENT's faults. [29]

KING DISCRIMINATION: That Bewilderment distinguishes himself as a fool. He is recorded as a receptacle of sin. I am not preoccupied, but rather, I feel angry for the offence. He obtained the fruit of his past deeds: for his own previous bad actions he was reborn evil. [30]

See, this brahman is unchanging, one. An ocean of bliss, his greatness manifold. Whose form has an incomparable radiance, his son is King Mind. He obliterated this greatness of his and made himself miserable. It is indescribable! Many births and deaths, multiple kinds of suffering. He floats, then drowns in an ocean of doubt. This is not Mind's true nature. It is the effect of the company of Great Bewilderment. Mind will not realise "I am that"[7] if Bewilderment harms him. In proximity to Mind this fact is evident: the nectar of immortality always flows, day and night. Mind left it for the company of Bewilderment. Now he drinks poison, not satisfied by objects of sense. [31]

The venerable Bond-with-Viṣṇu, compassionate, was favourable to me.
Now, uprooting the evil Bewilderment, I will make the root of nectar flow. [32]

He has one very powerful warrior, in whose hands there is a bow and flower-arrows. His name is Desire, together with his wife, who subdues many gods like Brahmā. He is Bewilderment's vanguard; hence he must be defeated first. If in my group there is a warrior who can conquer him; he is Investigation-into-Reality. *(He calls a staff-bearer, one called* KNOWLEDGE-OF-THE-VEDAS[8] *arrives.)* [33]

The king ordered her to bring INVESTIGATION-INTO-REALITY into his presence as soon as possible.

KNOWLEDGE-OF-THE-VEDAS bowed and left. [34]

INVESTIGATION-INTO-REALITY was backstage, preparing. KNOWLEDGE-OF-THE-VEDAS arrived and communicated the king's order. [35]

7 soha, a tadbhava of so'ham asmi ("I am he") from *Īṣa Upaniṣad* 16.

8 This staff-bearer is called either Veda Vidyā or Vidyā Veda. I chose the former in translating the meaning of the compound. The Sanskrit PC also presents many versions of this name in its different recensions. See Kapstein 2009: 306.

INVESTIGATION-INTO-REALITY *(angrily)*: That Desire is a major cause of suffering. There is no cheater worse than him, who by cheating, deceived the whole world. This king of sorrow is very base. He always does what pleases him. Taking a woman, the blackness of hell, he keeps painting everyone's face black. [36]

Even the most expert man, who knows all the notions of the world,
becomes a fool after seeing a woman because of Desire's crookedness. [37]

They are fools, in the power of Desire, as if their vision were restricted. Watching, they say a woman is beautiful – they do not ponder at all. Lovely limbs, uncovered, adorned with beautiful gold jewels. Multicoloured clothes on the body, full of good perfume and flowers. And all the other ornaments! Seeing just this, they describe a woman. [38]

If only they saw a woman's natural body, without ornaments and clothes,
as a receptacle of bones, skin, marrow, blood, and bad smells. [39]

Seeing a woman, resplendent for the jewels and clothes, a man falls in love with her.
He does not consider her shortcomings; being won, he burns as a moth attracted to light. [40]

A human, indeed, needs to dissect another human. By thinking, he ascertains the essence. The human body has to be recognised as not the essence, otherwise one is like an animal without a tail and horns. A multicoloured picture, which no one grasps by observation. Wise people know objects of the senses as such, they consider pondering over brahman as the essence. [41]

The essence is one, Hari's name, while worldly objects are without an essence,
vanishing quickly like dewdrops. [42]

All women of the world are made of Desire: without Desire they would stop, they would not oppress. They subjugate all people in their enjoyment and have bound them in an inauspicious union. People stubbornly insist on what is easily attainable, not understanding, the fools, that the dagger upon their head is sharp. There is no one who has understood this and can talk about it. How, in fact, can they grasp a good thought? [43]

The womanly body, for which much attraction is felt, is made of up of blood, flesh, and bones. When a shiny beauty, grace, and figure are visible, those with this appearance are considered good women. [44]

When a body is brilliant love arises but,
what if without body one is but dust? Reflect upon this. [45]

Having spoken in this way to KNOWLEDGE-OF-THE-VEDAS, INVESTIGATION-INTO-REALITY approached DISCRIMINATION and bowed. [46]

INVESTIGATION-INTO-REALITY *(saluting the king)*: What task, O king, has come to your mind today? Give me, O lord, an order and do not worry for, Your Majesty, I will execute it instantly. [47]

KING DISCRIMINATION: All the tasks could be easily accomplished by you. A war between Bewilderment and myself has commenced. Therefore, I summoned you at once. Apart from him, there is another powerful man, Desire, and you have to fight him. Tell, then, what weapon do you have by which the war can be won? [48]

INVESTIGATION-INTO-REALITY *(after listening to the king's order, dances, elated)*: Your Majesty, how many words I possess! Even if Desire has plenty of sayings and holds in his hand a bow and arrows of flowers, nevertheless, I – your servant – do not possess any weapon. But I swear by the king that when I find Desire, during war, when I catch sight of him, in that moment – consider it true – I will recognise that destroyer. [49]

I will knock him to the ground, after making him fly around.
Then I will go to find the traces of Desire's splendour. [50]

He has a great strength: woman, and through her he subdued the world, as if with a rope. I know woman well; she is a receptacle of vices, a vessel of dirt. Bone, flesh, marrow, and skin, a vessel of strong stench and impure. With her there is enjoyment, beauty, and pleasure, but she will become like a cloud of darkness. [51]

My strength is a fiery bull, a woman is a moth of delight and passion.
It will not take a long time for him [Desire] to fly, how much time will it take
him to become again bodiless? [52]

KING DISCRIMINATION *(pleased)*: You have to show me that you will win the battle with Desire. He is a fraud, an evil trickster who deceived all the sages. He made Nārada and others stumble and dispelled Śiva's intense absorption. Wine, springtime, a beautiful woman, and the sound of a vīnā; dance, lights, gardens, the voice of peacocks and cuckoos. As many ponds, etc., exist, these all are his aids. Never doubt, otherwise you will fall into his snare. [53]

INVESTIGATION-INTO-REALITY: By your majesty, O king, all the tasks are easy to accomplish. Bringing under control Desire's group, I will take the initiative of reducing them to smoke. Please give me the order now and I shall go, as swift as the wind. After having destroyed Desire, I will return and bow my head at your feet. [54]

The king rejoiced. INVESTIGATION-INTO-REALITY saluted him, bringing his head to the king's feet. [55]

KING DISCRIMINATION *(pondering awhile, then summoning the staff-bearer)*: Call Patience, she shall go to fight Anger. I've ascertained that without her, Anger cannot be defeated.

The staff-bearer left and went to PATIENCE. He told her to hurry since the king had summoned her for the task of defeating ANGER. [56]

PATIENCE *(laughing)*: Defeating Anger is already too many words!
If the king commands, I will destroy him in a second. [57]

Look, where Anger stays, there is obscurity night and day. Even if one person watches with both eyes and the world is manifest in front of his eyes, all the same he cannot perceive anything. Straying, he stumbles here and there. When one is dispirited because nothing is fixed, he falls down. [58]

If I inhabit that place, I will dispel darkness in an instant. The whole world is shining, a cool, beautiful concentrate of joy. All unbearable, evil words, etc., I will send them to hell. I will hit the extreme obscurity of that trickster, but without committing a sin or injury. All of this does not pervade me, I am like a mountain unshaken by the wind. [59]

The waters of lakes, rain, rivers, being collected, go to the ocean,
which is not so preoccupied that some river will destroy it. [60]

I stay absorbed in the ocean of my essence; I do not leave correct behaviour even in dreams. Where I manage to settle, there I will destroy Anger easily. The dust present in a storm – if someone keeps his eyes open, it fills the eyes, then one is in pain and suffers. If one stays there with the eyes closed and the head bent, then the storm will go away by itself. [61]

If a river obtains rain, it makes fall even a high tree.
A creeper stays with its tip bent – it grows, blossoms, and bears fruit. [62]

The actor playing PATIENCE said these words while arriving in the hall. She danced, then saluted the king. [63]

KING DISCRIMINATION: May you be happy, please, take your seat.

PATIENCE: For what task have you summoned me?

KING DISCRIMINATION: Sever Anger's head.

PATIENCE: Just give me the order.

King Discrimination: Kill him or subdue him. [64]

Patience: Thanks to your majesty, I will reduce him to dust immediately.
Just as Durgā killed Mahiṣāsura for the sake of the gods. [65]

King Discrimination: Anger has many weapons and the fool has many words.
Tell, with what weapons will you strike him? [66]

Patience: Forbearance is my shield; a gentle voice is my sharp sword. Even if someone, approaching me, would abuse or assault me unjustly, I would keep carrying out good deeds, resolutely. The actor who will kill Anger has arrived. Silence is my impenetrable armour; my power, lord, is due to your majesty. This is the situation of my weapons, O master: when they are in someone's hand, then no one can win them, their power is valid in the three worlds. Thanks to your majesty, O mighty king, today I will kill Anger as a good hawk kills a partridge.[9] [67]

In the moment I will go and kill Anger, thanks to the power of your majesty,
then Violence, Envy, Pride, and Intoxication, being agitated, will die
spontaneously as well. [68]

Discrimination *(surveying* Patience*'s qualities and might)*: Return with Anger, and may you triumph over all the others. *(He gives her a betel leaf.* Patience *bows her head. He dismisses her.)* [69]

Patience left. Discrimination sent a messenger to fetch Contentment. [70].

Knowledge-of-the-Vedas *(holding a staff, approaches* Contentment*)*: Quickly! The king has summoned you, he wants to send you to fight Greed.

Contentment *(standing up, pleased)*: Greed's so-called baseness has started spreading. There is no other cheat like him. He makes all the world's intellects falter. By his power, people are subjugated: whether they be of low or high status – he does not respect anyone. He praises vain wealth and makes the truth a lie in many ways. Similar to Indra, to a wish-fulfilling tree, to Kubera, Dadhīci left the sacrifice for the sake of Śiva, it is its proof.[10] Greedy people do not look at it in a good way,

9 This exchange between Discrimination and Patience is present in the KPC as well. There, however, the means at Patience's disposal are not made to correspond explicitly to her war equipment. See Kṛṣṇamiśra 2009: 165–166. From a literary point of view, the description of the PcN recalls Rāma's reply to Vibhīṣaṇa in RcM 6.79.2–6. Rāvaṇa's brother, worried, interrogates the Raghu hero about his weapons to defy the ten-headed king of Laṅka. Rāma answers with an outline of his chariot of dharma, where each component of the vehicle corresponds to a virtue. See subsection 3.2.1.
10 This is not entirely clear. I took it as a reference to the fact that Dadhīci left Dakṣa's sacrifice when he saw that Śiva had not been invited.

but, frowning, they say it is dull. They accept Greed's power, intent on some false blame. People taken by Greed's power approach the kings, they wish not only for wealth, but for all the grandeur. They stay in front of the king with folded hands, they do everything he bids – low or high. If gaining wealth is then at risk, first of all they gamble their own body. Every person who wanders, pleading, as beggars, some beg appropriately, others extort. Entreating, the latter person whines – I know this is the wound of Greed. Then, he slanders, troublesome. In this way Greed disrespects people. In this way he makes up multiple disguises and devises multiple means. He makes one waste one's own life by running here and there for money, if there is something to be gained, to be sure of obtaining it. [71]

Where wretched people exist,
subject to Greed, they float and die, like a fish in difficult waters.[11] [72]

In the hearts I inhabit, the light of innate bliss shines. Whatever obtainment occurs with time, one is always happy because of that, however much. When all kinds of obtainment are the cause of happiness, that happiness, always natural, is my banner. Where Greed dwells there is no happiness, even if Kubera's wealth were to be obtained. Those who have this innate nature of mine in them Greed cannot subdue, even its shadow indeed. Thus they do not consider themselves poor; they always consider themselves fulfilled. [73]

Those always content, why should they feel poor?
Neither happy for richness nor unhappy for misfortune, they feel spontaneous extreme joy. [74]

Tell, how can there be joy from Greed, when fulfilment is never obtained? In his company one accepts so much exertion, like a monkey dancing a thousand ways. The enjoyment and pleasures never leave, but taste for them wanes easily. The money gathered, being greedy, stay put, wherever one keeps them. That is why one who does not work – but squanders – whines: "How is that they decrease?" [75]

A man goes from house to house, asking for something, falling at many people's feet.
But if you beg for your gain, Discrimination will suffer. [76]

The many desires of one's heart are just as many sorrows, if you think about it. If you ponder appropriately, you will see that what you gain for the pleasure of the

11 The term used is kruvenī, which I could not attest elsewhere in itself. The word venī means "the flow of a water body or the confluence point of two or more bodies of water" (HSS 4599); the prefix kru- may mean "be rough or raw". See Macdonell 1929: 76. I thus chose to translate it as "difficult waters".

senses is not for your sake. If one goes around with a swollen belly, how can he be someone who is extremely poor, if he does not show his teeth for hunger a second time?[12] One who wants his liberation effortlessly – why doesn't he always take up the bond with Hari? [77]

The multiple fruits and roots that God gives are your food. Take some grass for your seat on the ground; it is needed just enough for the feet to stretch. A pond and a river soothe the heat, sun and fire ward off the cold. Why do you, fool, stubbornly stretch out your hand, base glutton? [78]

Leaving all worldly expectations, set in the heart joy and peace. Leave faith only in this life: live in the world as if you were dead. What happened up to now is past; now may the fluctuations of the mind[13] be directed towards discrimination. Concentrate on knowledge of yourself: to whom do you belong? Who are you? Where are you? [79]

All people are lost, worried, by always looking back or forward. No thinking being was saved in any place – water, ground, sky, or lower regions. Observe one instant and say, whose hope are you looking at? How can you find a meaning in it and prevent a meaningless human birth thus? [80]

You are engrossed in the intoxication of the sense of possession, exerting great effort you built many houses. You stayed dependent on greed, root of sin, the errors of the mind controlled you. You have been cheated, you count your last breaths, just as a partridge about to be overpowered by a hawk. Why don't you worship Nanda's son, who always tends to his devotees? [81]

After his long speech, CONTENTMENT, jubilant, arrived in the hall and danced spontaneously.

(*Approaching the king, he addresses the staff-bearer*) For what task has the king remembered Contentment? He has arrived, so what is the order for him? [82]

The king was delighted and called him close. Bowing his head, CONTENTMENT saluted the king. [83]

KING DISCRIMINATION *(rejoicing)*: Dear, Greed causes pain to the people, go at once and kill him. Without you that wicked man will not be defeated. This task is up to you; I am sure of this. [84]

---

12 To bare one's teeth is also associated with ascetics in the *Pañcatantra*. See Olivelle 2009: 119, v. 48.
13 The sense is to direct the activity of the mind towards viveka. The term employed here is citaceta, which I have translated as cittavṛtti, "fluctuations of the mind", following HSS 1576.

CONTENTMENT: Absolutely, O lord, I want to do it. Up until this moment, the king has not given me any orders. Otherwise, O lord, no matter how many orders you would have given, I would have beheaded him in a second. Now that the king has commanded – please know that it is as if Greed has already been eradicated from the root. [85]

Just as the wise Kānha knocked off the demon Keśī, in that way,
by your majesty, I will kill Greed instantly. [86]

Rejoicing, the king gave him a betel leaf. CONTENTMENT paid homage to the king, touching his feet, and departed. [87]

ASTROLOGER *(arriving in the hall and approaching the king holding a leaf in his hand)*: This is the appropriate time for victory. Depart, after praying to Gaṇeśa. [88]

KING DISCRIMINATION *(at the* ASTROLOGER*'s words, he addresses the herald* GOOD UNDERSTANDING*)*: Give the news to my army, all the soldiers must get ready. [89]

Many troops for the war came and stood to attention before the king. One SOLDIER stepped out.

SOLDIER *(loudly)*: All the army, be ready! King Discrimination has mounted his horse. He wants to hunt Bewilderment. Therefore, all prepare the equipment for war, everyone shall stay with his own unit. Go and stand in the courtyard – take your weapons against Bewilderment! [90]

Balance, Self-Restraint, Observance, Moral Principle, etc., who are all Discrimination's heroes.
Riders, be ready! All powerful, steady in the battle! [91]

The heroes prepared their body armour, seizes their swords, took their shields, seized the power of bows and arrows. Brave and mighty, the actors were delightful and stood tall with great pride. [92]

KING DISCRIMINATION *(to a* CHARIOTEER*)*: Quick, bring a chariot. *(Promptly the* CHARIOTEER *returns with a beautiful chariot.)* [93]

Remembering Hari, DISCRIMINATION rose to his feet, having prayed to Ganeśa. Following the ASTROLOGER's order, he sat upon his chariot. [94]

The flywhisk and parasol shone near the king. At the sound of the conch, the infantry departed. At the call of the heralds, all the riders appeared. Becoming an army, they left. The kettledrums sounded out loudly; a small trumpet rang out. There were plenty of cries and, gathering at the battlefield, the soldiers sang. They told each other stories about the sentiment of war, which increased their enthusiasm. [95]

At that moment DISPASSION entered in the company of the helper RELINQUISHMENT, and approaching, they saluted the king, who enquired of their identity. [96]

DISPASSION: We are Dispassion and Relinquishment, two riders.
Bond-with-Hari, calling us to help, sent us to you. [97]

KING DISCRIMINATION *(happy, embraces them respectfully)*: I had heard about the greatness of Bond-with-Hari, which increased by her being merciful to me. If Bond-with-Hari is my supporter, then the burden of Bewilderment's faction will not fall upon me. Now I have the conviction that my victory is doubtless. *(He rejoices again and leads the army forward.)* [98]

DISCRIMINATION's army went to where Kāśī was set in the hall.

CHARIOTEER *(to the king)*: Great king, this is the city of Kāśī. Pristine, a dagger which removes sin. Here live many knowing men that are at your orders. Bewilderment now wants to stay here: he wants to impose his rule. He entrusted the place to Deceit at first, who caused much harm. Even the people who stay home all live in fear of Deceit. [99]

KING DISCRIMINATION: If we have the support of Bond-with-Hari,
then, today, all the wicked men will earn their punishment. [100]

They approached a temple close to the riverbank, a huge shining and beautiful construction. At the door, a group of men stood, with gleaming bodies.

(To the CHARIOTEER*)*: What is this place, an abode of joy, indeed? [101]

CHARIOTEER: Great king, the venerable god Bindu Mādhava is shining. With his vision all sins are removed, the body is pure. Feeling a delightful bond with the god, touching the feet of the deity, going in the water, a man serves the deity. Those who, having faith, remember the name of the god, obtain the fulfilment of their wishes.

Listening to these words, DISCRIMINATION rose from the chariot and, diving into the waters, swam to the temple. Rejoicing, he touched the beautiful feet of Śrī Bindu Mādhava.

KING DISCRIMINATION *(worshipping with love)*: Please grant me the boon by which the war against Bewilderment will be won. [102]

DISCRIMINATION worshipped Mādhava in this way, then returned and mounted the chariot. A SPY entered.

SPY: Hearing about the arrival of the great king's army, Bewilderment was taken by agitation. Thus, please know that they have fled. Please, seize this occasion. [103]

KING DISCRIMINATION *(to the charioteer)*: Advance, hurry!
I must know if Bewilderment, cause of suffering, fleeing, is still alive. [104]

DISCRIMINATION exited Kīratibrahma's hall. The FIRST ACTOR and his troupe stood and began to dance. [105]

Thus concludes Act IV, entitled "Description of Discrimination's Army", of the "Drama of the Rise of the Wisdom Moon" in Brajbhāṣā.

ACT V

Kṛṣṇadās Bhaṭṭ: In this way this scene of the drama proved Discrimination's strategy.

After that, the First Actor arrived and addressed King Kīratibrahma. [1]

First Actor: The venerable Bond-with-Hari had sent Faith in help with a message.

Therefore, Faith stayed in the good company of Discrimination at the time of war. [2]

Faith came to report to Bond-with-Hari information

about how many fighting soldiers were in Bewilderment's service during the war. [3]

Even if Desire, etc., were Discrimination's enemies,

all the same Faith felt sadness, because of the destruction of family relations. [4]

Kṛṣṇadās's Disciple *(keeping his head on the ground)*: For what reason, my master,

did the actors prepare as soldiers? [5]

Kṛṣṇadās Bhaṭṭ: Against the actors who, having prepared themselves, fought the war, there were many other actors.

Murders take place, that is why Discrimination talked about war. [6]

Meanwhile, the actor impersonating Faith entered the hall, sobbing. [7]

Faith: God! The deeds I saw cannot be described. The lineage of Engagement-with-Action was destroyed in one second. Beautiful greenery, bushes, lovely flowers, but then winter affects even the root. Even if Desire, etc., are your enemies, all the same the love between father and son is not erased. That beauty, churning the heart, is extremely savoury. When the eyes see it, they fill with tears.[1] Anger, Greed, Intoxication, and Envy, wicked, were all ready. Fighting, they committed many actions in the war today. Seeing their beauty, majesty, and qualities, how could there be calm in the heart, full of suffering? One does not forget them, but affection for them is forgotten. Affection is a major cause of sorrow in the world. [8]

If it is observed through reason, then nothing is permanent.

The world is produced and destroyed like a water drop. [9]

---

1 jala lona, literally "water-salt", which I have translated as "tears".

https://doi.org/10.1515/9783111063836-014

Whatever one may desire, indeed, goes away, moves towards an end; what does affection fulfil? The body is destroyed as well. The sun, the moon, the stars, the gods, and the anti-gods, the seven seas, the earth also, and the sky – please, have trust – they will be destroyed too. From Brahmā to insects, all appear to be perishable; if one is dependent on his own sense of ego then he, indeed, will not live. With whom to accept a relationship, with whom to be kind or unkind? See, if there is sorrow, please verify who feels it. [10]

In this way I consider in my heart that the world is a false measure of reality.
All the same, thinking about Bewilderment's group's qualities, my heart does not find peace. [11]

Bond-with-Hari told me: "Go and tell Discrimination to fight Bewilderment tenaciously; give this message promptly." [12]

When Bond-with-Hari bid me farewell, the venerable leader added: "From this moment I will live in my Kāśī, where Violence is manifest now. The Auspicious Shrine of the Wheel, a propitious name, is a pure, unique place. I will reside there with Peace." Thus, she gave me this indication. [13]

Therefore, I am going to the abode of joy where Bond-with-Hari stays.
Without her vision I find no solace. [14]

Saying such words FAITH entered the hall. Then the FIRST ACTOR set in place the Auspicious Wheel. [15]

Calling the actors playing BOND-WITH-VIṢṆU and PEACE, he invited them sit. The two started conversing between themselves. [16]

PEACE *(holding* BOND-WITH-HARI*'s feet, sweetly)*: You are an ocean of compassion, O master Bond-with-Hari. For the people who take refuge at your feet the ocean of suffering is destroyed. You are always filled with bliss, a receptacle of joy, consisting of joy, bestower of good fortune. Why are you gripped by agitation today? Tell, O merciful one, what is the task to be carried out? [17]

BOND-WITH-HARI *(affectionately)*: Listen, O delicate daughter, agitation never touches me. However, I'm always preoccupied with the truth-knowers. The kings Discrimination and Bewilderment are great enemies. I came to know that a ferocious battle has started between them. I am worried about who will win. [18]

PEACE: Your mercy is beneficial to wretched people. Whosoever, in charge, obtains your help, there will never be defeat for him, I'm sure of this in my heart. The glory of victory is always with King Discrimination. Why is your heart agitated? [19]

BOND-WITH-HARI: Peace, what you say is true. However, whoever has a loving heart wants the dear ones to be good, immediately. This innate affection is eternal. I sent Faith with such task to Discrimination, but she has not returned until now. [20]

FAITH *(approaching* BOND-WITH-HARI *and touching her feet)*: Congratulations for the victory! [21]

BOND-WITH-HARI *(rejoicing and with a sweet and lovely voice)*: Tell me if the clever Discrimination has obtained victory. You, as well, after going there, were you fine? Among the armies of both kings, did you gain or are you in pain?

FAITH: O lady, thanks to your favour all places now are prosperous. I obtain extreme joy now by seeing your feet, the hook of all sufferings is destroyed. Discrimination obtained good fortune in victory only when you looked at him with mercy.

PEACE rejoiced at seeing her mother and, running, touched her feet. FAITH embraced her and sat next to BOND-WITH-HARI contently. [22]

BOND-WITH-HARI *(joyful)*: Tell me about Bewilderment.
Was he killed or did he escape? [23]

FAITH: Those opposed to you, O compassionate, are unfortunate.
They are like living corpses, they are all in death's clutches. [24]

When you left to come here [Kāśī], you bid me farewell and ordered me to go to the king's feet. As soon as I referred your order to Discrimination, he prepared a large army. Before dawn the nagāḍās sounded, his warriors were all there by early morning. When the sun rose, he departed with the army, a mass of strength. Opposite on the battlefield, Bewilderment marched with his own powerful troops. Approaching from the two sides, the armies stood still. The martial nagāḍās sounded on both sides, heroic warriors, heroes, marched forward with zealous hearts. The śaṅkha, the śahnāi, the bherī, and the nafīrī sounded; listening, the brilliance of the light became unstable.[2] The bards sung the raga mārū to the warriors. They sang the praises of the warriors, whom the war songs fortified. From both sides such enmity became intense, those alive at that time became deaf. [25]

First of all, King Discrimination, making of Science-of-Reasoning a messenger,
sent him by Bewilderment, with this message. [26]

FAITH recalled the scene.

2 [. . .] | bhaī kādarana kī sunata jyoti pīrī | |. This sentence is not clear, in particular the meaning of kādarana, which I interpreted as derived from the adjective kādara (HSS 898).

KING DISCRIMINATION: This is the city of salvation, the auspicious Kāśī. The ascetics living here now are all suffering because of you. Now, first, leave them to me. Then, as many other places of merit exist, where virtuous people live, do not attack them, for they belong to me. Stay away from the auspicious sites. The men who are not aware of the bond with Hari and the teacher are wicked and ignorant. Where there is my support, stay away from the city of their hearts. [27]

Otherwise today, on the battleground, your head – with the sword
    I will cut it separate from its possessor, as a jackal eats a dog. [28]
And the pride of your army for their body,
    it will vanish with a look from my warriors. [29]

SCIENCE-OF-REASONING approached BEWILDERMENT and relayed DISCRIMINATION's order to all.

KING BEWILDERMENT *(upon hearing it, with arrogance and anger jumped up, standing)*: Go away, messenger, for I am not killing you here and now. What should I do? Killing a messenger does not befit me. Go and, please, convey to everyone the answer to one issue: a work is always to be accomplished with a dagger, never with words. If the discussion about battle has escalated, then, how can it be subdued with a lie?[3] [30]

While talking to SCIENCE-OF-REASONING, BEWILDERMENT looked at his own army. He summoned the HERETICAL DOCTRINES and the MATERIALIST. [31]

Make this messenger understand my power.
    Drag him out of this army and tell him to go to Discrimination. [32]

FAITH: Upon being told, Science-of-Reasoning returned to his own king.
    He explained everything King Bewilderment told him. [33]

After that, King Bewilderment then, with his companions,
    marched, then stopped: he had come because of war. [34]

On this side, all of the great King Discrimination's warriors became alert.
    From the two directions cannonballs and arrows were fired. [35]

Then Bewilderment's warriors came out: Materialist, Sophistries, the heretical Jaina Scriptures, and the others, in whose doctrines hell is found. Here, from the part of Discrimination, the Mahābhārata and Science-of-Reasoning advanced. There were

3 He would lie by saying that he will leave.

the Traditional Texts, Bhāskara,[4] and the Six Philosophies, starting with Hermeneutics. They met there and the dialogue among them began. Equipped with weapons, from both directions came voices invoking Hari. [36]

PEACE: O mother, this is truly astonishing. These Philosophies, Traditional Texts, etc. – it is accepted that they are always in reciprocal opposition. One describes one god, the other worships a different god. They stay separated from each other, these doctrines were all in sin. Please, resolve this doubt. Please, clear up my uncertainty. [37]

FAITH: O daughter, crux of my life, you should not have any doubt in this matter. These Six Philosophies that are sung as distinct, all are born from the one Veda. If some enemy arrives from somewhere else, opposing their father, then all surely want to face and fight against him for their father. Even if they keep fighting each other, still, for the sake of their father, they all want to unite. Moreover, the philosophies which stay separate and say their doctrines differ: some say they are concerned primarily with ritual actions, some sing fate as superior. Traditional texts and philosophies describe worship of the sun, some maintain worship of Hari as superior. Consider them all as brahman's forms. No doubt should arise in the heart regarding this. If someone worships another god, know that in brahman there is the cause of other forms. Thus these are all worship of brahman, there is no other god. [38]

Moreover, know that who worships gods different from brahman,
those sinners are opposed to Hari's feet, they are dim-witted, very ignorant. [39]

And those who worship gods not different from brahman in their hearts,
they attain brahman, and I will describe such happiness.[5] [40]

If someone wants to see the ocean, he plunges into some river. He floats in that thus, and undoubtedly reaches the ocean.[6] [41]

In the same way is Hari, receptacle of bliss, the ocean that is brahman, unmanifest, imperishable. The living beings who worship him with a form, they know its manifest image. Those who worship him without egoistic desire, they obtain salvation,

4 The eighth/ninth-century Bhedābheda Vedānta philosopher, author of a commentary on the BS. See the concluding chapter of this book.

5 See BhG 9.23.

6 The same image, of waters coming from distinct rivers but which reach the one ocean, was employed by Madhusūdana Sarasvatī in his *Prasthānabheda*. Its meaning, however, has been interpreted in differing ways. See Hanneder 1999.

mine of all happiness. When they venerate the single brahman, being affectionate, they obtain Hari's feet, very fortunate. [42]

In such way the points of view of the philosophies appear to be distinct,
but they are all precisely one – be aware of this, Peace. [43]

BOND-WITH-VIṢṆU: The warriors of the two sides, tell,
who advanced first, who was brave in the fight? [44]

FAITH: One day, once, this happened: bringing the swords, spears, and daggers, the two armies became one. Weapons were used, here and there were cries of "Strike, strike!". The mighty warriors shouted and threatened, one or the other heroes advanced. Someone fell to the ground, someone fought to kill, brandishing his sword. Someone, taking shelter behind his shield, was saved and fired his arrows at the other. Heads started falling to the ground, from the headless bodies a flowing river of blood. A fine rain of blood was released, red as colouring powder. Imagine bloody soldiers, someone without a hand, someone else without a foot. They were as a bouquet of red palāśa flowers and the war was the forest. Some soldiers fell in line in the mud, and others, overpowered by death, did not move anymore. Someone, falling in the flow of the bloody river, sunk in it. Imagine that no one distinguished the enemies or his own companions, all with one. Not understanding anything, at that time the warriors were agitated. [45]

In one moment the war became infinitely dreadful:
Desire, among Bewilderment's numerous soldiers, arrived. [46]

Then, all the heretics, meeting, gathered in one place.
Forming a circle, they advanced towards the teacher of the Vedas. [47]

Then the teacher of the Vedas, very powerful, stood up, roaring in their direction. Upon hearing, the heretics fell to the ground trembling, all confused. Some was breathless, some dropped their weapons. One drowned in the Ganges, another stood and fled. Seeing this, the heretic Jaina, nervous, immediately ran away. With some trick he misled all the people here. Then Śaṅkara, the teacher, hurried and, in a second, dispelling the trick, destroyed the Jaina group. [48]

Some dead, sunk, others fled south.
No trace was left of them, they vanished instantly. [49]

Just as a strong wind carries off dried-up leaves,
in that way all the heretics were destroyed at once. [50]

Listening to his aims of human life, which are the best, BOND-WITH-VIṢṆU praised ŚAṄKARA. [51]

BOND-WITH-VIṢṆU *(to* FAITH*)*: And how was the war? Tell me in detail. [52]

FAITH: Then the Sophistries and Veda-Deniers stood, Bewilderment's warriors.

From Discrimination's side, Science-of-Reasoning and Hermeneutics, united, went to face them, brave. [53]

An extraordinary battle took place between them, the two sides fought fiercely. Taking the weapons in their hands in order to hurl them, the two factions called: "Kill!". They advanced with frightening swords and spears. Dreadful iron bars and sticks; hammers, scaling ladders, and axes; arrows, discs, and swords. If a weapon were to hit someone, he would have been reduced to pieces. Someone hit a mortal spot and blocked a blow with their shield, taking advantage of the situation. In this way a day of violent confrontation came to an end. Science-of-Reasoning, extremely brave, killed many excellent warriors with arrows. Their bodies motionless, they did not utter a word. The bodies of each and every person were bloody; they could not be brave anymore. [54]

Weak, agitated, and exhausted, they were no longer courageous.
Then Sophistries and Veda-Deniers, the extreme cowards, fled. [55]

Even if Bewilderment sent them back, many did not return to face us.
We could not find out where the people went, the search for them all is even now ongoing. [56]

FAITH, again, recalled the scene.

KING BEWILDERMENT *(angry, summons* DESIRE*)*: You are looking at someone whose soldiers have all fled. Your bow, arrows, and wife – will they ever be of use again? Will you go and launch an attack for which Discrimination will flee? [57]

DESIRE *(angered but bows his head)*: Lord, look today at the service your servant pays you.

FAITH *(resuming)*: He gathered all of his soldiers, putting forward Spring. Blooming trees, a lotus garden blossoms, as if they came out of Desire's quiver. A bee stayed upon them, longing after their nectar, brilliant like the dark notches of his arrows. Everywhere cuckoos sang, as heralds again prepared the army. In the sky the voices of the birds rang out, the sound spread, as if the noise of the army had increased. Dust rose into the air, the wind shook violently, as if it had risen in the four directions. The grace of a beautiful woman was like the radiance of a lightning, like the sparkle of Desire's sword. In this way Desire, enraged, ran. He arrived in front of

King Discrimination. Twanging his bow, a loud sound echoed. Upon hearing it Discrimination's army trembled. [58]

The Mind-stirring Desire, knower of magic formulas, shot five arrows.
Of the five released arrows, many in the enemy army fell. [59]

Several arrows rained; everyone was frightened. When the shower of arrows started, all lost consciousness. They could not stay calm, agitated by the pain caused by the arrows. The good heroes moaned and swayed. Someone, throwing their weapons, exclaimed, "Today we are defeated." Assembling together, they fled, overwhelmed by profound fear. Balance and the others hid. Being subdued, they retreated. If someone looked calm, in himself he was afraid. [60]

Investigation-into-Reality saw that Discrimination's army had become divided.
Previously brave faces now became panicked, seeing the fearsome rain of arrows from Desire's bow. [61]

Then, furious in his heart, the great, brave and powerful Investigation-into-Reality acted. Without any weapon, without companions, just like Hanumān with his diamond-like body, he came before Desire, threatening him. Springing, he leaped on the intoxicating Desire. Even if Desire, the god of loving recollection, hurled many arrows, they did not hit the mark, because Investigation-into-Reality shunned them. Destroying Desire, the intoxicator, and the whole of his army, he reduced them all to dust in a second. Leaping forward, he seized Desire and, lifting him, Investigation-into-Reality knocked him to the ground. He was unconscious for some time, then became bodiless at once. [62]

Thus Investigation-into-Reality killed the visible Desire,
and the army raised a cry of "Hail to Discrimination! Hail to Discrimination!" [63]

As soon as he heard the cries, Anger, outraged, moved, having assumed a huge form, resembling many fires. He took with him Violence, powerful and fearful; he held various weapons, as if a fire had taken him. Swooping, like fire, he ran straight at the opposing army for the attack. Arriving at Discrimination's army with great clamour, he stopped. There was the night of the day, a vast obscurity, one could not locate his own stretched hand. When one is not able to perceive, fear hits. How many were burned in Anger's fire! Worried, they all said: "How do we stay alive?" [64]

Then Patience, like a deep pond of cool water,
a flow of gentle words, directed herself towards Anger and fell upon him. [65]

Extinguishing him immediately, she blackened his face.
Then she took him away, until no trace was left. [66]

Then Greed assumed a powerful form, together with the formidable Thirst. He stretched his body out like a mountain cave, his stomach like a tank full of water. His long arms were like huge snakes; dreadful, he was like Kumbhakarṇa, Rāvaṇa's brother. He came in front of the army spreading his arms, enfolding everyone at once. It was clear at that time that Greed had become death for all.[7] [67]

Then, your majesty, Contentment, joining the arrows of satisfaction and eagerness,
piercing Greed's face with the bow, killed him easily. [68]

He killed him with only one arrow, and Greed breathed his last. After that, Envy arrived and was seized by Generosity. Imprisoning him, Generosity brought him in the presence of the king. The news spread in Bewilderment's faction: Desire, Anger, and Greed were defeated, and Envy was ungraciously imprisoned. Upon hearing this, Pride and Intoxication died, and fear pervaded the hearts of the rest of the army. The women, Passion, Violence, Thirst, etc., having prepared funeral pyres, burned themselves together with their husbands. Those soldiers who were scared fled secretly, like clouds disappearing unexpectedly.

Upon seeing all this, Great King Bewilderment said to himself: "All the warriors of the family are defeated; I am without an army." He was afraid, terrorised. [69]

And other people, such as those performing magic tricks,[8] were scared by the arrows.
He [Bewilderment] ran away with them; he could not be brave anymore. [70]

Now everyone is looking for where, indeed, he went into hiding.
Even though they have kept searching for some time, the king's people have not
found any trace of him. [71]

King Discrimination obtained victory, thanks to your favour.
He celebrated by striking the nagāḍā, then blowing a large conch. [72]

BOND-WITH-HARI: He did not do good, letting Bewilderment live. While there is life in Bewilderment's body, there is no ease of mind. If a leader is saved in war, it is not difficult for his army to assemble once again. Gathering together an army once more, in little time the enemy will return and, united, will not withdraw a second time. [73]

---

7 samujhi parī aisī tehi kālā | bhayo lobha aba saba ko kālā ||. Brajvāsīdās plays here with the double meaning of kāla as time/death.

8 The powers of the "false" ascetics/yogis are defined as ceṭaka nāṭaka, a display of illusion or magic here and, previously, in 2.13.

Tell, Faith, presently,
after having heard about Bewilderment's defeat, what is Mind's condition? [74]

FAITH: Alas, bestower of joy! Mind's condition cannot be described. His first wife, Engagement-with-Action, was his favourite – he felt most love for her. Once she heard that Desire and the others had been vanquished, she died because of her love for them. Thus Mind's dejection has increased. Now Mind wants to die as well! [75]

BOND-WITH-HARI: What is better than Mind dying?
All tasks then would be accomplished easily, without my effort. [76]

And Mind, indeed, released from extreme misery, would experience his true form.
He would be absorbed in brahman, that is joy, peerless in all respects. [77]

Is there such a means, Faith, tell,
by which Mind, abandoning the body, will reach a state of extinction? [78]

FAITH: If Upaniṣad came close to Discrimination and unite with him lovingly, then their son Wisdom would be generated from her womb. As soon as he is born, Mind will abandon the condition of the body. [79]

BOND-WITH-HARI *(mercifully)*: Now Mind is suffering, let's send Sarasvatī to make him understand. In such way the suffering will disappear, and he will find some courage in his heart. [80]

SARASVATĪ *(to* BOND-WITH-HARI*)*: Your command is pleasing.
After bowing, SARASVATĪ stood up to receive BOND-WITH-HARI's welcome.[9] [81]

The actor left the hall and went where MIND was staying in the company of two servants, INTENTION and DOUBT. [82]

MIND *(agitated, almost unconscious. He wails, cries, and calls his family's names. Restless, he falls to the ground)*: Where are all my sons? O Desire, Anger, and Greed! I feel distress for Envy, Intoxication, and Pride. Should not they meet their father today? Where did they go, leaving their old father? Throwing me into an ocean of pain, you went to the other bank, but your father is sinking. Alas! Passion, Desire's dear bride, and the other women: Thirst, Violence, etc. Where did they go, together with their husbands? They left me without protection. I am deprived of support in my old age, today the walking stick has fallen from my hand. Grasping her sons'

9 I have amended suvāgasa to suvāgata (Skt. svāgata).

fingers, the good woman Engagement went away, she left the house, deprived of them. Whom shall I make my support? *(Saying this, he fell unconscious.)* [83]

INTENTION and DOUBT lifted him up and tried to convince him with words and reflections. [84]

INTENTION: King Mind, you are wise. Why do you lament like an ignorant? Son, wife, who are they? Whose are they, O lord? This is but temporary company.[10] Just as the grass which flows in water, O wise one, it gathers somewhere for some time. Then such a current arrives and sweeps away as much grass as was there. Such is all a play of the Creator, why, indeed, does he cause separation and somewhere else a suitable union? Who depends on whom? This is all in the Creator's hands. Therefore, discard this sharp pain. Take refuge in Hari, please be calm. [85]

Wife and relatives are endowed with bodies: a bond of affection, for whose sake Mind, his intellect constrained, laments. Consider just the body: when the body will be destroyed in one second, upon destruction one becomes bodiless. Then where can you find the possessors of the bodies? Consider a house, some courtyard or a road – if someone moves close to a funeral pyre, then he is gone [i.e. is dead]. This is, indeed, the concern of those attached to a temporary existence, but, in the end, who is your companion? King Rāma.[11] [86]

MIND *(looking at* INTENTION *and* DOUBT*)*: Tell, where is the beautiful-eyed Engagement-with-Action, my delightful companion? [87]

INTENTION: Listen, O king, Engagement suffered a significant grief. How many were born from her womb, were all destroyed. The fire of pain was too consuming for her. She, too delicate, could not bear the suffering. Listening to affection, she abandoned her own body; now forsake, O king, any love for Engagement-with-Action.

Hearing this, MIND became agitated and, swooning, again fell to the ground.

---

10 [. . .] | yaha saba hai dasa dina ko sātha | |. Literally, "a company of just ten days"; a similar expression also appears in 6.86.

11 [. . .] anta ke samaya ko tero saṅgī rāmarāya hai | |. This half can also be translated in two other ways: the first is: "[but], at the end, is king Rāma your companion?"; the second is to translate the sentence not as a question but to treat ko as a postposition: "[but], in the end, King Rāma is your companion". I chose an intermediate version: on the one hand, I retained the interrogative aspect, in agreement with the tone of verse 86, whose objective is to unsettle Mind's convictions; on the other, I kept the affirmative aspect with "King Rāma" as a reply, highlighting the potential higher goal of one's existence.

MIND *(lamenting)*: Alas, Engagement-with-Action, you left me too! Now my life has unsteady foundation. There is no more serious trouble than this. Please, come and make a pyre of wood. After kindling it, place me on it. Without Engagement-with-Action, for me living is just useless . . . Hurry! There is no other way to stop suffering. [88]

MIND continued to mourn and wail. SARASVATĪ, sent by BOND-WITH-VIṢṆU, arrived. [89]

SARASVATĪ *(approaching)*: Son, after seeing you today, I have manifested here. Why do you feel such suffering, anguish, and crying today? You understand everything, dear, in the world about who and whose are your loved ones. God gave you clear thought, through your intellect. You have ascertained the nature of the world in all ways, by studying and pondering: what originates, assuming a body, is destroyed necessarily – be it small or big: Brahmā, Viṣṇu, Śiva, the gods, demons, sentient and insentient beings. All you can perceive with your eyes will be destroyed. One should be aware of one's own duty – why are you vainly sad? [90]

MIND: O mother, your words are true.
However, my heart is confused because of the great grief, that is why I cannot understand. [91]

SARASVATĪ: O son, this grief that comes to your heart has become intense.
Its cause is the excess of affection, know this. [92]

If you want affliction not to pervade you in the world,
then never feel affection for anybody, be removed from sorrow and love. [93]

Just as too much food is undigested, causing pain,
in that way, excess of affection is painful for you, O Mind. [94]

Love for all: mother, father, siblings, children, wife, servants, and disciples. You stayed with them, considering it love, always wandering together, like a swarm of bees. You suffer because of their sorrows – this is but affection for self-concern. Alive, they made you burn, dear, and they will make you burn even when they are dead. [95]

MIND: Receptacle of mercy, mother, all you say is true. Even if it is as you say, my desire for Engagement-with-Action does not vanish. By the fire of grief for the separation from her, my body is reduced to ashes, O merciful. You did good by coming, being compassionate, so that I could see your feet at the time of death. [96]

SARASVATĪ: Oh all! Why do you accept sorrow and affliction?
Being greatly distressed, how can the heart become calm, endowed with respect and compassion? [97]

Those you keep thinking yours – your children, your wife – are full of the sorrow of affection. The body is scorched by the fire of separation from them. If you reflect properly about them, you should not love them. From now on think that those you protected, being defeated, have died. Desire will not come by you. Now you are extremely grieved for their loss. If someone arrived, would that make you happy? [98]

Tell, from where did you yourself, and those relations of yours, come? Where have they all gone, and you, dying, where and how will you go? You dance, as in a puppet show, at the hands of the puppeteer, through the strings of the world. What you stubbornly call yours, how can it be really yours? Look at it with reason, how can you lament? [99]

MIND: Mother, even if what you say is true, and I consider it true, nevertheless all my dear ones, I protected them. I made all grow from children into adults, no? They were never apart from me. Meeting with them, going around with them, their' conversations, their looks and tricks, their sport and frolics! Now these loved ones do not slip away from my heart, but my heart[12] slips away at every moment. [100]

SARASVATĪ: If you, dear, have stubbornly accepted them as your kin,
they, your own, make your body suffer. [101]

As if you are nourishing a parrot, you protect it, being egoistic. If in your house there were a mouse as well, you would not be egoistical. If you ate the parrot, it would be dead, then you would not be pervaded by sorrow, O king. On the contrary, if a cat took away the parrot, then your pain would increase. Your egoism causes sorrow to you. The parrot and the mouse are the same, both will perish. [102]

If you forsake "I" and "mine", you will obliterate the self.[13]
Then suffering will not fill you, looking at all this as if a dream. [103]

You may say, "My sons are born from my body,
why are they compared to a parrot and a mouse?" [104]

In that case, sons are always born from the body, but a bedbug also is born from one's body.
One dandles a son, taking him on the knees, but becomes an enemy to the bedbug. [105]

---

12 The word employed is karejā, that is MSH kalejā (HSS 823).

13 Intended as the individual self as opposed to the Supreme Self.

Dear, consider brother, father, and mother, an enemy and a friend, son and commensal group. Consider yourself and another, consider man and woman, consider sorrow and happiness, day and night. Consider hell and heaven, sin and merit, loss and gain, they appear distinct. The world is entirely a lie, know it as the sun's rays[14] and accept the truth that all are one thing. [106]

MIND: Mother, please, now tell me about some means,

by which my heart, bound by the chains of affection, can become free from them. [107]

SARASVATĪ: O son, I tell you an accessible way, if you listen to it with the ears of intellect and concentrate. If you wish to sever the chains of affection, practise it and from then on, sorrow will not engulf you, for that way will always be joyful. First of all, leave behind your taste for the world: know the taste for worldly pleasure to be dull. It is all a deceiving decoration, like a sorceress's trick. Look at your whole family, ponder in your heart: they existed and where have they gone, then? They were not good acquaintances. Forsaking the way of affection for the deceiving ones, practise the way of affection for the truthful ones. Keep always contemplating a single essence, focus on it. Set in your heart this means, practise it, then you will be released from the noose of affection. You will not feel sorrow anymore, you will be a receptacle of bliss. [108]

MIND: Merciful mother, I feel some joy listening to your words. Pierced by the arrows of anguish, my heart was a burden. Now, thanks to your mercy, they are removed. Nevertheless, the bruise of the wound still hurts. Mother, now explain the effort for it: please, tell me that remedy by which all this suffering will dissolve, concentrating on it in my heart. [109]

SARASVATĪ: O son, I am going to explain to you an extremely beneficial cure:

the remedy of indifference for dear ones, removing the suffering of care at once. [110]

MIND: O mother, what you say about indifference as a beneficial remedy. Even I know this fact, that indifference procures much happiness. However, even if I treasure indifference in my heart, and I try to discard care in many ways, all the same the affliction of care is too strong: neither it exits the heart by itself nor am I able to discard it. [111]

---

14 The sun's rays are false if considered as independent from the sun; they are ultimately the sun. See subsection 4.1.2.

SARASVATĪ: The cause is the absence of Peace, dear. Because of it, you are not able to extirpate care from your heart. If you will make Peace your own, then you will extirpate care easily. [112]

MIND *(joining his hands)*: O mother, I request you humbly: tell, who is Peace? How should I find her? Describe her own form, tell me her characteristics and the means for knowing her. Now, quickly, tell me the means for which the affliction of care will leave me. [113]

SARASVATĪ: Listen, son, this is a highly secret matter.
It cannot be told to whomever without entitlement, dear. [114]

However, if someone distressed, deeply anguished, seeks the truth,
this secret is revealed to him and upon hearing it, he will find joy. [115]

MIND: I'm begging you, mother. There is no one more anguished than me, living in the world. Tell me the truth, I humbly request you, holding your feet. Now, grant me your support completely. Please tell, at once, after thinking about it, the means for which the burden of the pain of care will vanish. [116]

SARASVATĪ: Dear, if you would understand this matter, by knowing it, there would not be anyone more fortunate than you in the world. I will enlighten for you Peace's form. Listening to it, all sufferings indeed, like care and so on, will vanish. She is always absorbed in one blissful essence; joy, dismay, etc., never pervade her. Cold and hot, fear and error, all flee frightened as soon as Peace's happiness enters someone's heart. [117]

Peace comes to your heart when you concentrate on brahman. He is one: existence, consciousness, and bliss. Whose small part is the infinite egg of Brahmā [i.e. the universe]. Always a receptacle of bliss, of a single essence. Supreme light, imperceptible,[15] indestructible. Make the inner of your heart his abode, keep thinking about him. Spend your time in this way. Dear, with the quiet Peace arisen, you will obtain happiness. I will tell you more, listen, dear. A highly secret but extremely accessible good teaching. [118]

Brahman, pervading, a single essence, without distinguishing marks, beginningless, formless.
He has two forms: without qualities and with qualities – he is invisible, unfathomable, unique. [119]

---

15 alakha (Skt. alakṣya), "without marks", that is, imperceptible. It refers to the formless Absolute, used equally by the Nāth Yogis and Sants. See Thiel-Horstmann 1983: 70–71 (poem 8, 35, line 5). See also Schomer 1987: 70–71.

Unqualified, formless, faultless, he, indeed, out of the strength of the love [for/of] those bound to him, is always qualified. He does not enter the heart without form, unqualified. Therefore, one concentrates on the feet of the qualified form. I will tell you a description of that beautiful form. Having known his pleasing form, concentrate on it in the abode of your heart. How many autumn moons feel ashamed looking at his face? The body shines like a lotus, a blue sapphire, a cloud. The splendid beauty of the nose, lower lip, teeth, and chin. Who would not envy that sweet smile? The eyes attractive as fresh red lotus petals, the glance beautiful – at how much would it be sold? The auspicious mark on the forehead and the beauty of the eyebrows are charming. When one looks at this sweetest beauty, one desires it. Just as a snake-woman, he has curly, dark locks of hair. Devoid of the poison of the self, holding ambrosia, lovely. When his vision bites someone, he becomes immortal. If one shall make a comparison with him – how can someone else be similar? The pearl earrings at his ears and the crown on his head sparkle. Who can describe the magnificence in a gathering? A garland of forest flowers shines on the neck, the kaustubha jewel is radiant on the chest. The great beauty of the śrivatsa mark on the heart. The two arms beautiful and strong, the adornment of jewels is perfect and divine. [120]

A yellow band around the waist, the belly covered by a garment, attractive and lovely.
The beautiful navel is like a swirl on a boundless ocean of beauty. [121]

Beautiful lotus-feet, auspicious, cool, removing the threefold suffering,
pearls filled with pollen – dear Mind, hover on them like a black bee! [122][16]

He assumes such a form for the sake of the bond with him, sporting in many ways. Sometimes, holding the flute at his lips, he plays it: a mass of bliss who, taking pleasure, produces the bliss of joy. Sometimes he holds in his hand bow and arrows, killing the enemies, removes their weight from the earth. Sometimes he assumes four arms, extremely beautiful, in his hands the conch, the wheel, the mace, and the lotus are radiant. Such a god with a form with qualities, serve him with love, in many ways. Concentrate on his feet without egoistic desire, be pleased with a vision without other fruits. [123]

By worship of the form of brahman with qualities in nine ways,
the form without qualities, incomparable, then naturally will enter your heart. [124]

16 This head-to-toe description of Viṣṇu/Kṛṣṇa in verses 120–122 adaptively reuses one of Rāma in the RcM. For an analysis and comparison, see subsection 4.1.1.

When Peace will manifest, then any suffering due to false worldly existence will not pervade you.
With birth and death destroyed; all afflictions will vanish. [125]

Listening to this teaching, MIND was liberated from the cycle of worldly existence. His distress removed, his afflictions burned. He grasped SARASVATĪ's feet. [126]

MIND *(paying homage and joining his hands)*: O mother! By your mercy now I have a protector. [127]

Moving, in the ocean of worldly existence, I was absorbed in it, a burden of suffering.
Making me mount the vessel of your words, you, mother, made me cross that ocean. [128]

Mother, thanks to your mercy, grief is gone. Thanks to the blessing of your feet, now I have become happy. Now, merciful, please tell again by what means error will not occur anew. Suffering and anguish will not pervade me again. Considering such a state of error, my heart trembles.

As MIND became thus humble, SARASVATĪ rejoiced. [129]

SARASVATĪ: Now, son, if this teaching of mine seemed good to you,
then, by abandoning inclination for worldly pleasures, the heart has become a good place. [130]

DISPASSION arrived, extremely elated.

DISPASSION: Four heads do possess four intelligences. See how he [Brahmā] made the beautiful human body:[17] mixing bones, flesh, veins, blood, pus, and marrow. He made different qualities in each limb for thinking, moving, talking. He covered the body with skin with truly wonderful colours. If there were not the skin of the body, the flies would eat all the flesh, until what point could one, then, frighten away kite and crow, that will take its pieces away? Through that body one relishes sensual enjoyment, a rope trapping one in this body, and from which man never enjoys freedom, even for a second. [131]

One must see that everyone, indeed, is caught in the pleasure of this body.
Making one's own the condition of the body, one sees but is not seen.[18] [132]

---

17 dekhahu nara tana kahā raco sundara ko mabahara | hāḍa māsa nasa rudhira pība majjā milāya kara ||. I was not able to find a meaning or propose an emendation for mabahara that rhymes with kara.

18 The meaning would be that when one identifies with his/her own body, one does not possess the needed detachment to observe his/her own self in order to ascertain his/her nature.

SARASVATĪ *(to* MIND*)*: You are very fortunate!
Without even calling him, Dispassion has come to you by himself. [133]

MIND *(looking at* DISPASSION, *lovingly)*: Son, Dispassion, come to me quickly, embrace me. [134]

Son, your arrival was auspicious, a cause of great happiness to me. All sorrow and fatigue have vanished, my despair has gone away. Now the destruction of inner and outer darkness has taken place. Embracing you, dear, my body is soothed. Being eradicated, suffering and affliction are obliterated. *(He embraces* DISPASSION.*)* [135]

DISPASSION: It is heard that you, father, are extraordinarily skilful; for what reason did you bear sorrow and affliction? How many the dear ones may be: mother, father, son, relatives, they are as travellers on a highway, meeting briefly and then parting. Many voyagers meet in the middle of the way and take rest at one good spot. Just as at night, the conjunction result of past actions dissolves, and all proceed on their own path in the new day. [136]

As men who, climbing on a single boat to cross a river,
having crossed, land and find different paths. [137]

And what else is seen by the eye, it is not to be considered permanent. In the way a tree, on the bank of a watercourse, is swept away in the water one day, in the same way all, indeed, are dry food for Time: some munched for good, others – already munched – munch again.[19] Such is the norm of the dream that is the world. All joy and sadness are meaningless. [138]

Living in the world without comprehension should be understood, O father.
Dead, one lives like in dreams; awaken and the error vanishes. [139]

Without knowing, one doubts, like in the case of the rope mistaken for a serpent.
By listening to scriptures, one knows and, then, the world vanishes, there is no doubt. [140]

How could Mind keep forgetting this? He was carried away night and day in the enjoyment of sense objects. The world is a lie, an abode of smoke, the deer desires good water because of thirst. Keeping on running, the deer died out of exhaus-

19 [. . .] | kachu cabe kachu cābai cabai ||. The verse is not completely clear in sense. See Vaudeville 1974: 249, 16a, for a similar image from Kabīr's sākhīs.

tion, only one hand was left.[20] Thanks to understanding,[21] abandoning egoism, he will experience an ocean of equanimity, happiness, and bliss only today. [141]

MIND *(to* SARASVATĪ*)*: O mother, benevolent, the impurity in my heart was removed by listening to Dispassion's words. The moon of peace has risen inside me, the obscurity of care has faded. The moonlight of indifference has appeared, bestowing calmness and joy, destroyer of the threefold affliction. [142]

SARASVATĪ: Listen, dear, you are first King Mind, a householder saddled with responsibility. The way of living and conduct of the brave Dispassion are different from those of a householder. By keeping Dispassion's teaching in your heart and outwardly staying detached, like a cloud, as a first thing, forgo the objects of enjoyment and the cure of kingly duty will spread. In whose heart Dispassion dwells, he is always happy in himself: joy for the possession of wealth, affliction for any misfortune do not pervade him. The deceit of the world does not attach to him; he stays in the abode of the forest. For him, union with or separation from the objects of enjoyment are the same, as if he were bodiless on earth. For this, you have a second beautiful wife, the intelligent Disengagement-with-Action. You left her, wretched, she who accomplishes the highest good of her husband-god. Now, showing love and comfort, call for her, keep her at home. If you serve her, you will be pleased. Then summon Friendship, Compassion, Gladness, and Sympathy – by Bond-with-Viṣṇu's command the four sisters should stay with you. And the sons, like Balance, Self-Restraint, Observance, etc., are your virtuous teachers. Give them an order to stay next to you, reading the Vedas. Then Discrimination, your elder son, is worthy, young, and urbane: order him to stay in your house with his wife Upaniṣad. Make him crown prince, give him the burden of kingly duty from today. In such a way make, O king, the three worlds your decoration. [143]

MIND *(standing then falling at* SARASVATĪ*'s feet, humbly)*: Your orders are borne atop my head. [144]

Many actors entered, beginning with BALANCE and DISCRIMINATION. They arrived in the hall, joyful, and bowed to MIND. [145]

---

20 dhāvata dhāvata dhāya marau śramahīṃ ika kevala hātha rahai |. The meaning of ika kevala hātha rahai is not clear but could be an idiom.

21 ceta is employed here, which I have translated as "understanding". The image of the deer who wants to assuage its thirst by a mirage is employed by early modern authors and amply commented on during kathā sessions. See Thiel-Horstmann 1989: 155–157.

MIND, joyful and embarrassed, embraced his son DISPASSION. SARASVATĪ, observing, rejoiced and blessed MIND. [146]

SARASVATĪ: Now, son, sit on the king's throne and make of them your ministers. Be the king of the three worlds, dear. Give happiness to everyone in this way. Individual Self, your pure father, son of the Supreme Self, highly revered. You caused much suffering to him, being dependent on affection. You abandoned him, because of Bewilderment. He lived disappointed by you; he suffered a million difficulties. [147]

Give to Discrimination all the burden of kingly duties!
Love Individual Self, remove your father's sorrow! [148]

Take in consideration Individual Self's good by abandoning the fickleness that is intrinsic to your nature.
If you see proper conduct as foremost, Supreme Self will come in your heart. [149]

MIND, listening to SARASVATĪ's words, then kept DISPASSION's words in his heart. He abandoned his own nature, became steady, virtuous. He was purified, affectionate to his father's feet. [150]

SARASVATĪ *(observing* MIND, *who had become pure with the destruction of all sorrow)*: Son, now leave. Do your ablutions properly in the river of the gods. Perform the last rites for those who died in the war, become pure by removing suffering and anguish. [151]

SARASVATĪ, MIND, and the rest exited the hall, talking among themselves. The stringed instruments started. [152]

Thus concludes Act V of the "Drama of the Rise of the Wisdom Moon" in Brajbhāṣā.

ACT VI

For some time auspicious songs were sung and the townsmen danced in front of the king. Then the actor playing PEACE arrived from offstage. [1]

PEACE: Today King Discrimination sent me with this order: [2]

"O daughter, you got to know everything of how I fought Bewilderment. The crafty schemers Desire, Anger, Greed, and so on, I fought all and the wicked mingled with the dust [i.e. they died]. Bewilderment, fallen, left the battlefield and fled. It is not known where he is hiding. They are all conmen, devious, disgraced, engrossed in enmity, deceiver of the universe. [3]

Like grass is completely burned to ashes by a little light.
All, indeed, obtained the fruit of their actions for good. [4]

And Mind, in the company of Bewilderment, in the power of his own misjudgement, was like milk mixed with water: now, in effect, he is separated from Bewilderment. He has renounced his affection for him, because of whom the world had become his own possession.[1] Forsaking the enjoyment of worldly objects, removing any hope for pleasure, he has given up egoism. He has become stable, abandoning fickleness, and has found a single essence, equanimity. All opposites vanishing, he felt love again for Individual Self. Having seen the greatness of Individual Self, he has now become pristine, having destroyed the stain of egoism. Now, after making the moon of Wisdom appear, he wants to meet Supreme Self. [5]

Therefore, be quick and approach Upaniṣad.
Solving all her doubts, make her come to me, respectfully." [6]

After PEACE had reported in this way, FAITH entered the hall, delighted. PEACE addressed her.

PEACE: Mother, today you seem ecstatic, tell me the reason for this joy.

FAITH: O daughter, our enemies, who gave us great pain, oppressive: with the favour of Bond-with-Hari now they are destroyed, Discrimination killed them. Now all of King Discrimination's community has become untroubled, the double rule[2] has ceased to exist. The excellent truth-knowers are now all happy, meditating at

1 aba tyāga kari anurāga tina soṁ jaga kachū āpuna mayo | [. . .]. I have emended mayo to bhayo, since mayo (Skt. maya) is attached in general to nouns. Here, however, there is the reflexive pronoun āpuna; therefore, I think bhayo is more probable and grammatically appropriate.
2 durājū can also be translated as "the evil rule" (HSS 2320).

https://doi.org/10.1515/9783111063836-015

Hari's feet with the attitudes of Bond-with-Hari. They now walk on the path of Bond-with-Hari, in the middle of happiness; the thorns and mud of error and fear wiped away. [7]

All these the reasons of happiness, with Bond-with-Hari gaining strength.
That is why the bliss in my heart grows at every second. [8]

PEACE: The things told by my mother happened truly only for the favour of Bond-with-Hari. She has manifested, root of bliss. The pangs of confusion have disappeared. [9]

Now tell, how does Individual Self act these days?
How does he show favour and affection for King Mind? [10]

FAITH: Listen, daughter, this matter is quite extraordinary. Individual Self, in the company of Mind, experienced great pain at his hands. Some of those pains are indescribable, who can talk about so many unbearable difficulties? Understand that he is suffering, hence he knows him [i.e. Mind] as an enemy. Now he wants to kill him. Tell, are there any favour and affection left? [11]

PEACE: In that case, I had to know this, mother.
If Individual Self appropriates sovereignty, is Mind's lordship lost? [12]

FAITH: Listen, young girl. One should know that such is Individual Self's conduct: even if Mind now is dependent on Individual Self, and lives considering his own benefit as impure, nevertheless Individual Self does not even look at him, for he considers kingly duties to be his responsibility. Thinking about the many previous troubles, he does not trust Mind. Now, remembering his father's lordship, he wishes to meet him immediately. [13]

PEACE: I had to know, mother, that Mind was deprived of sovereignty.
The Supreme Self was angered and assumed it on himself. [14]

Mother, tell, I have to know everything:
How does brahman love Illusion now? [15]

At that time, seeing Material Nature's qualities, he showed incredible favour.
Now, please tell, how is his behaviour today, compared with the past? [16]

FAITH: Supreme Self is an ocean of joy, incomparable in all respects, cultured. He acknowledged Illusion as misleading, studded with deceit from head to toe: "As many individual selves are in the world from my own self, all stay unconscious by her hand. She does not let them move towards me. They have forgotten everything; she does not let them recognise me. Therefore, she is an extremely astute and cruel,[3] she is to be kept captive." He pondered thus. [17]

PEACE: With great fortune Individual Self's Highest Self got to know.
Now, leaving Illusion, he wants to take your company. [18]

Be merciful, give me your own opinion.
Now I would like to know if the king has created an eternal reign. [19]

Please tell, mother, how now is Discrimination, together with the family?
Does he fulfil his kingly duty, the manifold decorations of a king? [20]

FAITH: Daughter, you are completely acquainted with Discrimination's eternal skilfulness and greatness. Pleasures, like the body and so on, are temporary and false, made by Material Nature and permitted by the Highest Goal. Individual Self made his dwelling there [i.e. in the bodies], accepted pleasure like in a dream, being in the position of a king. [21]

King Discrimination now wants to illuminate him;
he wants the joy that is brahman, always a single essence, to inhabit that region. [22]

Where there is no gain or loss, no one praises or blames. Where there is no death or retribution for past actions, enemy or friend, the light of supreme bliss is pure. He wants to establish his capital in such a secure place, according to the scriptures. [23]

And Discrimination has much love for Dispassion.
He does not carry out any royal duty without asking him. [24]

Compassion, Sympathy, Friendship, Gladness, the four sisters, Duty's daughters, they stay in the king's service. As soon as they receive an order, they execute it. The ministers, like Balance, Self-Restraint, and so on, devotedly attend to the king, with the army and with Observance. Advising him day and night, how could I

3 I have taken the meaning of kharī khoṭī from BHK 289, where it corresponds to kaḍavī-kasailī. This is a composite adjective indicating something like "bitter and astringent; adverse" (OHD 160). However, it could also have a sense which is not completely negative but paradoxical: she is ambiguous, in the sense of both good and evil. This reading is supported by HSS 1184, where khoṭā kharā is uttama aur nikṛṣṭa. I have chosen the first option because the context does not seem to leave space for a positive side of Māyā, being focused on the drawbacks of her influence on individual selves. The context differs from her presentation in 1.140 and 1.146, where she is described mostly in apophatic or positive terms because she is introduced as brahman's śakti.

describe their proximity to the king? Of the rest of the enemies, no one is alive, the king has no doubt. [25]

Selfishness, Bewilderment,[4] and the doorkeeper False Company,
Intention and Doubt, two of Mind's servants, are killed. [26]

PEACE: It was important to know all about the enemy and his faction.
He [Discrimination] rules Individual Self's reign, then he meets the Supreme Brahman. [27]

Tell me everything, mother: Individual Self, Duty, and King Mind,
how do they stay together today, compared with yesterday? [28]

FAITH: Today King Mind has love only for Dispassion, having abandoned any hope of auspicious or inauspicious fruits attached to Action.[5] He does not love Duty much, because he considers everything to be impermanent. He believes Dispassion and Relinquishment to be his highest good. [29]

That is why King Mind, upon seeing Duty, was enraged.
Stopping his efforts, he isolated himself, hiding with Action. [30]

PEACE: Oh mother! I know about this. Please, explain to me: Great King Bewilderment deserted the battlefield, where did he flee with the Supernatural Powers? Where, indeed, did he manifest? Is there any news of him?

FAITH: O daughter, you do not know about Bewilderment? That wicked conman, shrewd deceiver, you are not aware of his deceits? The honeyed Supernatural Power, called Knowledge of Indra's Net – he adorned her in many ways, like a golden doll, a fine lady. He told her to approach King Mind with sidelong glances and many womanly flirtatious gestures; to please him in whatever way, to persuade and enchant King Mind. [31]

FAITH recounted the scene.

KING BEWILDERMENT: If Mind is pleased by your qualities and shows affection for you,
then nothing will happen between Discrimination and Upaniṣad. [32]

PEACE: At hearing such things, my body starts shaking.
Please, go ahead, what happened next, O mother? [33]

4 There seems to be an inconsistency here in the text as Bewilderment is mentioned as being in hiding in 6.31.
5 karman, the actions associated with dharma as kingly duty.

FAITH: Honeyed Knowledge listened to Bewilderment's command and approached Mind, rejoicing. Extremely deceitful, clever – how many sweet words! Just like an enchantress who subjugates the world. At seeing her, she was like a well-made, beautiful court, a knowledgeable city of ignorance. The *Mahābhārata*, the Vedas, the Purāṇas, the *Bhagavadgītā* – as if she knew by heart all the pure treatises! The many profound tales in the world, she was knowledgeable with the course of earth, heaven, and hell. That astonishing fancy went where King Mind was hidden. [34]

The Supernatural Power, with the form of a supernatural woman, presented
herself to Mind.
Right as in a magic show, she composed the Vedas. [35]

HONEYED KNOWLEDGE *(smiling, to* MIND*)*: Wise king,
for the sake of what salvation are you, learned, unhappy and enduring sorrows? [36]

Considering it, bestower of pleasure, look at heaven. Where there are all kinds of gratifications, day and night. No troubles, no one is sad: the afflictions of people's hearts have no place there. There is no anguish, king, but always merriment. All pleasures, as wish-fulfilling trees are spread. No misfortune, but a reservoir of nectar. It is a lovely place filled with beauty: the women of the gods sing; they cheer the hearts. No bodily diseases, no pain; ultimate enjoyment, all favourable unions. Whatever desired is obtained, without effort. [37]

Such an auspicious good place, everywhere there is the fulfilment of all desires.
Bestower of extreme repose: regard it as your own auspicious abode! [38]

FAITH continued.

Bewilderment's company somehow resided in his heart again.
Mind, listening about the pleasures of heaven, looked greedily at the
Supernatural Powers. [39]

ILLUSION *(explaining to* MIND*)*: These powers are good. Learn, talk, and practise what is pleasant to you.

INTENTION and DOUBT, too, began speaking of the manifestation of the pleasures of heaven. [40]

FAITH resumed.

Hearing Intention and Doubt's words and Illusion's advice,
Mind returned again to the cycle of existence with a body fit to enjoy the
Supernatural Powers, abandoning previous guidance. [41]

PEACE: Alas, alas, King Mind! How does an ignorant spoil his efforts?! So far, indeed, he has not rejected impurity: the thoughts of an unfortunate elephant cannot be erased.[6] He considers them good, time after time, he does not know they cause suffering. All these are killers of the Highest Truth. The aspirant to salvation shall not gain Supernatural Powers! [42]

For this reason Individual Self wants the sovereignty for himself, after killing him.
He is fickle, a liar, you cannot trust him. [43]

FAITH: O daughter, in that moment
Science-of-Reasoning stood next to Mind, bowing her head. [44]

SCIENCE-OF-REASONING *(with folded hands)*: Oh! Listen to my auspicious supplication! At first, Lord, you were in their company, like a deer enamoured by a sound. Caught in the snare of the whirlpool of existences, the doubts of salvation were many. But, broken the fetters through yoga, luckily, you have been somehow saved. [45]

FAITH: Mind, listening to Science-of-Reasoning's pure words, was ashamed.
He sent away the Supernatural Powers and started blaming them. [46]

Then, summoning Dispassion, he took his counsel that he will make him meet Individual Self. "Please come meet me" – Individual Self's royal order, from which extreme happiness will arise. [47]

PEACE: It was good, mother, that Mind was saved.
What a miserable situation, indeed, if Mind had fallen in their power! [48]

If King Mind understands this and accepts Individual Self's royal order to go meet him; if Mind stays in Individual Self's power, then there will not be any Bewilderment again. Tell me, mother, where are you going today? Tell me the reason, what is your task?

FAITH: Beautiful daughter, the great king Individual Self sent me. [49]

He told me to bring Discrimination to him, as soon as possible.
Hence, I am going to summon him. Now tell, where are you going? [50]

PEACE: King Discrimination ordered me to summon Upaniṣad.
He yearns for Wisdom's birth. [51]

6 The mind is often compared to an intoxicated elephant, which lacks control. See e.g. Kabīr's verses in Dharwadker 2003, specifically "Moth" and "Shalok 58" from the *Ādi Granth*.

FAITH and PEACE rejoiced and after dancing for some time, they exited the hall, joyful. [52]

INDIVIDUAL SELF *(while entering the hall from backstage)*: This cycle of existence is an ocean of doubt, filled with the water of Illusion. The waves of Thirst rise high, the whirlpool of Greed is like a harsh drought. Its flame, Violence, rises, destroying the mare-fire beneath the sea.[7] The group of sons, wives, friends, and enemies are fishes and huge snakes. The five organs of sense, moving like huge crocodiles, draw one under deep water. No one can traverse it, immense, impenetrable, profound. Deceit, Pride, and Intoxication are like blasts of wind raised by Bewilderment, they have no equal. There is no support, no near or other side to be seen. I was sinking in such a painful, frightening, vast ocean, desperate; nevertheless, obtaining the boat of Bond-with-Viṣṇu's favour, I have found the other bank: hope that is the Lord. [53]

Now, all my doubts cleared, I have found the bank of bliss.

Climbing the boat of Bond-with-Hari's favour, the other shore of the ocean of existence appears before me. [54]

The actors playing PEACE and QUEEN UPANIṢAD entered, talking with each other as they came. [55]

UPANIṢAD *(teary-eyed, hesitantly)*: My husband, King Discrimination, has completely forgotten me. He sent me away from home, he does not think of me anymore. Now I am enraged, I do not want to see Discrimination's face anymore. [56]

PEACE: O virtuous, wise queen, what you said does not befit you. Does a faithful wife blame her husband? Now please narrate your own story nicely: what sufferings did you endure, O queen? Where did you pass all these days? How did you spend them without the king? [57]

UPANIṢAD: When the king forgot me and did send for news of me, oblivious, destiny created such a condition for me, that I fell in the hands of ignorant people. I endured animals but for tail and horns, similar to flies in milk. They stole all my ornaments, they spoiled my beauty. They wanted to make me their slave, they did not ponder about what has a sense and what is senseless. [58]

In this way I experienced innumerable sufferings among stupid people.

The cause of it is that Discrimination forgot about me. [59]

7 For the mare-fire image (often associated with a woman) in literature, see Doniger 1980: 213–237.

PEACE: This, virtuous queen, is not Discrimination's fault. You are Discrimination's beloved life, he does not forget your face, even for an instant. What should have he done? His enemies were burdensome. They, starting with Bewilderment, were performing many tricks. When the king wanted to meet you, they caused him great trouble. He did not have even the occasion of a moment free from them that he could call you to him. Be blessed, you are virtuous and wise. Your intelligence is praiseworthy, you are fortunate, O queen. The duty appropriate for a woman, you carried it out rightly amid uncountable difficulties. Now do not worry, all your enemies are destroyed by your majesty. Now approach Discrimination quickly, meet him happily, with no grief.

UPANIṢAD: O friend, when I left my refuge, I met Bhagavad-Gītā on the way and she lovingly instructed me: [60]

"If Discrimination and Individual Self, both benevolent, sit together and,
meeting you, they ask you something, do not hesitate to reply. [61]
When Individual Self, rejoicing, will ask, then tell him about your separation from Discrimination."
But I am afraid, I am still preoccupied.[8] How could I utter any word, approaching the lord? [62]

PEACE: Queen, the instruction Gītā gave,
please accept it definitely, consider it your highest good. [63]

UPANIṢAD: O Peace, if what you say is just as Gītā told me,
then I come with you and whatever happens, let it happen. [64]

KING DISCRIMINATION *(to* FAITH*)*: Did Peace go to fetch Upaniṣad? We have been separated for a long time; any meeting and union between us was not possible. Now only God knows if Peace has brought her or not.

FAITH *(joining her hands)*: Listen, O lord of the world. Peace has surely brought her, there is no doubt. The mount Mandara, where the worship of Hari takes place in many beautiful ways, is distant. There dwelled Upaniṣad, together with Gītā, out of fear of Sophistries. Bond-with-Hari told me that Faith is bringing her. King, when Upaniṣad arrives, she will tell you all about her condition. Now Individual Self sent me to summon you; please come, O king. [65]

8 [. . .] | mopai lā na jāta nivārī | |. The text appears to be missing a part after the lā, which cannot stand by itself. Since metrically it is a caupāī, we are only missing one syllable to arrive at the total count of sixteen morae. I surmise the syllable to be -ja, so that the word would be lāja, which also makes sense with the character expressing her insecurity.

DISCRIMINATION stood and approached INDIVIDUAL SELF, bowing many times. [66]

INDIVIDUAL SELF *(takes his hand and embraces him, then invites him to sit)*: O son, anchor of my life, even if I am your senior in matters of age, you are superior to me in terms of insight. One should not consider superiority a question of age, for those who have greater insight are superior. [67]

As they talked, PEACE arrived with UPANIṢAD. [68]

UPANIṢAD fell at INDIVIDUAL SELF's feet, then paid homage at the lotus-feet of her husband and sat, bowing her head. [69]

INDIVIDUAL SELF (*looking at* UPANIṢAD, *praises her*): You paid homage at my feet. Who does what is appropriate is very fortunate. However, you are like my mother, of pure conduct, clever. Tell, under what kind of constraint did you leave the company of your husband? Where have you been until now? How did you spend all these days? How did you fulfil your duty? [70]

UPANIṢAD: O receptacle of qualities, teacher of the world, god! How could I narrate my own misfortune, god? Every day was spent like an era, wise god! [71]

At first I met several people attached to the views of the Sophistries,
whose doctrines are all like an amorphous crystal of a single, white colour.[9] [72]

The line originated from me, consisting of qualities, they deformed it more and more. Even if they listened to the doctrine of the Vedas, all the same they did not reflect upon it well. A blind thought, without eyes – they did not truly understand me. Even if I expounded them my doctrine and tried to make them understand in many ways, all the same, they misconstrued my humble words, extreme wickedness in their hearts and thoughts. [73]

Then I saw by myself that the afflictions resided there.
Knowing that the fulfilment of my duty was not possible in this way, I left them. [74]

Then a place came into view, fully equipped for the worship of various deities. Seeing it, I thought that I could take refuge there for some days. They [people] are extremely wicked, ignorant, of Bad Reasoning, opposed to their own good. Their own form, always a receptacle of bliss, they forget that it is unborn and indestructible. They roam to forget themselves; they stubbornly yearn for something temporary to have continuity. They do not know at all the secret of my doctrine and because of this I became sad, O god. [75]

9 I have translated asphaṭika as "amorphous", following BHK 110.

Individual Self: O daughter, you met numerous men here and there.
Was there someone among them who knew your doctrine? [76]

Upaniṣad: O god! I did not meet anyone who would listen attentively to my teaching, or make of it his authority. If someone listens to my instructions, pondering well, then the impurities of dividedness vanish – no duality is left. [77]

Individual Self: After leaving their company, where did you go?
Narrate all of your story. [78]

Upaniṣad: From there, O god, I proceeded in a distant direction. How could I describe the extreme toil of the way? There I saw an enclosure, beautiful, clean, and spacious. There dwelled the precepts of Sacrificial Science, where she carries out numerous sacrifices. She kills animals, like the horse and so on, from which she thinks she obtains the desired result. I saw all those who live there, ensnared by she who maintains that ritual action is pre-eminent. They carry out many kinds of rituals and observances, follow the orders of Sacrificial Science. Seeing them, I thought that I could spend some days there and that, seeing me, maybe they will acknowledge me in their hearts. Maybe, among them, someone will know about me. With this hope in my heart, I went to Sacrificial Science. Approaching her, I said my name and requested to stop there. [79]

Upaniṣad recalled the scene.

Sacrificial Science: I do not know your actions.
Clarify, what are your conduct and observances. [80]

Upaniṣad: Then I said that this is my conduct: brahman without qualities, unique, without external appearance, who is found in everyone, and everyone, indeed, is found inside him. Who is called always one, a single essence, in one bliss.[10] Without origin or end, unmanifested. Seeing that there is nothing else but him, I always sing his glorious fame. [81]

Sacrificial Science *(agitated and harsh)*: What actions does Self carry out? In the way in which he carries out auspicious and inauspicious actions, according to it he will obtain the fruit. If you do not believe my words, be aware that the ritual section of the Vedas knows all desires. You did not call Self one who carries out

10 [. . .] | sadā eka rasa eka eka ānanda meṁ kahiye | |. It can also be translated as "always one essence, one in one bliss".

actions, you said that Self is not an agent.[11] One might suppose, then, that you are opposed to my point of view. Therefore, I am not pleased. [82]

If you wish to stay here, then praise my doctrine!
Worship Self as an agent otherwise leave! [83]

King Discrimination *(standing next to* Individual Self, *listening to* Upaniṣad's *words, smiling)*: Understand Sacrificial Science's thought! [84]

Her eyes, confused, are blinded by the haze coming from the smoke of the sacrifice.
That is why she says wrong words, which should not be reflected upon. [85]

Supreme Self, in the bending of whose eyebrows Material Nature creates the world. In one second destroys then creates, in this way she dances in front of him.[12] He is the Self and has two forms: one is Individual Self, lord of action; the other is the Supreme Self, immaculate, inactive, the Self that is Lord of the universe. [86]

The inactive Self is like a magnet, the active is like immovable iron, it does not move. Only if the magnet wills does it set it in motion. The power of the inactive is present in the active, that is why the active carries out actions. As with a wooden puppet: it dances, but it is the player who, after making it, makes it dance. The inactive Supreme Self, all-pervading; know him as the puppeteer. The active Individual Self stays in his power, like a puppet. Everyone can see the active puppet; no one can glimpse the puppeteer. [87]

In that way the inactive Self enters by his power in duality.
Sacrificial Science does not perceive this fact, hence how could she make you stay? [88]

This cycle of existence is a deep well. You are the water of misconception, nescience. Rites are the plants in the well, one form in all the beings trapped in the cycle. They are extremely frenzied because of the rain of sense objects, and, fallen in the well, they talk nonsense in many ways. They are happy in that well, who can shed light on darkness? [89]

The light in the sky, outside the well, incomparable mercy,
they do not know it, the dull-witted, fallen in the well of darkness. [90]

---

11 In the ritual context, the puruṣa has to be a kartā, a "doer", or "active" in order to enjoy the fruits of his actions. She ignores the distinction among the embodied puruṣa and the Highest Puruṣa, as will be shown.

12 The image of the dancer to describe Prakṛiti's activity is found in Iśvarakṛṣṇa's *Sāṅkhya Kārikā* 42.

In this way Sacrificial Science commits many faults.
    (To INDIVIDUAL SELF*)* Are you silent? [91]

INDIVIDUAL SELF *(to* UPANIṢAD*)*: O my well-wisher,
    what else did Sacrificial Science tell you? Explain it. [92]

UPANIṢAD: Then, pensive, Sacrificial Science spoke to me: “Upon closer scrutiny, your doctrine of obtaining salvation is an absolute elixir: there is no rebirth and death, hell and heaven, pain and joy are destroyed. There is always one essence, bliss, no other mood. If those who believe in sacrifice get to know about it, they will become detached and give up rites. Therefore, I will become useless, as no one would carry out sacrificial actions. Hence, it is not good that you stay here. Go wherever you may like, quickly!”

Upon hearing Sacrificial Science’s words, I left, worried. [93]

Then, proceeding on the way, I saw a matchless place.
    There, with other people, sat someone whose name is Hermeneutics. [94]

She classifies different kinds of utterances,[13]
    she explains her doctrine as giving authority to the words of the Vedas. [95]

I approached her with folded hands, saying beseeching words, very humble. I asked her if, being merciful, she could allow me to take refuge there. She enquired: “Tell me, what are your rites and prescribed conduct?” Then I told her about my way, describing it. [96]

Endless, one in this and the other shore. Not visible, without a form, without a beginning, the waves of the ocean are the multiple worlds. The multiple worlds, like the waves of the ocean, rise. Then, they are absorbed again in the ocean; they do not separate from it. Pure, unfluctuating, profound, his essence a single essence. The ocean of bliss of brahman, whose glory the Vedas sing. [97]

Meditation always takes place in that ocean of bliss in the heart,
    the tongue sings only his glory day and night. [98]

HERMENEUTICS *(listened to my words and told her people)*: She, in fact, worships a Self who carries out sacrifices. She says she remembers and sings about him. That is why may she stay here happily and constantly sing the glory of an active Self.

---

13 bhāṁti bhāṁti kī uktī kari āni vividhi anumāna | [. . .]. Literally, “Bringing various deductions with respects to different kinds of utterances.”

KUMĀRALA *(the master who goes into the subtle points of* HERMENEUTICS, *getting up)*: Listen, O merciful, if you talk in such way, then all is wrong. The god she believes in, praising him, is inactive. She does not bring him in the rites. It is of this Self she sings the pure glory.

MAN *(sitting next to* HERMENEUTICS*)*: Oh! The active Self is one, is there someone else? The worldly Self, who is the individual self, progressively becomes a god. When he improves by himself, then he becomes the lord of the three worlds. [99]

KUMĀRALA *(smiling)*: I am going to tell how in such doctrine there are two kinds of Self. [100]

One Self is all-pervading, dwells in all places, having assumed the form of observer. He does not carry out any action by himself, he does not move, does not walk, does not leave, does not grasp. Always a single essence, extremely luminous, treasury of joy, receptacle of joy in joy. Invisible, complete, unsurpassed, endless, illuminating everything, separated from everything. [101]

The second Self, knowledgeable, shining, is pure.
He carries out actions and has attachment for Illusion, Bewilderment, Pride. [102]

The one who carries out actions, he can be excellent or the worst. What auspicious or inauspicious actions he likes to carry out, he obtains the fruit according to that. Hell, heaven, all conform to his will, he acts and then returns to the test of the actions in this way. Therefore, the active one accepts happiness and sorrow. [103]

The inactive Self, the stain of action does not attach to him,
auspicious and inauspicious, the two are equal for him, therefore he is untarnished. [104]

DISCRIMINATION, upon discovering such detail from UPANIṢAD, rejoiced, praising the master KUMĀRALA. [105]

KING DISCRIMINATION: Good, master Kumārala, you are a sage, may you live long with such deep intelligence! What master Kumārala said, the Vedas maintain it as well. Talking of this cycle of existence as a tree, two birds sit on its branch. One bird eats many kinds of fruit, but it always stays weak. It bears several difficulties and griefs, sad, impure, miserable. The second bird, next to it, sees everything with the form of an observer. Sorrow and happiness happen to it and it observes them, but they do not affect it. It does not eat nor drink, stays next to it [i.e. the other bird], but without touching it. Even though it does not eat nor drink, it is always nourished and content. Delighted, full of supreme joy: one who knows, just knows! [106]

He continued to thus sing the praises of master KUMĀRALA, then fell silent, standing next to INDIVIDUAL SELF. [107]

INDIVIDUAL SELF *(benevolently, to* UPANIṢAD*)*: Tell, returning from that place, where did you travel? [108]

UPANIṢAD: Science-of-Reasoning's abode was nearby.
Several agreeable people sat there, discussing.[14] [109]

The numerous instruments of inquiry, embellished with various examples,
I saw they were practising the multifarious science of debate. [110]

There someone discusses, someone practises the points of defeat, someone sticks to dispute and trivial objection; someone spreads intentional misinterpretation and incorrect rejoinder. These are said to be six categories of reasoning, they all keep fighting through them. The characteristics of the six categories I am going to describe, for all skilled people know them. [111]

Two scholars sit in a place, they reflect together upon reality.[15] They deliberate about the possessor of the body and the body, they ponder upon what is permanent and impermanent. When one upholds verbal testimony[16] which is impartial, this is called Discussion in the Science-of-Reasoning. [112]

Where two people, very self-conceited, do not understand the words of authoritative testimony, they are intent only on themselves, the stubborn fools. They are not sad for others, they are sorrowful only for themselves. Who likes only fighting, this is called Point-of-Defeat in the Science-of-Reasoning. [113]

Two scholars discuss, they fix the mind on the desire for one's own victory. Being devoid of an essence, they do not know the taste of brahman's essence. This is called Dispute in the Science-of-Reasoning. [114]

When there are people with little knowledge, two behaving arrogantly with one another, they are intent on big discussions with one another but do not know anything true. They consider themselves to be the best scholar, this is called Trivial Objection in the Science-of-Reasoning. [115]

---

14 For an analysis of these categories in relation to classical Nyāya texts and observations about Nyāya expressed in Brajbhāṣa, see Pastore 2023.

15 *Tarkabhāṣā* 135: tatvabubhutsvoḥ kathā vādaḥ, "Discussion is argument between two persons or parties desirous of arriving at the truth (of the topic under discussion)"; text and translation in Iyer 1979: 230. Taking my cue from that, I translated tatva in the meaning of "existing", i.e. reality.

16 That is, "The instructive assertion of a reliable person", according to *Nyāya Sūtra* 1.1.7.

Joining meaning and lack of meaning, to what extent do they make up something which does not have the same meaning of what has been said? Afterwards, they blame the other person; this is called Intentional Misinterpretation in the Science-of-Reasoning. [116]

An excellent person asks a question, then someone answers him, but he is not satisfied. On the contrary, listening to the reply, he becomes even sadder. Who knows this as Incorrect Rejoinder in the Science-of-Reasoning thus recognises the six categories.[17] [117]

I have ascertained properly that these six are baseless.
    By constantly reflecting, I could not find any joy in them. [118]

They discussed among themselves in the following way: "Material Nature is the seed of the world. Know that from her I-maker is produced, that is then called threefold.[18] I know the five gross elements, five subtle elements, ten organs, five of perception, five of action. Putting them together in this way, I describe these as the twenty-four elements. The Supreme Brahman, uniting with Material Nature, made her create the cosmos. He himself stays separated from all, he is not contained in that." I saw that their way of acting is such, thinking that I should meet all. I approached them and I have observed all these behaviours. [119]

Seeing me, they asked about my thought and conduct:
    "Who are you? What's your doctrine? What your way of acting?" [120]

Then I told them about my way, as I did with the others.
    I said that I concentrate on that pure single brahman. [121]

Without external appearance, he himself is everywhere. He is always pervading, as from the seed one obtains the name of the tree. When the name of the tree exists, it is said that in that flowers and fruits exist already. But in the absence of a seed, how can there be branches, leaves, roots and a big tree? The tree exists, indeed, in the seed; the seed, in fact, is the true being of the tree. I always praise the form of this all-pervading brahman. [122]

17 See Jonardon Ganeri on jati: "A sophistical rejoinder (*jati*) is a debating tactic in which the opponent tries unsuccessfully to prove a counter-argument, an argument designed to prove the opposite thesis. It is sophistical because the counter-argument is based on a false or superficial resemblance." Ganeri 2001: 31–32.

18 This ahaṃkāra should correspond to the antaḥkaraṇa, which includes buddhi, ahaṃkāra, and manas. In this way we have a total of twenty-four tattvas.

As soon as they heard these words from me, they all stood up, angry. All of them, furious, shouted: "Now get out from here!" Without understanding, they thus spoke: "You made the Lord of the world become the world. But the world has a form of suffering, always impermanent; on the other hand, the Supreme God is an ocean of happiness, joy." Who told them this idea? The Supreme God is the form of the universe. The production of the world occurs always in this way, it is not by those twenty-four elements. [123]

UPANIṢAD resumed reporting their arguments.

"This world arose from Material Nature who brought together the twenty-four elements.
Brahman always stays separate from it, the world is impermanent and will vanish. [124]

The world arose from atoms, God is distinct. He is the efficient cause, the former are material causes.
The efficient cause is like a cane, the material a thick particle." [125]

KING DISCRIMINATION *(angrily)*: May Science-of-Reasoning's thought be cursed! She does not even know this! Material Nature, and the atoms called material cause, are false. The universe is like a mirage of water due to the sun, which, by mistake, appears as not false. If you can see the origin and destruction of something, then how can this be called real? Until the individual self is awakened, this universe is what is perceived. Like the error of mistaking a rope for a serpent or of perceiving silver in a seashell. The confusion dispelled, there is no serpent nor silver, only the pure true being is perceived. Material Nature and the atoms are all conformable to Supreme Self: like waves and water, ornaments and gold, they are just one thing, incomparable. Always a brilliant form, unborn, imperishable, creating, sustaining, and withdrawing the universe – no change occurs to him! Different masses of rain-bearing clouds can be seen as they disperse, then they gather again, yet there is no modification in the sky!

INDIVIDUAL SELF *(as the king explains, he becomes happy)*: I like this way of thinking! [126]

(To UPANIṢAD *again)* Then what happened to you?
O mother, tell everything essential of your story. [127]

UPANIṢAD: Then those stupid tricksters, without thinking, ran behind me to kill me. They were saying: "Who is she, where has she come from? She will not survive. To call salvation the destruction of the world, this is but a world-denier's thought!" Overwhelmed by fear, I ran to the Daṇḍaka forest. My garments torn, my jewels broken, they spoiled all my clothes. In that forest, there was a temple on a mountain, a temple of Hari in a beautiful place. Since my daughter Gītā stays there, I knew and reached the place, extremely frightened. [128]

INDIVIDUAL SELF: When, fleeing, you arrived at Gītā's hermitage.
How was it there? Tell me. [129]

UPANIṢAD: Hari's devotees hastened from his temple, they all rushed to protect me. Holding beautiful maces in their hands, they drove away those evil people. As soon as I saw them, I felt relieved. [130]

KING DISCRIMINATION: Viṣṇu could not bear to see you attacked. You are a vessel abounding with Hari's favour. How would you say were Hari's devotees as protectors? [131]

UPANIṢAD: Seeing me agitated, terrified, Gītā came at once, running. Meeting me, she paid respect and reassured me in many ways. She told me to live there.

UPANIṢAD recalled.

GĪTĀ: Do not be afraid, there is no fear here, mother. [132]

Mother, you surely are an authority. Who does not consider you an authority expounds knowledge by saying what they fancy. All these thoughts are evil, foolish people are carried away by them. God will punish them; in the end they will dwell in hell. God, indeed, and the Vedas affirm this, and it is known as true in the three worlds: those who reject the revealed Word and transgress duty, the angry, the cruel, and the evil-doers, those men will obtain an evil rebirth, then will stay in hell.[19] [133]

In this way UPANIṢAD narrated the doctrines of all scriptures and her own story. [134]

After that, INDIVIDUAL SELF was purified of ILLUSION's stain, and a clear intention of knowing his own true form arose. [135]

INDIVIDUAL SELF: O pleasing mother,
who is it that you call God? Please tell me. [136]

UPANIṢAD *(smiling)*: What a surprise!
You do not know your own name, how to explain it? [137]

INDIVIDUAL SELF: And you say this smiling! You consider that I am God, how about the world? How can I deem this true? O mother, stop smiling and please tell me! [138]

---

19 This is a paraphrased quotation from BhG 16.19.

Upaniṣad: Understand this fact: you, dear, belong to the eternal Supreme Self. You are not separate from him, who is known to be two due to Illusion. As the sun is but one, water, snow, and hailstones are not distinct either. In this way you are a god; hence I call you God! [139]

Individual Self, astonished after listening to Upaniṣad's authoritative words, addressed King Discrimination. [140]

Individual Self: Are you listening, dear, to what Upaniṣad has been saying? I cannot understand; what is she saying, is she making it up? What is described as the Supreme Brahman, she tells me, "That you are.[20] Being, consciousness, and bliss, incomparable, that is your form."[21] I am the Individual Self, helpless, poor, dependent on eternal rebirth. Without essence, suffering, old age, etc.[22] – I keep being oppressed by their weight. When, indeed, have I become brahman? This is an unfathomable secret to me. [141]

King Discrimination: You do not understand the meaning of what it refers to; that is why your heart is full of confusion. If you stop being attached to Illusion, then you will recognise your pure form. By destroying your sense of ego[23] you will find your own self, then you will not be different from God. [142]

Individual Self: O dear, I am not able to understand this. Therefore, having mercy, describe it in a way I can figure out its meaning. How can I shed the sense of ego? How can I realise my own form? [143]

King Discrimination: First, distinguish your "I" from your own self,[24] ponder it deeply. One talks and hears, but who am I? Realise who is seeing and thinking. Then reflect on the words of the Vedas, its appeal "Neither this, nor that." You are not soil, water, fire, space, and wind. You are not the I-maker, nor the great element – there is no reality of yours in them. Now, taking those elements, place them: take them as if they were reversed. Put the earth in the smell, then place the smell in the water; the water in taste, then bring together taste with fire; unite fire with form, then form in wind. Afterwards, place together wind and touch, and then, touch and space. Space will vanish into sound; sound will unite with the triple I-maker. I-maker mingles with the great element, then absorbs the great element in Material

20 so tū āhi, paraphrasing tat tvam asi of CU 6.8.7.
21 sata citānanda anūpa | so ahai tero rūpa | |. This sentence is like a sort of commentary, specifying the meaning of the so tū āhi.
22 dukha jarā ādi asāra | [. . .]. I took asāra as an adjective, following HSS 379, but we would need a noun to go with the ādi and to link with suffering and old age.
23 Āpā as a synonym for ahaṃkāra. See HSS 450.
24 Lit. rūpa, "form".

Nature. Material Nature will merge with nothing else but what you are. You are a droplet of that ocean – not a god nor a man, not an evil creature nor an insect. This world is false, a well of darkness – by eradicating it, see your own nature! Someone self-luminous, a treasury of joy, "That you are", thus the Vedas affirm. [144]

As King Discrimination made Individual Self aware with such speech, the latter, having comprehended the real meaning, became supremely blissful. [145]

He started thinking about it afresh and felt happy again and again. He examined his ego and the confusion in his self dissipated. [146]

Deep Meditation (*arriving, dancing happily*): Bond-with-Viṣṇu, who is the essence of the Blessed Lord, treasury of mercy, commanded me to go to Upaniṣad, who sits next to King Discrimination saying agreeable words. I have to explain to her Bond-with-Viṣṇu's profound purpose. Then, I have to go to dwell in Individual Self's heart: if I stay there with Love, it will become spotless, pure. [147]

Deep Meditation approached Upaniṣad and respectfully addressed her. [148]

Listen, O right-minded mother, the venerable, beautiful Bond-with-Viṣṇu sent a message to you. I will report it to you as she told me. [149]

Bond-with-Viṣṇu: You have loved King Discrimination and I have concentrated on a pregnancy. Your body is not made up of the five elements, and so the foetus was generated in the union.[25] I got to know about this auspicious association through my power[26] and now you, believing it, are confused. In your womb there are two children and I have orders for them. There is one daughter, called Knowledge. Pulling her out of your womb,[27] I will place her in Mind, kindling love for her: in this way, I will dispel the darkness in his heart. Then there is a son, Wisdom Moon, whom I will place in Individual Self's heart. Come to me, together with Discrimination. Come now, do not delay further. [150]

---

25 It is not a worldly pregnancy and birth: not made up of the five elements means not related to materiality, therefore, not subject to disease, old age, death, etc. It may also hint at vidyā and prabodha as permanent attainments.

26 maiṁ suyoga bala karikai jānī | [. . .]. This could also be translated as: "I consider the auspicious union as a power."

27 The saṅkarṣaṇa vidyā is a kind of knowledge that allows one to transfer a foetus from a woman's womb to another's. It is employed in the case of the seventh child of Devakī, Balarāma, who was transferred to Rohiṇī's womb. For this reason, he is also called Saṅkarṣaṇa. In the BhP, Viṣṇu asks his śakti, Yogamāyā, to transfer the foetus. See BhP 10.2.7–8 and 13. According to Nelson, Jñānadeva seems to have been the first to identify bhakti with a śakti. See B. P. Bahirat quoted in Nelson 2004: 347n5.

UPANIṢAD: Yes, I am certainly pregnant. Whatever task the eminent Bond-with-Viṣṇu instructs, I will carry it out. *(She joins her husband* DISCRIMINATION *and they leave the hall to where* BOND-WITH-VIṢṆU, *sentiment of joy, dwells.)* [151]

The two actors playing UPANIṢAD and DISCRIMINATION exited the hall swiftly. [152]

DEEP MEDITATION, rejoicing, went to dwell in INDIVIDUAL SELF's heart. The latter fell into inward meditation. [153]

KNOWLEDGE arrived, with shining radiance. That mass of obscurity MISCONCEPTION drifted apart from MIND. KNOWLEDGE showed her form and destroyed it immediately. She annihilated the whole extent of ILLUSION's manifestation. Then, together with MIND, she became absorbed in the SELF. Like wood burned by fire, even the own self will merge into the SELF. [154]

Then, having arrived from backstage, the lustre of the actor playing WISDOM MOON spread. As he rose, all desire for the three worlds was obliterated. He appeared, like a brave hero.

WISDOM MOON *(looking here and there)*: O Individual Self, look at me! I am Wisdom and I bow to you. Whatever you command me, I will accomplish at once.

INDIVIDUAL SELF, hearing WISDOM MOON's voice, got unsteadily to his feet, rejoicing and feeling happy. Running, he embraced WISDOM MOON and felt blissful. [155]

Some trace of darkness was still in the abode of INDIVIDUAL SELF's heart. Facing WISDOM MOON, it vanished, and there was only light. [156]

INDIVIDUAL SELF *(to himself)*: Oh my fortune is beyond words! I meet Wisdom Moon, bestower of happiness, a cool ray destroying the fear of darkness. I had fallen and stayed in the company of evil people, forgetting my pure form. This is the favour of Bond-with-Viṣṇu who, compassionate, saved me. Now I will not love anyone else, I will never forget the single eternal brahman. I will not trust evil people for I have experienced a lot of pain from them. I cannot see the different regions,[28] I have become pacified, having obliterated the confusion of worldly existence. There was the union with the being of Viṣṇu, with the help of Bond-with-Viṣṇu.[29] [157]

---

28 Or "I do not look at the various directions". Diśi and bidiśa mean opposite quarters or regions, which are not perceived by one who, liberated, no longer sees distinctions.

29 [. . .] | viṣṇu hona ko yoga bhayo aba viṣṇubhakti kī kṛpā sahāya || viṣṇu hona ko yoga bhayo is not easily translatable. I kept yoga as the subject and interpreted it in its literal meaning of "yoking, joining" (MW 856) and hona as a verbal substantive of ho-. See Strnad 2013: 448–451.

Individual Self thus untainted, the stain of Illusion was wiped away. Then the venerable Bond-With-Viṣṇu, consisting of the essence, spread all auspicious qualities and splendour.

Bond-with-Viṣṇu *(approaching* Individual Self *with extreme brilliance and majesty)*: It has been long time, son, that you have been separated from this bliss. Now all your tasks have been accomplished and my desire, too, has been fulfilled today. It was good that as many of your enemies were, they obtained destruction. I have seen you happy after that, that is why I, myself, felt happy. [158]

Individual Self, listening to her words, attained paramount bliss. Hastily, he fell at Bond-With-Hari's feet. [159]

Individual Self *(praising her with folded hands)*: Mother, you are a receptacle of all happiness. What is difficult to obtain if one has your favour? You are the fulfilment of all desires. By your majesty[30] I have now found a great bliss, erasing sorrow. Your infinite greatness, endless qualities, purifying deeds – how can they be expressed? [160]

Bond-with-Viṣṇu *(satisfied)*: Bless you, bless you! Now, if you have a wish, reveal it, for I will fulfil it immediately. Consider me extremely pleased and ask for your desired boon.

Individual Self: O lady, my tasks have been fulfilled today. Destroying the confusion of worldly existence, I have become stable, obtaining extreme peacefulness in happiness. All King Discrimination's enemies were destroyed. You settled me in a pure, fearless, and blissful place. Now, even if I have no desire left, all the same I will make a request of you. The sorrowful individual selves in the world – they wander in the whirlpools of the ocean of worldly existence, carried away in the wave of the sense of possession they die, sinking, they lose courage – make them cross, being merciful. O lady, please, grant me this boon!

Bond-with-Viṣṇu *(wisely)*: May prosperity be with all!

All the people exited the hall with the actors; they felt now that all had attained brahman. [161]

The First Actor danced beautifully. King Kīratibrahma, having seen the performance, became content. [162]

30 tuma pratāpa. Tuma is interpreted as a possessive. The same expression is repeated in 2.146 and 5.68.

The scum of the stain of the sense of possession vanished,
the First Actor had fulfilled all wishes with respect and reverence. [163]

When Kṛṣṇadās Bhaṭṭ had narrated this book to his pupil,
he, with a pure understanding, fell at his guru's feet. [164]

The forms of and inclination towards the objects of the senses became dull and disappeared from consciousness.
He became discriminating, learned, with clear understanding, and incomparable. [165]

Whoever will read, listen, understand, or ponder over this book,
the knot of misconception and ego will be untied from their hearts. [166]

There will be a bond with the Blessed Lord and the effulgence of discrimination.
Brajvāsīdās says: "Without a bond with God the three worlds are distressed. [167]

O servants of he who dwells in Braj (Kṛṣṇa), if you wish for eternal bliss,
then forsake desire for the objects of the senses and share a bond with Hari." [168]

Thus concludes Act VI of the "Drama of the Rise of the Wisdom Moon" in Brajbhāṣā.

Bibliography

Brajvāsīdās: Primary Sources

Lithographic Editions

Brajvāsīdās (1873): *Brajavilāsa (Dvitiyāvṛtti)*. Bombay: Nirnay Sagar Press.

Brajvāsīdās (1874): *Prabodhacandrodaya Nāṭaka*. Benares: Benares Light Press.

Brajvāsīdās (1875): *Prabodhacandrodaya Nāṭaka*. Lucknow: Naval Kishore Press. http://dbooks.bodleian.ox.ac.uk/books/PDFs/N13445609.pdf

Manuscripts

MS 455 (VS 1878): *Brajavilāsa*. 428 folios. Allahabad: Allahabad Museum.

MS 14008 (VS 1908): *Prabodhacandrodaya Nāṭaka*. Vrindavan: Vrindavan Research Institute.

MS 17098 (n.d.): *Prabodhacandrodaya Nāṭaka*. Vrindavan: Vrindavan Research Institute.

Additional Primary Sources

Ādiśeṣa (1980): *The Essence of Supreme Truth* (*Paramārthasāra*): *Sanskrit Text with Translation and Notes*. Translated by Henry Danielson. Leiden: Brill.

Banādās (1980): *Mahātmā Banādās viracit Ubhay Prabodhak Rāmāyaṇ*. Edited by Bhagvati Prasad Singh. Allahabad: Lokbharati Prakashan.

Dādū Dayāl (n.d.): *Śrī Dādūvāṇī*. Edited by Swami Narayandas. Jaipur: Shri Dadu Dayalu Mahasabha.

Dhauṁkal Miśra (1962): "*Prabodhacandrodaya Nāṭaka*". In: *Prabodhacandrodaya aur uskī hindī Paramparā*. Edited by Saroj Agrawal. Prayag: Hindi Sahitya Sammelan, 402–407.

Dīkṣit, Rāmadās (1835): "Prakāśa (Scholien)". In: *Prabodha Chandrodaya. Krishna Misri Comoedia*. Edited by Hermann Brockhaus. Lipsiae: Brockhaus, 3–106.

Gandhi, Aarti (2017): *Bihārī Satasaī ke Śreṣṭh Dohoṁ kā Saṅkalan*. New Delhi: Diamond Pocket Books.

Govarddhandās (1898): *Brajavilāsa Sārāvalī*. Kanpur: Naval Kishore Press.

Gulābsiṃh (1905): *Prabodhacandrodaya Nāṭaka*. Edited by Udasi Guruprasad. Bombay: Venkateshwar Steam Press.

Hariścandra, Bhāratendu (1987): "Nāṭak". In: *Bhāratendu Samagra*. Edited by Hemant Sharma. Varanasi: Hindi Pracarak Samsthan, 555–579.

Kabīr (1950): *Kabīr Sahāb kā Bījaka*. Edited by Hansdas Shashtri and Mahabir Prasad. Barabanki: Kabir Granth Prakashan Samiti.

Kabīr (1968): *Kabīr Granthāvalī (Śyāmsundardās Dvārā Sampādit Kabīr Granthāvalī ke Anusār)*. Commentary by S. Shukla and R. Caturvedi. Lucknow: Prakashan Kendra.

Kabīr (2013): *Mūlabījakaṭīkāsahita*. Mumbai: Khemraj Krishnadas.

Kṛṣṇamiśra (2009): *The Rise of the Wisdom Moon*. Edited by Matthew Kapstein. New York: New York University Press.

Malha kavi (1962): "*Prabodhacandrodaya Nāṭaka*". In: *Prabodhacandrodaya aur uskī hindī Paramparā*. Edited by Saroj Agrawal. Prayag: Hindi Sahitya Sammelan, 381–383.

https://doi.org/10.1515/9783111063836-016

Nānakdās (1962): "*Prabodhacandrodaya Nāṭaka*". In: *Prabodhacandrodaya aur uskī hindī Paramparā*. Edited by Saroj Agrawal. Prayag: Hindi Sahitya Sammelan, 397–401.

Olivelle, Patrick (trans. and ed.) (1998): *The Early Upaniṣads: Annotated Text and Translation*. Oxford: Oxford University Press.

Shastri, J. L. (ed.) (1983): *Bhāgavata Purāṇa of Kṛṣṇa Dvaipāyana Vyāsa with Sanskrit Commentary Bhāvārthabodhinī of Śrīdhara Svāmin*. Delhi: Motilal Banarsidass.

Siṃh, Jasvant (1972): "*Prabodhacandrodaya Nāṭaka*". In: *Jasvantsiṃh Granthāvalī*. Edited by Vishvanathprasad Mishra. Varanasi: Nagari Pracharini Sabha, 81–113.

Śaṅkara (2009): *Brahma-Sūtra-Bhāṣya of Śrī Śaṅkarācārya*. 10th impr. Translated by Swami Gambhirananda. Kolkata: Advaita Ashrama.

Tulsīdās (1962): *Vinaya-patrikā. Devadīpikāṭīkāsamalaṅkṛtā*. 2nd edn. Edited by Devnarayan Dvivedi. Varanasi: Jnanmandal Ltd.

Tulsīdās (1972): *Śrī Rāmacaritamānasa or the Mānasa Lake Brimming Over with the Exploits of Śrī Rāma*. 2nd edn. Edited by Hanumanprasad Poddar. Gorakhpur: Gita Press.

Tulsīdās (2006): *Śrī Kṛṣṇagītāvalī*. Gorakhpur: Gita Press.

Tulsīdās (2016–2023): *The Epic of Ram*. Vols 1–7. Translated by Philip Lutgendorf. Cambridge, MA: Harvard University Press.

Vālmīki (1986): *The Rāmāyaṇa of Vālmīki: An Epic of Ancient India*. Vol. 2, *Ayodhyākāṇḍa*. Edited by Robert P. Goldman. Translated by Sheldon Pollock. Princeton, NJ: Princeton University Press.

Viśvakarmā, Bihārīlāl (1972): "*Kauśalendra Kautuka*". In: *Tulsīdāsottara Hindī Rām-sāhitya*. Edited by Ram L. Pandey. Allahabad: Abhinava Bharati, 54–55.

Vivekananda, Swami (1957): "Paper on Hinduism". In: *Complete Works*, vol. 1,. 10th enl. edn. Calcutta: Advaita Ashrama, 6–20. 1st edn published in 1907.

Dictionaries

Callewaert, Winand M / Sharma, Swapna (2009): *Dictionary of Bhakti*, New Delhi: D.K. Printworld.

Das, Shyamsundar (1965–1975): *Hindī Śabdasāgar*, Kashi: Nagari Pracarini Sabha.

Gupta, Dindayal / Tandan, Premnarayan (1974): *Brajbhāṣā Sūr-Koś*. Lucknow: Lucknow Vishvavidyalaya.

Kulkarni, Krushnaji Pandurang (1993): *Marathi VyutpattiKkosh: Aitihasik va Taulanik*. Pune: Shubhada-Sarasvata Prakashan.

McGregor, Ronald S. (1993): *The Oxford Hindi-English Dictionary*. Oxford; Delhi: Oxford University Press.

MacDonell, Arthur A. (1929): *A Practical Sanskrit Dictionary with Transliteration, Accentuation, and Etymological Analysis Throughout*. London: Oxford University Press.

Monier-Williams, Monier (1899): *A Sanskrit-English Dictionary*. Oxford: Clarendon Press.

Platts, John T. (1884): *A Dictionary of Urdu, Classical Hindi, and English*. London: W. H. Allen & Co.

Prasad, Kalika (1956): *Bṛhat Hindī Koś*. Varanasi: Prakashak Gyanmandal.

Tiwari, Bholanath (1954): *Tulsī-Śabdasāgar*. Ilahabad: Hindustani Ekedamy.

Secondary Sources

Agrawal, Purushottam (2018): "*Nirguṇa* and *Saguṇa*". In: *Brill's Encyclopedia of Hinduism Online*. Edited by Knut A. Jacobsen, Helene Basu, Angelika Malinar, and Vasudha Narayanan. http://dx.doi.org/10.1163/2212-5019-beh-all (13/01/2021).

Agrawal, Saroj (1962): *Prabodhacandrodaya aur uskī hindī Paramparā*. Prayag: Hindi Sahitya Sammelan.

Alam, Muzaffar (1998): "The Pursuit of Persian: Language in Mughal Politics". *Modern Asian Studies* 32.2: 317–349.

Alam, Muzaffar (2010): *The Languages of Political Islam in India c. 1200–1800*. Ranikhet: Permanent Black.

Alam, Muzaffar / Subrahmanyam, Sanjay (2004): "The Making of a Munshi". *Comparative Studies of South Asia, Africa and the Middle East* 24.2: 61–72.

Ali, Daud (2004): *Courtly Culture and Political Life in Early Medieval India*. Cambridge: Cambridge University Press.

Ali, Daud (2011a): "Rethinking the History of the *Kāma* World in Early India". *Journal of Indian Philosophy* 39.1: 1–13. https://doi.org/10.1007/s10781-010-9115-7 (16/09/2020).

Ali, Daud (2011b): "Padmaśrī's *Nāgarasarvasva* and the World of Medieval Kāmaśāstra". *Journal of Indian Philosophy* 39.1: 41–62.

Alfaro, María J. M. (1996): "Intertextuality: Origins and Development of the Concept". *Atlantis* 18.1–2: 268–285.

Allchin, Frank R. (1966): *Tulsī Dās: The Petition to Rām*. London: George Allen & Unwin.

Allen, Graham (2011): *Intertextuality*. 2nd edn. London: Routledge.

Allen, Michael S. (2013): "*The Ocean of Inquiry*: A Neglected Classic of Late Advaita Vedānta". Unpublished PhD dissertation, Cambridge, MA, Harvard University.

Allen, Michael S. (2016): "Dueling Dramas, Dueling Doxographies: The *Prabodhacandrodaya* and *Saṃkalpasūryodaya*". *Journal of Hindu Studies* 9.3: 273–297.

Allen, Michael S. (2017): "Greater Advaita Vedānta: The Case of Niścaldās". *International Journal of Hindu Studies* 21.3: 275–297.

Allen, Michael S. (2020): "Greater Advaita Vedānta: The Case of Sundardās". *Journal of Indian Philosophy* 48.1: 49–78.

Allen, Michael S. / Venkatkrishnan, Anand (2017): "Introduction". *New Directions in the Study of Advaita Vedānta*. Special issue of *International Journal of Hindu Studies* 21.3: 271–274.

Bailey, Greg (2011): "'Him I Call a Brahmin': Further Instances of Intertextuality". *Studia Orientalia Electronica* 110: 3–19.

Bader, Jonathan (2000): *Conquest of the Four Quarters: Traditional Accounts of the Life of Śaṅkara*. New Delhi: Aditya Prakashan.

Bangha, Imre (2007): "Courtly and Religious Communities as Centres of Literary Activity in Eighteenth-Century India: Ānandghan's Contacts with the Princely Court of Kishangarh-Rupnagar and with the Maṭh of the Nimbārka Sampradāy in Salemabad". In: *Indian Languages and Texts Through the Ages: Essays of Hungarian Indologists in Honour of Prof. Csaba Tottossy*. Edited by Csaba Dezso. Delhi: Manohar, 307–354.

Bangha, Imre (2013): *It's a City-Showman's Show! Transcendental Songs of Anandghan*. London: Penguin.

Bangha, Imre (2014): *Scorpion in The Hand: Brajbhasha Court Poetry from Central India around 1800; A Critical Edition of Ṭhākur's Kabittas*. New Delhi: Manohar.

Bangha, Imre (2015): "A Curious King, a Psychic Leper, and the Workings of Karma: Bajid's Entertaining Narratives". In: *Tellings and Texts: Music, Literature and Performance in North India*. Edited by Francesca Orsini and Katherine Butler Schofield. Cambridge: OpenBook Publishers, 359–381.

Bangha, Imre (2019): "Shifts in Kabīr Contexts and Texts from Mughal to Modern Times". In: *Early Modern India: Literatures and Images, Texts and Languages*. Edited by Maya Burger and Nadia Cattoni. Heidelberg: CrossAsia, 125–172.

Bartley, Christopher J. (2002): *The Theology of Rāmānuja: Realism and Religion*. London: Routledge.

Barz, Richard (1976): *The Bhakti Sect of Vallabhācārya*. Faridabad: Thomson Press India.

Barz, Richard (1994): "The *Caurāsī Vaiṣṇavan kī Vārtā* and the Hagiography of the Puṣṭimārg". In: *According to Tradition: Hagiographical Writing in India*. Edited by Winand M. Callewaert and Rupert Snell. Wiesbaden: Harrassowitz, 43–64.

Beck, Guy L. (2012): *Sonic Liturgy: Ritual and Music in Hindu Tradition*. Columbia: University of South Carolina Press.

Benjamin, Walter (2012): "The Translator's Task". In: *The Translation Studies Reader*. 3rd edn. Edited by Lawrence Venuti. London: Routledge.

Bharadwaj, Ramdat (1979): *The Philosophy of Tulsidas*. New Delhi: Munshiram Manoharlal.

Bhattacarya, Siddheshwara (1960): *The Philosophy of the Śrīmad-Bhāgavata*. 2 vols. Santiniketan: Visva-Bharati.

Biardeau, Madeleine (1978): "Études de Mythologie Hindoue (V): Bhakti et Avatāra". *Bulletin de L'École française d'Extrême-Orient* 65.1: 87–238.

Bilimoria, Purushottama (1989) "The Idea of Authorless Revelation (Apauruṣeya) in the Mīmāṃsā". In: *Indian Philosophy of Religion*. Edited by Roy W. Perrett. Dordrecht: Martinus Nijhoff, 143–166.

Birch, Jason (2018): In: "Chapter 3: The Proliferation of Āsanas in Late-Medieval Yoga Texts". In: *Yoga in Transformation: Historical and Contemporary Perspectives*. Edited by Karl Baier, Philipp A. Maas, and Karin Preisendanz. Vienna: Vienna University Press, 101–179.

Birch, Jason / Hargreaves, Jacqueline (2016a): "The Yamas and Niyamas: Patanjali's View". *Yoga Scotland Magazine* 49: 32–35.

Birch, Jason / Hargreaves, Jacqueline (2016b): "The Yamas and Niyamas: Medieval and Modern Views". *Yoga Scotland Magazine* 50: 1–10.

Black, Brian (2015): "Dialogue and Difference: Encountering the Other in Indian Religious and Philosophical Sources". In: *Dialogue in Early South Asian Religions: Hindu, Buddhist, and Jain Traditions*. Edited by Brian Black and Laurie Patton. Burlington, VT: Ashgate, 243–257.

Bouthillette, Karl-Stéphan (2020): *Dialogue and Doxography in Indian Philosophy: Points of View in Buddhist, Jaina, and Advaita Vedānta Traditions*. Abingdon; New York: Routledge.

Bouy, Christian (1994): *Les Nātha-yogin et les Upaniṣads: étude d'histoire de la littérature hindoue*. Paris: De Boccard.

Brajratnadas (1944): *Hindī Nāṭya Sāhitya*. Benares: Hindi Sahitya Kutira.

Brockhaus, Hermann (1835): *Prabodha Chandrodaya. Krishna Misri Comoedia*. Lipsiae: Brockhaus.

Brockington, John L. (2004): "The Concept of 'Dharma' in the *Rāmāyaṇa*". *Journal of Indian Philosophy* 32.5/6: 655–670.

Brockington, John L. (2005): "The Epics in the Bhakti Tradition". In: *The Intimate Other: Love Divine in Indic Religions*. Edited by John L. Brockington and Anna S. King. Hyderabad: Orient Longman, 31–53.

Bryant, Kenneth / Hawley, John S. (2015): *Sur's Ocean: Poems from the Early Tradition*. Cambridge, MA: Harvard University Press.

Buchta, David (2016): "Devotion and Karmic Extirpation in Late Vedānta: Viṭṭhalanātha and Baladeva Vidyābhūṣaṇa on *Brahmasūtra* 4.1.13–19". *Journal of Hindu Studies* 9.1: 29–55.
Buitenen, Johannes A. B. van (1964): "The Large Ātman". *History of Religions* 4.1: 103–114.
Buitenen, Johannes A. B. van (1979): "'Ānanda' or All Desires Fulfilled". *History of Religions* 19.1: 27–36.
Bulcke, Camille (1999): *Ram Kathā: Utpatti aur Vikās*. 6th edn. Prayag: Hindi Sahitya Parishad.
Burchett, Patton (2013): "Bitten by the Snake: Early Modern Devotional Critiques of Tantra-Mantra". *Journal of Hindu Studies* 6.1: 1–20.
Burchett, Patton (2019): A Genealogy of Devotion: Bhakti, Tantra, Yoga, and Sufism in North India. New York: Columbia University Press.
Burger, Maya (2014a): "La *Sarvāṅgayogapradīpikā* de Sundardās : une classification des chemins du yoga au 17e siècle". *Asiatische Studien – Études Asiatiques* 68.3: 683–708.
Burger, Maya (2014b): "Jayatrām's *Jogpradīpakā*: Between Haṭha-yoga and Bhakti in the Eighteenth Century". In: *Bhakti Beyond the Forest: Current Research on Early Modern Literatures in North India, 2003–2009*. Edited by Imre Bangha. New Delhi: Manohar, 177–190.
Busch, Allison (2010): "Riti and Register: Lexical Variation in Courtly Braj Bhasha Texts". In: *Before the Divide: Hindi and Urdu Literary Culture*. Edited by Francesca Orsini. New Delhi: Orient Blackswan, 84–120.
Busch, Allison (2011): *Poetry of Kings: The Classical Hindi Literature of Mughal India*. New York: Oxford University Press.
Busch, Allison (2015): "Listening for the Context: Tuning in to the Reception of *Riti* Poetry". In: *Tellings and Texts: Music, Literature and Performance in North India*. Edited by Francesca Orsini and Katherine Butler Schofield. Cambridge: OpenBook Publishers, 249–282.
Busch, Allison (2018): "The Poetics of History in Padmakar's *Himmatbahādurvirudāvalī*". In: *Text and Tradition in Early Modern North India*. Edited by Tyler Walker Williams, Anshu Malhotra, and John Stratton Hawley. Delhi: Oxford University Press, 260–281.
Callewaert, Winand M. (2000): *The Hagiographies of Anantadās: The Bhakti Poets of North India*. London: Curzon.
Callewaert, Winand M. (2018): "Sants". In: *Brill's Encyclopedia of Hinduism Online*. Edited by Knut A. Jacobsen, Helene Basu, Angelika Malinar, and Vasudha Narayanan. http://dx.doi.org/10.1163/2212-5019-beh-all (13/01/2021).
Callewaert, Winand M. / Snell, Rupert (eds) (1994): *According to Tradition: Hagiographical Writing in India*. Wiesbaden: Harrassowitz.
Cappello, Giuseppe (2019): "*Il Gulzār-i ḥāl* by Banwālī dās: Un testimone storico della letteratura indo-persiana del periodo di Dārā Šikūh (XVII secolo)". Unpublished PhD dissertation, Naples, Università degli Studi di Napoli "L'Orientale".
Cappello, Giuseppe (2021): "Sufi-Vedāntic Interactions and Literary Interconnections in the *Gulzār-i ḥāl* by Banwālīdās". *Eurasian Studies* 18.2: 255–298.
Cappello, Giuseppe (forthcoming): "The *Gulzār-i ḥāl* by Banvālīdās: Two Possible Prefaces of an Indo-Persian Text". In: *Early Modern Literary Cultures in North India: Current Research*. Edited by Dantua Stasik and Imre Bangha. Oxford: Oxford University Press.
Caracchi, Pinuccia (2010): "Alcune considerazioni sulla percezione della figura di Sītā ieri e oggi". In: *Voci e conflitti*. Edited by Alessandra Consolaro. Alessandria: Edizioni dell'Orso, 1–22.
Carlisle, Clare / Ganeri, Jonardon (eds) (2010): *Philosophy as Therapeia*. Cambridge: Cambridge University Press.
Carman, John B. (1974): *The Theology of Rāmānuja: An Essay in Interreligious Understanding*. New Haven, CT; London: Yale University Press.

Cattoni, Nadia (2020): *Dev, l'artisan-poète du 18ème siècle et la nāyikā dans le Rasavilāsa: circulation et échanges, intertextualité et transformations*. Berlin: De Gruyter.

Cavaliere, Stefania (2020): "*Dhārmik* Kings in Courtly Agendas: The Figure of Rāma in the Works of Keśavdās". In: *Oral-Written-Performed: The Rāmāyaṇa Narratives in Indian Literature and Arts*. Edited by Danuta Stasik. Heidelberg; Berlin: CrossAsia, 53–75.

Chakrabarti, Samiran C. (1996): "*Ākāśa*". In: *Kalātattvakośa: A Lexicon of Fundamental Concepts of the Indian Arts*, vol. 3. Edited by Bettina Bäumer. New Delhi: Indira Gandhi National Centre for the Arts, 103–141.

Chand, Tara / Abidi Amir H. (eds) (1961): *Banvālī Dās Gulzār-ī ḥāl yā Ṭulū'-i qamar-i ma'rifat*. Aligarh: Aligarh Muslim University.

Chattopadyaya, Debiprasad / Gangopadhyaya, Mrinalkanti (eds) (1990): *Cārvāka/Lokāyata: An Anthology of Source Materials and Some Recent Studies*. Delhi: Indian Council of Philosophical Research.

Chaturvedi, Parashuram (1953): *Mānas ki Rāmkathā*. Allahabad: Kitab Mahal.

Chaudhuri, Roma (2012): *Ten Schools of the Vedānta*. Kolkata: Rabindra Bharati University.

Clark, Matthew J. (2006): *The Daśanāmī-Saṃnyāsīs: The Integration of Ascetic Lineages into an Order*. Leiden: Brill.

Clooney, Francis X. (2007): "Ramanuja and the Meaning of Krishna's Descent and Embodiment on This Earth". In: *Krishna: A Sourcebook*. Edited by Edwin F. Bryant. New York: Oxford University Press, 329–356.

Clooney, Francis X. (2017–2018): "Rāmānuja's *Nityam*: A Neglected Key to His Theology". In: *Śrī Rāmānuja's Sahasrābdi*. Special issue of *Brahmavidya: The Adyar Library Bulletin* 81/82: 231–264.

Clooney, Francis X. (2021): "Rāmānuja's *Nityagrantham* ('Manual of Daily Worship'): A Translation". *International Journal of Hindu Studies* 24.3: 345–380.

Clooney, Francis X. / Stewart, Tony K. (2004): "Vaiṣṇava". In: *The Hindu World*. Edited by Sushil Mittal and Gene Thursby. New York: Routledge, 165–166.

Cohen, Signe (2008): "The *Śvetāśvatara Upaniṣad*". In: *Text and Authority in the Older Upaniṣads*. Leiden: Brill, 213–245.

Comans, Michael (1993): "The Question of the Importance of Samādhi in Modern and Classical Advaita Vedānta". *Philosophy East and West* 43.1: 19–38.

Corcoran, Maura (1995): *Vṛndāvana in Vaiṣṇava Literature: History-Mythology-Symbolism*. Vrindavan: Vrindavan Research Institute; New Delhi: D.K. Printworld.

Cort, John E. (2015): "Making it Vernacular in Agra: The Practice of Translation by Seventeenth-Century Digambar Jains". In: *Tellings and Texts: Music, Literature and Performance in North India*. Edited by Francesca Orsini and Katherine Butler Schofield. Cambridge: OpenBook Publishers, 61–105.

Couture, André (2001): "From Viṣṇu's Deeds to Viṣṇu's Play, or Observations on the Word Avatāra as a Designation for the Manifestations of Viṣṇu". *Journal of Indian Philosophy* 29: 313–326.

Cummer, Katie / DiStefano, Lynne D. (eds) (2021): *Asian Revitalization: Adaptive Reuse in Hong Kong, Shanghai, and Singapore*. Hong Kong: Hong Kong University Press.

D'Onofrio, Svevo (2010): "A Persian Commentary to the *Upaniṣads*: Dārā Šikōh's *Sirr-i akbar*". In: *Muslim Cultures in the Indo-Iranian World during the Early-Modern and Modern Periods*. Edited by Denis Hermann and Fabrizio Speziale. Berlin: Klaus Schwarz, 533–564.

Dahl, Eystein / Stroński, Krzysztof (2016): "Ergativity in Indo-Aryan and beyond". In: *Indo-Aryan Ergativity in Typological and Diachronic Perspective*. Edited by Eystein Dahl and Krzysztof Stroński. Amsterdam; Philadelphia, PA: John Benjamins, 1–37.

Dalal, Neil (2014): "Contemplative Grammars: Śaṅkara's Distinction of *Upāsana* and *Nididhyāsana*". *Journal of Indian Philosophy* 44.1: 179–206.
Dalal, Neil (2020): "Contemplating Nonduality: The Method of *Nididhyāsana* in Śaṅkara's Advaita Vedānta". In: *The Bloomsbury Research Handbook of Vedānta*. Edited by Ayon Maharaj. London: Sydney Bloomsbury Academic, 45–74.
Dalmia, Vasudha (2006): *Poetics, Plays, and Performances: The Politics of Modern Indian Theatre*. Delhi: Oxford University Press.
Dalmia, Vasudha (2017): *The Nationalisation of Hindu Traditions: Bhāratendu Hariśchandra and Nineteenth-Century Banaras*. 3rd impr. Ranikhet: Permanent Black.
Dalmia, Vasudha / Faruqui, Munis D. (eds) (2014): *Religious Interactions in Mughal India*. New Delhi: Oxford University Press.
Dasa, Satyanarayana / Edelmann, Jonathan B. (2013): "Agency in the Gauḍīya Vaiṣṇava Tradition". In: *Free Will, Agency, and Selfhood in Indian Philosophy*. Edited by Matthew R. Dasti and Edwin F. Bryant. Oxford: Oxford University Press, 279–307.
Dasgupta, R. K. (1977): "Theology of Tulasīdāsa". In: *Tulasidasa: His Mind and Art*. Edited by Nagendra. New Delhi: National Publishing House, 117–124.
Dasgupta, Surendranath (1922): *A History of Indian Philosophy*. Vols 1–5. Cambridge: Cambridge University Press.
Dasgupta, Surendranath / De, Sushil Kumar (2017): *A History of Sanskrit Literature: Classical Period*. 1st MLBD edn. Delhi: Motilal Banarsidass.
de Bruijn, Thomas (2010): "Dialogism in a Medieval Genre: The Case of the Avadhi Epics". In: *Before the Divide: Hindi and Urdu Literary Culture*. Edited by Francesca Orsini. New Delhi: Orient Blackswan, 121–141.
Deussen, Paul (1906): *The Philosophy of the Upanishads*. Translated by Alfred Shenington Geden. Edinburgh: T. & T. Clark.
Deussen, Paul (1907): *Outlines of Indian Philosophy*. Berlin: Karl Curtius.
Deutsch, Eliot / Dalvi, Rohit (eds) (2004): *The Essential Vedānta: A New Source Book of Advaita Vedānta*. New Delhi: New Age Books.
Devanandan, Paul D. (1954): *The Concept of Māyā: An Essay in Historical Survey of the Hindu Theory of the World, with Special Reference to the Vedānta*. London: Lutterworth Press.
Dharwadker, Vinay (2003): *Kabīr: The Weaver's Songs*. New Delhi; New York: Penguin Books.
Dimock, Edward C. (1966): "Doctrine and Practice among the Vaiṣṇavas of Bengal". In: *Krishna: Myths, Rites, and Attitudes*. Edited by Milton B. Singer. Honolulu: East-West Center Provinces, 41–63.
Divanji, P. C. (1948): "Brahman-Ākāśa Equation: Its Origin and Development". *Bharatiya Vidya Bhavan* 9: 148–173.
Djurdjevic, Gordan / Singh, Shukdev (trans.) (2019): *Sayings of Gorakhnath: Annotated Translation of the Gorakh Bānī*. New York: Oxford University Press.
Doniger, Wendy (1980): *Women, Androgynes, and Other Mythical Beasts*. Chicago; London: University of Chicago Press.
Drocco, Andrea (2016): "The Restoration of the Ergative Case Marking of 'A' in Perfective Clauses in New Indo-Aryan: The Case of the Braj Language". In: *Indo-Aryan Ergativity in Typological and Diachronic Perspective*. Edited by Eystein Dahl and Krzysztof Stroński. Amsterdam; Philadelphia, PA: John Benjamins, 201–236.
Dwyer, Rachel (2006): *Filming the Gods: Religion and Indian Cinema*. London: Routledge.
Eco, Umberto (2020): *Dire quasi la stessa cosa. Esperienze di traduzione*. 2nd repr. Milan: Giunti/Bompiani. 1st edn published in 1992.
Entwistle, Alan W. (1987): *Braj: Centre of Krishna Pilgrimage*. Groningen: Egbert Forsten.

Ernst, Carl (2003): "Muslim Studies of Hinduism? A Reconsideration of Arabic and Persian Translations from Indian Languages". *Iranian Studies* 36.2: 173–195.

Faruqui Munir D. (2012): *The Princes of the Mughal Empire 1504–1719*. New York: Cambridge University Press.

Fitzgerald, James L. (2017a): "A Semantic Profile of Early Sanskrit '*Buddhi*'". *Journal of Indian Philosophy* 45.4: 669–709.

Fitzgerald, James L. (2017b): "The *Buddhi* in Early Epic *Adhyātma* Discourse (The Dialog of Manu and Bṛhaspati)". *Journal of Indian Philosophy* 45.4: 767–816.

Fort, Andrew O. (1984): "The Concept of Sākṣin in Advaita Vedānta". *Journal of Indian Philosophy* 12.3: 277–290.

Fort, Andrew O. (1998): *Jīvanmukti in Transformation: Embodied Liberation in Advaita and Neo-Vedanta*. Albany: State University of New York Press.

Fort, Andrew O. (2000): "Reflections on Reflection: *Kūṭastha, Cidābhāsa* and *Vṛttis* in the *Pañcadaśī*". *Journal of Indian Philosophy* 28.5/6: 497–510.

Frazier, Jessica (2018): "Philosophy and Religion in India". In: *History of Indian Philosophy*, vol. 1. Edited by Purushottama Bilimoria. London: Routledge, 59–70.

Freschi, Elisa (2015): "The Reuse of Texts in Indian Philosophy: Introduction". *Journal of Indian Philosophy* 43.2/3: 85–108.

Freschi, Elisa (2017–2018): "Bhakti in Rāmānuja: Continuities and Changes of Perspective". In: *Śrī Rāmānuja's Sahasrābdi*. Special issue of *Brahmavidya: The Adyar Library Bulletin* 81/82: 285–311.

Freschi, Elisa (2019): "We Resort to Reason: The Argumentative Structure in Veṅkaṭanātha's *Seśvaramīmāṃsā*". In: *In Dialogue with Classical Indian Traditions: Encounter, Transformation, and Interpretation*. Edited by Brian Black and Chakravarthi Ram-Prasad. Abingdon; New York: Routledge, 66–83.

Freschi, Elisa / Maas, Philipp A. (2017): "Introduction: Conceptual Reflections on Adaptive Reuse". In: *Adaptive Reuse: Aspects of Creativity in South Asian Cultural History*. Edited by Elisa Freschi and Philipp A. Maas. Wiesbaden: Harrassowitz, 11–25.

Friedlander, Peter G. (1996): *A Descriptive Catalogue of the Hindi Manuscripts in the Library of the Wellcome Institute for the History of Medicine*. London: Trustees of the Wellcome Trust.

Friedlander, Peter G. (2011): "Before Translation?". In: *Translation in Asia: Theories, Practices and Histories*. Edited by Ronit Ricci and Jan van der Putten. Manchester: St. Jerome Publishing, 45–56.

Fuchs, Martin (2019): "Self-Affirmation, Self-Transcendence and the Relationality of Selves: The Social Embedment of Individualisation in *Bhakti*". In: *Religious Individualisation: Historical Dimensions and Comparative Perspectives*, vol. 1. Edited by Martin Fuchs, Antje Linkenbach, Martin Mulsow, Rahul Bjørn Parson, and Jörg Rüpke. Berlin; Boston: De Gruyter, 257–288.

Gambhirananda, Swami (trans.) (1965): *Eight Upaniṣads with the Commentary of Śaṅkarācārya*. Vol. 2. Kolkata: Advaita Ashrama.

Gandhi, Supriya (2014): "The Prince and the *Muvaḥḥid*: Dārā Shikoh and Mughal Engagements with Vedānta". In: *Religious Interactions in Mughal India*. Edited by Vasudha Dalmia and Munis D. Faruqui. Delhi: Oxford University Press, 65–101.

Gandhi, Supriya (2020a): *The Emperor Who Never Was: Dara Shukoh in Mughal India*. Cambridge: Belknap Press of Harvard University Press.

Gandhi, Supriya (2020b): "The Persian Writings on Vedānta Attributed to Banwālīdās Walī". *Journal of Indian Philosophy* 48.1: 79–99.

Ganeri, Jonardon (2001): *Philosophy in Classical India: The Proper Work of Reason*. London; New York: Routledge.

Ganeri, Jonardon (2010): "A Return to the Self: Indians and Greeks on Life as Art and Philosophical Therapy". In: *Philosophy as Therapeia*. Edited by Clare Carlisle and Jonardon Ganeri. Cambridge: Cambridge University Press, 119–135.

Ganeri, Jonardon (2011): *The Lost Age of Reason: Philosophy in Early Modern India, 1450–1700*. Oxford: Oxford University Press.

Ganeri, Jonardon (2014): "Philosophical Modernities: Polycentricity and Early Modernity in India". *Royal Institute of Philosophy Supplement* 74: 75–94.

Ganeri, Jonardon (2017): *The Oxford Handbook of Indian Philosophy*. New York: Oxford University Press.

Garcin de Tassy, Joseph H. (1839–1947): *Histoire de la littérature hindoui et hindoustani*. Paris: Printed under the auspices of the Oriental Translation Committee of Great Britain and Ireland.

Gaur, Krishnaprasad (ed.) (1964): *Hastalikhit Hindī Pustakoṃ kā Saṃkṣipt Vivaraṇ (Dvitīya Khaṇḍa)*. Varanasi: Nagari Pracharini Sabha.

Genette, Gérard (1982): *Palimpsestes. La littérature au second degré*. Paris: Editions du Seuil.

Gerow, Edwin (1971): *A Glossary of Indian Figures of Speech*. The Hague; Paris: Mouton.

Ghosh, Manmohan (trans.) (1951): *The Nāṭyaśāstra ascribed to Bharata-Muni*. Vol. 1. Calcutta: Asiatic Society.

Gokhale, Pradeep (1993): "The Cārvāka Theory of *Pramāṇas*: A Restatement". *Philosophy East and West* 43.4: 675–682.

Goldman, Robert P. (1986): "The Serpent and the Rope on Stage: Popular, Literary, and Philosophical Representations of Reality in Traditional India". *Journal of Indian Philosophy* 14.4: 349–369.

Goldstücker, Theodor (1842): *Prabodha-Chandrodaya oder die Geburt der Begriffs: ein theologisch-philosophisches Drama*. Königsberg: Theile.

Gonda, Jan (1968): "The Concept of a Personal God in Ancient Indian Religious Thought". *Studia Missionalia* 17: 111–136.

Gonda, Jan (1969): *Ancient Indian Kingship from the Religious Point of View*. 2nd edn. Leiden: Brill. 1st edn published in 1966.

Goudriaan, Teun (1978): *Māyā Divine and Human: A Study of Magic and its Religious Foundations in Sanskrit Texts*. Delhi: Motilal Banarsidass.

Goudriaan, Teun (1992): "The Pluriform Ātman from The Upaniṣads to the Svacchanda Tantra". *Wiener Zeitschrift für die Kunde Südasiens / Vienna Journal of South Asian Studies* 36: 163–186.

Goudriaan, Teun (1994): "Rejection of Duality: On the Ātman Conception in the *Viṣṇu Purāṇa*". *Wiener Zeitschrift für die Kunde Südasiens / Vienna Journal of South Asian Studies* 38: 435–449.

Grierson, George (1903): "Art. XVI. Tulasī Dāsa, Poet and Religious Reformer". *Journal of the Royal Asiatic Society* 35.3: 447–466.

Growse, Frederic S. (1883): *Mathura: A District Memoir*. 3rd rev. edn. [Allahabad?]: North Western and Oudh Government Press. 1st edn published in 1874,

Gupta, Bina (1998): *The Disinterested Witness: A Fragment of Advaita Vedānta Phenomenology*. Evanston, IL: Northwestern University Press.

Gupta, Gopal K. (2020): *Māyā in the Bhāgavata Purāṇa*. Oxford: Oxford University Press.

Gupta, Kishorilal (1967): *Saroj Sarvekṣaṇ*. Allahabad: Hindustani Academy.

Gupta, Ravi M. (2007): *The Caitanya Vaiṣṇava Vedānta of Jīva Gosvāmī: When Knowledge Meets Devotion*. London: Routledge.

Gupta, Ravi M. (2016a): "Where One is Forever Two: God and World in Jīva Gosvāmī's *Bhāgavata-sandarbha*". In: *Caitanya Vaiṣṇava Philosophy: Tradition Reason and Devotion*. Abingdon; New York: Routledge, 35–60. Volume first published in 2014 by Ashgate.

Gupta, Ravi M. (2016b): "A Pearl Necklace of Propositions: Baladeva Vidyābhūṣaṇa's *Prameyaratnāvalī*". In: *Caitanya Vaiṣṇava Philosophy: Tradition Reason and Devotion*. Abingdon; New York: Routledge, 113–134. Volume first published in 2014 by Ashgate.
Gupta, Ravi M. (2018): "Translators' Troubles: Seeking a Method that Suits the *Bhāgavata Purāṇa*". *International Journal of Hindu Studies* 22.1: 175–189.
Gupta, Sanjukta (2006): *Advaita Vedānta and Vaiṣṇavism: The Philosophy of Madhusūdana Sarasvatī*. London: Routledge.
Gupta, Somnath (1958): *Pūrv-Bhāratendu Nāṭak Sāhitya*. Allabhabad: Hindi Bhavan.
Hacker, Paul (1960): "Zur Entwicklung der Avatāralehre". *Wiener Zeitschrift für die Kunde Südasiens / Vienna Journal of South Asian Studies* 4: 47–70.
Hacker, Paul (1995): "Distinctive Features of the Doctrine and Terminology of Śaṅkara: Avidyā, Nāmarūpa, Māyā, Īśvara". In: *Philology and Confrontation: Paul Hacker on Traditional and Modern Vedānta*. Edited by Wilhelm Halbfass. Albany: State University of New York Press, 57–100.
Halbfass, Wilhelm (1983): *Studies in Kumārila and Śaṅkara*. Reinbek: I. Wezler Verlag für orientalistische Fachpublikationen.
Halbfass, Wilhelm (1991): "The Therapeutic Paradigm and the Search for Identity in Indian Philosophy". In: *Traditions and Reflection: Explorations in Indian Thought*. Albany: State University of New York Press, 243–263.
Hanneder, Jurgen (1999): "A Conservative Approach to Sanskrit Śāstras: Madhusūdana Sarasvatī's '*Prasthānabheda*'". *Journal of Indian Philosophy* 27: 575–581.
Hansen, Kathrin (1983): "Indian Folk Traditions and the Modern Theatre". *Asian Folklore Studies* 42.1: 77–89.
Hansen, Kathrin (1992): *Grounds for Play: The Nauṭaṅkī Theatre of North India*. Berkeley: University of California Press.
Hardy, Friedhelm (1979): "The Philosopher as Poet: A Study of Vedāntadeśika's *Dehalīśastuti*". *Journal of Indian Philosophy* 7.3: 277–325.
Hardy, Friedhelm (1983): *Viraha-Bhakti: The Early History of Kṛṣṇa Devotion in South India*. Delhi; Oxford: Oxford University Press.
Hawley, John S. (1988): "Author and Authority in the *Bhakti* Poetry of North India". *Journal of Asian Studies* 47.2: 269–290.
Hawley, John S. (1992): "Creative Enumeration in the *Vinaya* Poetry of Sūrdās". In: *Devotional Literature in South Asia: Current Research 1985–1988*. Edited by Ronald S. McGregor. Cambridge: University of Cambridge, 225–236.
Hawley, John S. (2005): *Three Bhakti Voices: Mirabai, Surdas, and Kabir in Their Time and Ours*. Delhi; Bombay: Oxford University Press.
Hawley, John S. (2009a): *The Memory of Love: Sūrdās Sings to Krishna*. Oxford: Oxford University Press.
Hawley, John S. (2009b): "The *Bhāgavata-Māhātmya* in Context". In: *Patronage and Popularisation, Pilgrimage and Procession: Channels of Transcultural Translation and Transmission in Early Modern South Asia; Papers in Honour of Monika Horstmann*. Edited by Heidi R. M. Pauwels. Wiesbaden: Harrassowitz, 81–100.
Hawley, John S. (2011): "The Four Sampradāys: Ordering the Religious Past in Mughal North India". *South Asian History and Culture* 2.2: 160–183.
Hawley, John S. (2016): "Can There Be a Vaishnava Kabir?". *Studies in History* 32.2: 147–161.
Hegewald, Julia A. B. / Mitra, Subrata K. (eds) (2012): *Re-Use: The Art and Politics of Integration and Anxiety*. New Delhi: Sage.
Holdrege, Barbara (2013): "The Gauḍīya Discourse of Embodiment: Re-visioning Jñāna and Yoga in the Embodied Aesthetics of Kṛṣṇa Bhakti". *Journal of Hindu Studies* 6.2: 154–197.

Holdrege, Barbara (2014): "Meditation as Devotional Practice in Jīva Gosvāmin's Philosophy of Education". *ISKCON Studies Journal* 2: 45–70.
Holdrege, Barbara (2015): *Bhakti and Embodiment: Fashioning Divine Bodies and Devotional Bodies in Kṛṣṇa Bhakti*. London: Routledge.
Hopkins, Steven P. (2002): *Singing the Body of God: The Hymns of Vedāntadeśika in Their South Indian Tradition*. New York: Oxford University Press.
Horstmann, Monika (2009): *Der Zusammenhalt der Welt: religiöse Herrschaftslegitimation und Religionspolitik Maharaja Savai Jaisinghs (1700–1743)*. Wiesbaden: Otto Harrassowitz.
Horstmann, Monika (2011): "Theology and Statecraft". *South Asian History and Culture* 2.2: 184–204.
Horstmann, Monika (2017): "Nāth and Dādūpanthī Critique of Jains". *International Journal of Jaina Studies* 13.1: 1–72.
Horstmann, Monika (2018): "Three Brajbhāṣā Versions of the *Bhāgavata Purāṇa*". *International Journal of Hindu Studies* 22.1: 123–174.
Horstmann, Monika / Mishra, Anand (2013): "Vaishnava Sampradayas on the Importance of Ritual: A Comparison of the Two Contemporaneous Approaches by Viṭṭhalanātha and Jīva Gosvāmī". In: *Bhakti Beyond the Forest: Current Research on Early Modern Literatures in North India, 2003–2009*. Edited by Imre Bangha. New Delhi: Manohar, 147–168.
Huart, Clément / Massignon, Louis (1926): "Les entretiens de Lahore [entre le prince impérial Dârâ Shikûh et l'ascète hindou Baba La'l Das]". *Journal Asiatique* 209: 285–334.
Hultzsch, E. (1892): "A Chandella Inscription from Mahobā". In: *Epigraphia Indica: A Collection of Inscriptions, Supplementary to the Corpus Inscriptionum Indicarum of the Archaeological Survey* (Archaeological Survey of India), vol. 1. Edited by Jas. Burgess. Calcutta: Printed and published by the Superintendent of Government Printing, India, 217–222.
Ingalls, Daniel H. (1967): "Bhāskara the Vedāntin". *Philosophy East and West* 17.1/4: 61–67.
Iyer, S. R. (ed. and trans.) (1979): *Tarkabhāṣā of Keśava Miśra*. Varanasi: Chaukhambha Orientalia.
Jhaveri, Indukala H. (1956): "The Concept of Ākāśa in Indian Philosophy". *Annals of the Bhandarkar Oriental Research Institute* 37.1/4: 300–307.
Johnson, Willard (1976): "On the ṚG Vedic Riddle of the Two Birds in the Fig Tree (RV 1.164.20–22), and the Discovery of the Vedic Speculative Symposium". *Journal of the American Oriental Society* 96.2: 248–258.
Kaplan, Stephen (2018): "Avidyā: The Hard Problem in Advaita Vedānta". In: *History of Indian Philosophy*, vol. 1. Edited by Purushottama Bilimoria. London: Routledge, 242–250.
Kapstein, Matthew (2009): *The Rise of the Wisdom Moon*. New York: New York University Press.
Kapur S. K. (2011): *Bihari Satsai: A Commentary*. New Delhi: Publications Division, Ministry of Information and Broadcasting.
Karmarkar, Ragunath D. (1919–2020): "Comparison of the Bhaṣyas of Śaṅkara, Ramanuja, Keśavakaśmirin and Vallabha on Some Crucial Sutras". *Annals of the Bhandarkar Oriental Research Institute* 1.2: 105–127.
Karmarkar, Raghunath D. (1953): *Gauḍapāda-kārikā*. Poona: Bhandarkar Oriental Research Institute.
Katre, S. L. (1963): "*Jñānadipikā*, a Genuine & Pre-*Rāmacaritamānasa* Work of Gosvāmin Tulasīdāsa, its Date–Saṁvat 1622 = 1565 A.C.". In: *Munshi Indological Felicitation Volume*. Edited by Jayantkrshna H. Dave, H. D. Velankar, and A. K. Majumdar. Bombay: Bharatiya Vidya Bhavan, 401–411.
Kaul, Mrinal (2020): "'Foreword' to 'Explorations in Indian Philosophy'". In: *Explorations in Indian Philosophy*. Edited by Rajendran Chettiarthodi. New Delhi: D.K. Printworld, v–xxx.
Keating, Malcolm (2020): "The Literal-Nonliteral Distinction in Classical Indian Philosophy". In: *The Stanford Encyclopedia of Philosophy*. Edited by Edward N. Zalta. https://plato.stanford.edu/entries/literal-nonliteral-india/ (20/04/2021).

Keith, A. Berriedale (1954): *The Sanskrit Drama in its Origin, Development, Theory & Practice*. London: Oxford University Press. Repr. of 1924 1st edn.

Kellogg, Samuel Henry (1989): *A Grammar of the Hindi Language*. 3rd edn. New Delhi; Madras: Asian Educational Services.

Kinra, Rajeev (2015): *Writing Self, Writing Empire: Chandar Bhan Brahman and the Cultural World of the Indo-Persian State Secretary*. Oakland: University of California Press.

Krishnamachariar, M. (1937): *History of Classical Sanskrit Literature*. Madras: Tirumalai-Tirupati Devasthanams Press.

Kristeva, Julia (1986): *The Kristeva Reader*. Edited by Toril Moi. New York: Columbia University Press.

Lalitacharan, Goswami (1957): *Śrī Hita Harivaṃśa Gosvāmī: Sampradāya aur Sāhitya*. Vrindavan: Venu Prakashan.

Lamb, Ramdas (2002): *Rapt in the Name: The Ramnamis, Ramnam, and Untouchable Religion in Central India*. Albany: State University of New York Press.

Larson, Gerald J. (1969): *Classical Sāṃkhya: An Interpretation of its History and Meaning*. Delhi: Motilal Banarsidass.

Larson, Gerald J. (2018): *Classical Yoga Philosophy and the Legacy of Sāṃkhya: With Sanskrit Text and English Translation of Pātañjala Yogasūtra-s, Vyāsa Bhāṣya and Tattvavaiśāradī of Vācaspatimiśra*. Delhi: Motilal Banarsidass.

Leclère, Basile (2013): *Le théâtre de l'Inde médiévale entre tradition et innovation : le Moharājaparājaya de Yaśaḥpāla*. Marburg: Indica et Tibetica.

Lipner, Julius (1986): *The Face of Truth: A Study of Meaning and Metaphysics in the Vedāntic Theology of Rāmānuja*. Houndmills; Basingstoke: Macmillan Press.

Lipner, Julius (2018): "Rāmānuja". In: *Brill's Encyclopedia of Hinduism Online*. Edited by Knut A. Jacobsen, Helene Basu, Angelika Malinar, and Vasudha Narayanan. http://dx.doi.org/10.1163/2212-5019-beh-all (19/01/2021).

Lutgendorf, Philip (1991): *The Life of a Text: Performing the Rāmcaritmānas of Tulsidas*. Berkeley: University of California Press.

Lutjeharms, Rembert (2016): "An Ocean of Emotion: Rasa and Religious Experience in Early Caitanya Vaiṣṇava Thought". In: *Caitanya Vaiṣṇava Philosophy: Tradition Reason and Devotion*. Edited by Ravi Gupta. Abingdon; New York: Routledge, 175–215. Volume first published in 2014 by Ashgate.

Maas, Philipp A. (2017): "From Theory to Poetry: The Reuse of Patañjali's Yogaśāstra in Māgha's *Śiśupālavadha*". In: *Adaptive Reuse: Aspects of Creativity in South Asian Cultural History*. Edited by Elisa Freschi and Philipp A. Maas. Wiesbaden: Harrassowitz, 29–62.

Maas, Philipp A. (2018): "Chapter 2: 'Sthirasukham Asanam': Posture and Performance in Classical Yoga and Beyond". In: *Yoga in Transformation: Historical and Contemporary Perspectives*. Edited by Karl Baier, Philipp A. Maas, and Karin Preisendanz. Vienna: Vienna University Press, 51–100.

Madaio, James (2018): "The Instability of Non-dual Knowing: Post-gnosis Sādhana in Vidyāraṇya's Advaita Vedānta". *Journal of Dharma Studies* 1.1: 11–30.

Mahadevan, T. M. P. (1969): *The Pañcadaśī of Bhāratītīrtha-Vidyāraṇya: An Interpretive Exposition*. University of Madras: Centre of Advances Study in Philosophy.

Maharaj, Ayon (Swami Medhananda) (ed.) (2020): *The Bloomsbury Research Handbook of Vedānta*. London: Sydney Bloomsbury Academic.

Malinar, Angelika (2007): *The Bhagavadgītā: Doctrines and Contexts*. Cambridge: Cambridge University Press.

Malinar, Angelika (2011): "Hindu Cosmologies". In: *The Continuum Companion to Hindu Studies*. Edited by Jessica Frazier. London: Continuum, 60–74.

Malinar, Angelika (2015): "Religious Pluralism and Processes of Individualisation in Hinduism". *Religion* 45.3: 386–408.
Malinar, Angelika (2017a): "Philosophy in the *Mahābhārata* and the History of Indian Philosophy". *Journal of Indian Philosophy* 45.4: 587–607.
Malinar, Angelika (2017b): "Narrating Sāṃkhya Philosophy: Bhīṣma, Janaka and Pañcaśikha at *Mahābhārata* 12.211–12". *Journal of Indian Philosophy* 45.4: 609–649.
Malinar, Angelika (2018): "Puruṣa". In: *Brill's Encyclopedia of Hinduism Online*. Edited by Knut A. Jacobsen, Helene Basu, Angelika Malinar, and Vasudha Narayanan. http://dx.doi.org/10.1163/2212-5019-beh-all (15/11/2020).
Mallinson, James (2014): "*Haṭhayoga*'s Philosophy: A Fortuitous Union of Non-Dualities". *Journal of Indian Philosophy* 42: 225–247.
Mallinson, James / Singleton, Mark (2017): *Roots of Yoga*. London: Penguin Books.
Mangraviti, Fabio (2019): "The Construction of a Hindu Reformer: A Diachronic Study of Tulsidas' Moral Reception in Hindi Literary Criticism and Historiography (19th to Mid 20th Century)". *Zeitschrift Für Indologie Und Südasienstudien* 36: 138–171.
Marfatia, Mridula (1967): *The Philosophy of Vallabhācarya*. New Delhi: Munshiram Manoharlal.
Matchett, Freda (2008): *Kṛṣṇa: Lord or Avatāra? The Relationship between Kṛṣṇa and Viṣṇu*. London: Routledge. First published in 2001 by Curzon.
McGregor, Ronald S. (1968): *The Language of Indrajit of Orchā: A Study of Early Braj Bhāṣā Prose*. Cambridge: Cambridge University Press.
McGregor, Ronald S. (1971): "Some Manuscripts Containing Nanddās's Version of the *Prabodhacandrodaya* Drama". *Journal of the American Oriental Society* 91.4: 487–493.
McGregor, Ronald S. (1976): "Tulsīdās' *Śrīkṛṣṇagītāvalī*". *Journal of the American Oriental Society* 96.4: 520–527.
McGregor, Ronald S. (1984): *Hindi Literature from its Beginnings to the Nineteenth Century*. Wiesbaden: Harrassowitz.
McGregor, Ronald S. (1986): "A Brajbhāṣā Adaptation of the Drama Prabodhacandrodaya by Nanddās of the Sect of Vallabha". In: *Perspectives on Indian Religion: Papers in Honour of Karel Werner*. Edited by Peter Connolly. Delhi: Sri Satguru Publications, 135–144.
McGregor, Ronald S. (2003): "The Progress of Hindi, Part 1: The Development of a Transregional Idiom". In: *Literary Cultures in History: Reconstructions from South Asia*. Edited by Sheldon Pollock. Berkeley: University of California Press, 912–957.
McHugh, James (2011): "The Incense Trees of the Land of Emeralds: The Exotic Material Culture of *Kāmaśāstra*". *Journal of Indian Philosophy* 39.1: 63–100.
Menezes, Walter (2017): *Exploring Ātman from the Perspective of the Vivekacūḍāmaṇi*. Cham: Palgrave Macmillan.
Miner, Allyn (1997): *Sitar and Sarod in the 18th and 19th Centuries*. Delhi: Motilal Banarsidass.
Minkowski, Christopher (2011): "Advaita Vedānta in Early Modern History". *South Asian History and Culture* 2.2: 205–231.
Minkowski, Christopher / O'Hanlon, Rosalind / Venkatkrishnan, Anand (2015): "Introduction: Social History in the Study of Indian Intellectual Cultures?". *South Asian History and Culture* 6.1: 1–9.
Mody, Sujata (2012): "Literary Self-Determination and the Disciplinary Boundaries of Hindi Literature in the Early Twentieth Century". *South Asia Research* 32.3: 233–256.
Mujtabai, Fatullah (1978): *Aspects of Hindu-Muslim Cultural Relations*. Delhi: National Book Bureau.
Myers, Michael W. (1998): "Śaṅkarācārya and Ānanda". *Philosophy East and West* 48.4: 553–567.
Nair, Shankar (2020): *Translating Wisdom: Hindu-Muslim Intellectual Interactions in Early Modern South Asia*. Oakland: University of California Press.

Nambiar, Sita Krishna (1971): *Prabodhacandrodaya of Kṛṣṇa Miśra*. Delhi: Motilal Banarsidass.
Nanddās (1973): *The Round Dance of Krishna and Uddhav's Message*. Translated by Ronald S. McGregor. London: Luzac.
Narain, K. (2003): *The Fundamentals of Advaita Vedānta*. Varanasi: Indological Research Centre.
Narain, K. (2004): *The Philosophy of the Vāllabha School of Vedānta*. Varanasi: Indological Research Centre.
Narayanan, Vasudha (1996): "Arcāvatāra: On Earth as He Is in Heaven". In: *Gods of Flesh, Gods of Stone: The Embodiment of Divinity in India*. Edited by Joanne Punzo Waghorne and Norman Cutler. New York: Columbia University Press, 53–68.
Nelson, Lance E. (2004): "The Ontology of *Bhakti*: Devotion as *Paramapuruṣārtha* in Gauḍīya Vaiṣṇavism and Madhusūdana Sarasvatī". *Journal of Indian Philosophy* 32.4: 345–392.
Nelson, Lance E. (2007): "Krishna in Advaita Vedanta: The Supreme Brahman in Human Form". In: *Krishna: A Sourcebook*. Edited by Edwin F. Bryant. New York: Oxford University Press, 309–328.
Nicholson, Andrew J. (2005): "Vijñānabhikṣu's Yoga: A Note on Doctrine and Identity in Late Medieval India". *Journal of Vaishnava Studies* 14.1: 43–63.
Nicholson, Andrew J. (2007): "Reconciling Dualism and Non-Dualism: Three Arguments in Vijñānabhikṣu's Bhedābheda Vedānta". *Journal of Indian Philosophy* 35.4: 371–403.
Nicholson, Andrew J. (2010): *Unifying Hinduism: Philosophy and Identity in Indian Intellectual History*. New York: Columbia University Press.
Norelius, Per-Johan (2016): "The Honey-Eating Birds and the Tree of Life: Notes on *Ṛgveda* 1.164.20–22". *Acta Orientalia* 77: 3–70.
Novetzke, Christian L. (2007) "Bhakti and its Public". *International Journal of Hindu Studies* 11.3: 255–272.
Novetzke, Christian L. (2015): "Note to Self: What Marathi *Kirtankars*' Notebooks Suggest about Literacy, Performance, and the Travelling Performer in Pre-Colonial Maharashtra". In: *Tellings and Texts: Music, Literature and Performance in North India*. Edited by Francesca Orsini and Katherine Butler Schofield. Cambridge: OpenBook Publishers, 169–184.
O'Connell, Joseph T. (2016): "Caitanya Vaiṣṇava Ethics in Relation to Devotional Community". In: *Caitanya Vaiṣṇava Philosophy: Tradition Reason and Devotion*. Edited by Ravi Gupta. Abingdon; New York: Routledge, 135–162. Volume first published in 2014 by Ashgate.
Ojha, Dasharath (1995): *Hindī Nāṭak: Udbhav aur Vikās*. Delhi: Rajpal & Sons.
Okita, Kiyokazu (2014): *Hindu Theology in Early Modern South Asia: The Rise of Devotionalism and the Politics of Genealogy*. Oxford: Oxford University Press.
Okita, Kiyokazu (2016): "A Vedāntic Refutation of Buddhism in 18th-Century North India: The *Tattvadīpikā* of Baladeva Vidyābhūṣaṇa". *Journal of Vaishnava Studies* 25.1: 153–162.
Okita, Kiyokazu (2020): "Rejecting Monism: Dvaita Vedānta's Engagement with the *Bhāgavatapurāṇa*". *Journal of Indian Philosophy* 48.3: 447–465.
Olivelle, Patrick (1997): "Orgasmic Rapture and Divine Ecstasy: The Semantic History of 'Ānanda'". *Journal of Indian Philosophy* 25.2: 153–180.
Olivelle, Patrick (2009): *Pañcatantra: The Book of India's Folk Wisdom*. Repr. of 1999 pbk edn. New York: Oxford University Press. 1st edn published in 1997.
Olivelle, Patrick (2013): *King, Governance, and Law in Ancient India: Kauṭilya's Arthaśāstra*. New York: Oxford University Press.
Orsini, Francesca (1998): "Tulsī Dās as a Classic". In: *Classics of Modern South Asian Literature*. Edited by Rupert Snell and Ian Raeside. Wiesbaden: Harrassowitz, 119–142.

Orsini, Francesca (2004): "Pandits, Printers and Others: Publishing in Nineteenth-Century Benares". In: *Print Areas: Book History in India*. Edited by Swapan Chakravorty and Abhijit Gupta. Delhi: Permanent Black, 103–138.

Orsini, Francesca / Butler Schofield, Katherine (eds) (2015): *Tellings and Texts: Music, Literature and Performance in North India*. Cambridge: OpenBook Publishers.

Pahlarai, Prem (2013): "Niścaldās and his *Vṛttiprabhākar*: Advaita Vedānta in the Vernacular". Unpublished PhD dissertation, Seattle, University of Washington.

Pandey, Chandrabhushan (1955): *Rājā Rudrapratāp Siṁh viracit Susiddhāntottam Rāmkhaṇḍ ke Rāmkathā: Sandarbhoṁ kā Ālocanātmak Viśleṣaṇ*. Allahabad: University of Allahabad.

Pandey, Indrajit / Mishra, Vidyadhar (ed.) (1993). *Rāmāśvamedh: Uttar Rāmāyaṇ*. Kolkata: Jivan Jyoti Nyas.

Pandey, R. C. (1977): "Philosophy of Tulasīdāsa". In: *Tulasidasa: His Mind and Art*. Edited by Nagendra. New Delhi: National Publishing House, 103–116.

Pandey, Ram L. (1972): *Tulsīdāsottara Hindī Rām-Sāhitya*. Allahabad: Abhinava Bharati.

Parikh, Dvarkadas (2009): *Do Sau Bāvan Vaiṣṇavan kī Vārtā: Tīn Janma kī Līlā Bhāvnā Vālīsahit*. Indore: Vaishnava Mitra Mandal Sarvananik Nyas.

Pastore, Rosina (2023): "Discussing Nyāya in Brajbhāṣā? On Six Categories of Reasoning in Brajvāsīdās's *Prabodhacandrodaya Nāṭaka*". In: *Puṣpikā: Proceedings of the 12th International Indology Graduate Research Symposium (Vienna, 2021)* (Puṣpikā – Tracing Ancient India Through Texts and Traditions: Contributions to Current Research in Indology, Vol. 6). Edited by Dominik Haas, Vitus Angermeier and Christian Ferstl. Heidelberg: Heidelberg Asian Studies Publishing, 149–168. Previous volumes in the series published by Oxbow Books.

Pastore, Rosina (forthcoming a): "Exploring the Relationship between Bhakti, Bhakta and Yoga in the *Prabodhacandrodaya Nāṭaka* by Brajvāsīdās". In: *Early Modern Literary Cultures in North India: Current Research*. Edited by Danuta Stasik and Imre Bangha. Oxford: Oxford University Press.

Pastore, Rosina (forthcoming b): "'For the Benefit of People': Bhuvadev Dube's (19th C.) Hindi Translation of Brajvāsīdās's Brajbhāṣā *Prabodhacandrodaya Nāṭaka* (1760)", In: *Proceedings of the 13th International Conference on Early Modern Literatures in North India (ICEMLNI)*. Edited by Hiroko Nagasaki, Kiyokazu Okita, and Heidi R. M. Pauwels. Heidelberg: Heidelberg Asian Studies Publishing.

Pauwels, Heidi R. M. (1996): *Kṛṣṇa's Round Dance Reconsidered: Harirām Vyās's Hindi Rās-pañcādhyāyī*. Richmond, Surrey: Curzon.

Pédraglio, Armelle (1974): *Le Prabodhacandrodaya de Kṛṣṇamiśra : un drame allégorique sanskrit*. Paris: E. de Boccard.

Pellegrini, Gianni (2015): "'Old is Gold!' Madhusūdana Sarasvatī's Way of Referring to Earlier Textual Tradition". *Journal of Indian Philosophy* 43.2: 277–334.

Pellegrini, Gianni (2018): ". . . And What about the *Vedānta Paribhāṣā*'s *Paribhāṣātva*? Some Hypotheses on the Use of *Paribhāṣā*s in Later Advaita Vedānta". *Asiatische Studien – Études Asiatiques* 72.2: 597–645.

Peterson, Jonathan R. (2020a): *Pluralizing the Non-dual: Multilingual Perspectives on Advaita Vedānta, 1560–1847*. Special issue of *Journal of Indian Philosophy* 48.1.

Peterson, Jonathan R. (2020b): "Pluralizing the Non-dual: Multilingual Perspectives on Advaita Vedānta, 1560–1847" [introductory paper to special issue of same name]. *Journal of Indian Philosophy* 48.1: 1–7.

Phillips, Robert (2014): "The Urdu-Language *Khushtar Ramayan*: Verbal- and Visual-Narrative Repertoires and 'Sense of Place'". *South Asia: Journal of South Asian Studies* 37.3: 454–473.

Pillai, Sohini S. (2021): "Blessed Beginnings: Invoking Viṣṇu, Kṛṣṇa, and Rāma in Two Regional Mahābhāratas". In: *Many Mahābhāratas*. Edited by Nell Shapiro Hawley and Sohini Sarah Pillai. Albany: State University of New York Press, 257–275.

Pinch, William R. (2006): *Warrior Ascetics and Indian Empires*. Cambridge: Cambridge University Press.

Plevoets, Bie / Van Cleempoel, Koenraad (2011): "Adaptive Reuse as a Strategy Towards Conservation of Cultural Heritage: A Literature Review". *Structural Studies, Repairs and Maintenance of Heritage Architecture XII* 118.12: 155–163.

Polizzotti, Mark (2018): "L'art de la traduction". *Aeon*, 9 October, https://tinyurl.com/fe6hjb62 (15/03/2022).

Pollock, Sheldon (1998): "Bhoja's *Śṛṅgāraprakāśa* and the Problem of *Rasa*: A Historical Introduction and Annotated Translation". *Asiatische Studien – Études Asiatiques* 52.1: 117–192.

Pollock, Sheldon (2011): "The Language of Science in Early Modern India". In: *Forms of Knowledge in Early Modern Asia*. Durham, NC: Duke University Press, 19–48.

Potter, Karl H. (ed.) (1970–2019): *Bibliography of Indian Philosophies*. Vol. 1 of *Encyclopedia of Indian Philosophies*. Delhi: Motilal Banarsidass. Reproduced at https://faculty.washington.edu/kpotter/ (no longer active, last accessed 28/09/2021).

Potter, Karl H. (1981): *Encyclopedia of Indian Philosophies: Advaita Vedānta up to Śaṃkara and His Pupils*. Princeton, NJ: Princeton University Press.

Potter, Karl H. (1999): *Presuppositions of India's Philosophies*. 2nd repr. Delhi: Motilal Banarsidass. 1st edn published in 1991.

Prasad, Jagannath "Bhānu" (1938): *Chandaḥ Prabhākar*. Bilaspur: Jagannath Press.

Rajpurohit, Dalpat (2018): "Making the War Come Alive: Ḍiṅgal Poetry and Padmakar's *Himmatbahādurvirudāvalī*". In: *Text and Tradition in Early Modern North India*. Edited by Tyler Walker Williams, Anshu Malhotra, and John Stratton Hawley. Delhi: Oxford University Press, 282–298.

Rajpurohit, Dalpat (2019): "Vaiṣṇava Models for Nirguṇa Devotion". *Journal of Vaishnava Studies* 28.1: 157–171.

Ram, Kanshi (1995): *Integral Non-Dualism: A Critical Exposition of Vijñānabhikṣu's System of Philosophy*. Delhi: Motilal Banarsidass.

Ramanujan, A. K. (1989): "Telling Tales". *Daedalus* 118.4: 238–261.

Ramanujan, A. K. (1993): "On Folk Mythologies and Purāṇas". In: *Purāṇa Perennis: Reciprocity and Transformation in Hindu and Jaina Texts*. Edited by Wendy Doniger. Albany: State University of New York Press, 101–120.

Ramanujan, A. K. (2004): "Three Hundred *Rāmāyaṇas*: Five Examples and Three Thoughts on Translation". In: *The Collected Essays of A. K. Ramanujan*. Edited by Vinay Dharwadker. Delhi: Oxford University Press, 131–160.

Ram-Prasad, Chakravarthi (2013): *Divine Self, Human Self: The Philosophy of Being in Two Gītā Commentaries*. New York: Bloomsbury.

Rao, Srinivasa (1998): *Perceptual Error: The Indian Theories*. Honolulu: University of Hawai'i Press.

Rao, Velcheru N. / Shulman, David (1998): *A Poem at the Right Moment: Remembered Verses from Premodern South India*. Berkeley: University of California Press.

Redington, James D. (2000): *The Grace of Lord Krishna: The Sixteen Verse-Treatises (Sodasagranthah) of Vallabhacharya*. New Delhi: Sri Satguru Publications.

Reinöhl, Uta (2016): *Grammaticalization and the Rise of Configurationality in Indo-Aryan*. Oxford: Oxford University Press.

Richman, Paula (1991): *Many Rāmāyaṇas: The Diversity of a Narrative Tradition in South Asia*. Berkeley: University of California Press.

Roebuck, Valerie (trans.) (2000): *The Upaniṣads*. eBook. New Delhi; London: Penguin Books.
Sastri, Sitarama (trans.) (1905): *The Isa, Kena and Mundaka Upanishads, and Sri Sankara's Commentary*. Madras: G. A. Natesan.
Saxena, Baburam (1938): *Evolution of Awadhi (A Branch of Hindi)*. Delhi: Motilal Banarsidass.
Schomer, Karine (1987): "The *Dohā* as a Vehicle of Sant Teachings". In: *The Sants: Studies in a Devotional Tradition of India*. Edited by Karine Schomer and W. H. McLeod. Delhi: Motilal Banarsidass, 61–90.
Schreiner, Peter (2013): *Viṣṇupurāṇa. Althergebrachte Kunde über Viṣṇu. Aus dem Sanskrit übersetzt und herausgegeben*. Berlin: Verlag der Weltreligionen.
Schwarz, Jason (2017): "Parabrahman Among the Yogins". *International Journal of Hindu Studies* 21.3: 345–389.
Semgar, Shivsingh (1926): *Śivsiṃh Saroj*. Edited by Rupnarayan Pandey. 7th edn. Lucknow: Naval Kishore Press. 1st edn published in 1878.
Sharma, Devendra (2020): "Community, Artistry, and Storytelling in the Cultural Confluence of Nautanki and Ramlila". *Asian Theatre Journal* 37.1: 107–132.
Sheridan, Daniel (1986): *The Advaitic Theism of the Bhāgavata Purāṇa*. Delhi: Motilal Banarsidass.
Shukla, Ramchandra (1942): *Hindī Sāhitya kā Itihās*. Rev. and enl. edn. Benares: Nagari Pracharini Sabha.
Shukla, Ramchandra (1973–1977): *Tulsī Granthāvalī*. 4 vols. Kashi: Nagari Pracharini Sabha.
Shukla, Ramchandra (1977): "Emotive Appeal in Tulasīdāsa". In: *Tulasidasa: His Mind and Art*. Edited by Nagendra. New Delhi: National Publishing House, 196–218.
Shulman, David (1991): "The Yogi's Human Self: Tāyumāṉavar in the Tamil Mystical Tradition". *Religion* 21.1: 51–72.
Shulman, David (1993): "From Author to Non-Author in Tamil Literary Legend". *Journal of the Institute of Asian Studies* 10.2: 1–23.
Shulman, David (2016): *Tamil: A Biography*. Cambridge; London: Belknap Press of Harvard University Press.
Singh, Bhagvati Prasad (1976): *Mahātmā Banādās: Jīvan aur Sāhitya*. Varanasi: Vishvavidyalaya Prakashan.
Singh, Bhagvati Prasad (1990): *Banādās Granthāvalī*. Gorakhpur: Avadh Sahitya Mandir.
Singh, Himanshu S. (1996): *Mahātmā Banādās Kā Nirguṇ Bhaktikāvya*. Varanasi: Nagari Pracarini Sabha.
Singh, Shukadev / Hess, Linda (2002): *The Bijak of Kabir*. Oxford; New York: Oxford University Press.
Singh, Naveen P. (2020): "*Prabodha Candrodaya Nāṭak* ke Hindī Rūpāntar aur Antaḥdhārmik Saṃvād". Unpublished PhD dissertation, Varanasi: Kashi Hindu Vishwavidyalaya.
Sinha, Maheshwari (1964): *The Historical Development of Mediaeval Hindi Prosody (Rāmānanda-Keśava 1400–1600 A.D.)*. Bhagalpur: Bhagalpur University.
Sitaram, Lala (1926): *Selections from Hindī Literature*. Calcutta: University of Calcutta Press.
Skinner, Quentin (2002): *Visions of Politics*. Vol. 1. Cambridge: Cambridge University Press.
Smith, Frederick M. (1998): "*Nirodha* and the *Nirodhalakṣaṇa* of Vallabhācārya". *Journal of Indian Philosophy* 26.6: 489–551.
Smith, Frederick M. (2004): "Vedic and Devotional Waters: The *Jalabheda* of Vallabhācārya". *International Journal of Hindu Studies* 8.1–3: 107–136.
Smith, Frederick M. (2011): "Predestination and Hierarchy: Vallabhācārya's Discourse on the Distinctions Between Blessed, Rule-Bound, Worldly, and Wayward Souls (the *Puṣṭipravāhamaryādābheda*)". *Journal of Indian Philosophy* 39.2: 173–227.

Smith, Frederick M. (2016): "The *Siddhāntamuktāvalī*: Vallabhācārya's 'Pearl Necklace of Correct Views' as an Epitome of his Vedānta". *Journal of Vaishnava Studies* 25.1: 123–151.
Snell, Rupert (1991): *Eighty Four Hymns of Hita Harivaṃśa: An Edition of the Caurāsī Pada*. Delhi: Motilal Banarsidass.
Söhnen-Thieme, Renate (1996): "The Ahalyā Story through the Ages". In: *Myth and Mythmaking: Continuous Evolution in Indian Tradition*. Edited by Julia Leslie. London: Curzon Press, 39–62.
Squarcini, Federico (2011): "*Pāṣaṇḍin, Vaitaṇḍika, Vedanindaka* and *Nāstika*: On Criticism, Dissenters and Polemics and the South Asian Struggle for the Semiotic Primacy of Veridiction". *Orientalia Suecana* 60: 101–115.
Srivastava, Surendra K. (2010): "Advaita Vedānta in Hindi". In: *Advaita Vedānta*. Edited by Ramachandran Balasubramaniam. Vol. 2, Part 2 of *History Of Science, Philosophy and Culture in Indian Civilization*. Edited by Debi Prasad Chattopadhyaya. 2nd repr. New Delhi: Munshiram Manoharlal, 582–599.
Stark, Silvia (1992): "Śvetāśvataropaniṣad 3.8 – 3.11 in the Light of Viśiṣṭādvaita (with Special Reference to the Interpretation of Vātsya Varadaguru)". *Wiener Zeitschrift Für Die Kunde Südasiens / Vienna Journal of South Asian Studies* 36: 187–194.
Stark, Ulrike (2004): "Hindi Publishing in the Heart of an Indo-Persian Cultural Metropolis: Lucknow's Newal Kishore Press (1858–1895)". In *India's Literary History: Essays on the Nineteenth Century*. Edited by Stuart H. Blackburn and Vasudha Dalmia. Delhi: Permanent Black, 251–279.
Stark, Ulrike (2007): *An Empire of Books: The Naval Kishore Press and the Diffusion of the Printed World in Colonial India*. Ranikhet: Permanent Black.
Stasik, Danuta (2009): *The Infinite Story: The Past and Present of the Rāmāyaṇas in Hindi*. New Delhi: Manohar Publishers.
Stasik, Danuta (2013): "Following the Path of One's Duty: Tulsīdās's *Rāmcaritmānas* as a Socio-Cultural Code". *Cracow Indological Studies* 15: 237–248.
Stoler Miller, Barbara (1982): "The Divine Duality of Rādhā and Krishna". In: *The Divine Consort: Rādhā and the Goddesses of India*. Edited by John Stratton Hawley and Donna Marie Wulff. Delhi: Motilal Banarsidass, 13–26.
Stoler Miller, Barbara (1996): *Yoga: Discipline of Freedom – The Yoga Sutra Attributed to Patanjali; A Translation of the Text, with Commentary, Introduction, and Glossary of Keywords*. Berkeley: University of California Press.
Strnad, Jaroslav (2013): *Morphology and Syntax of Old Hindi: Edition and Analysis of One Hundred Kabir Vani Poems from Rajasthan*. Leiden: Brill.
Sundaresan, Vidyasankar (2003): "Yoga in Śaṅkaran Advaita Vedānta". In: *Yoga: The Indian Tradition*. Edited by David Carpenter and Ian Whicher. New York: Routledge, 99–129.
Suthren Hirst, Jacqueline (1996): "Strategies of Interpretation: Śaṃkara's Commentary on *Bṛhadāraṇyakopaniṣad*". *Journal of the American Oriental Society* 116: 58–75.
Suthren Hirst, Jacqueline (2005): *Śaṃkara's Advaita Vedanta: A Way of Teaching*. London: Routledge.
Talbot, Cynthia (2009): "Becoming Turk the Rajput Way: Conversion and Identity in an Indian Warrior Narrative". *Modern Asian Studies* 43.1: 211–243.
Tam, Vivian W. / Hao, Jane J. (2019): "Adaptive Reuse in Sustainable Development". *International Journal of Construction Management* 19.6: 509–521.
Taylor, James (1893): *Prabodha Chandrodaya or Rise of the Moon of Intellect: A Spiritual Drama and Ātma Bodha or the Knowledge of the Self*. 2nd edn. Bombay: Tookaram Tatya. 1st edn published in 1811/1812.
Theodor, Ithamar (2016): "Introduction". In: *The "Fifth Veda" of Hinduism: Poetry, Philosophy and Devotion in the Bhāgavata Purāṇa*. London: I. B. Tauris, 1–8.

Thiel-Horstmann, Monika (1983): *Crossing the Ocean of Existence: Braj Bhāṣā Religious Poetry from Rajasthan*. Wiesbaden: Harrassowitz.

Thiel-Horstmann, Monika (1984): "The *Bhakti* Theology of the Dādūpanthī Sundardās". *Indologica Taurinensia* 12: 263–279.

Thiel-Horstmann, Monika (1989): "Dādūpanthī Sermons". In: *Living Texts from India*. Edited by Richard Barz and Monika Thiel-Horstmann. Wiesbaden: Harrassowitz, 141–183.

Timalsina, Staneshwar (2009): *Consciousness in Indian Philosophy: The Advaita Doctrine of "Awareness Only"*. London: Routledge.

Tiwari, Gopinath (1959): *Bhāratendukālīn Nāṭak Sāhitya (1850–1900)*. Allahabad: Hindi Bhavan.

Torella, Raffaele (2011): *The Philosophical Traditions of India: An Appraisal*. Varanasi: Indica Books.

Tripathi, R. K. (1977): "Tulasī's Concept of Bhakti". In: *Tulasidasa: His Mind and Art*. Edited by Nagendra. New Delhi: National Publishing House, 125–139.

Truschke, Audrey (2011): "The Mughal *Book of War*: A Persian Translation of the Sanskrit *Mahabharata*". *Comparative Studies of South Asia, Africa and the Middle East* 31.2: 506–520.

Tubb, Gary (2014): "Poetry and Play in Kavikarṇapūra's Play Within the Play". In: *Innovations and Turning Points: Toward a History of Kāvya Literature*. Edited by Yigal Bronner, David Shulman, and Gary Tubb. Oxford: Oxford University Press, 691–713.

Uskokov, Aleksandar (2018): "The Long and Short of It: Mahā-vākya from Mīmāṁsā to Jīva Gosvāmin, from the Veda to the Bhāgavata". *Journal of Hindu Studies* 11.1: 38–52.

Vatsyayan, Kapila (2007): *Traditional Indian Theatre: Multiple Streams*. 1st repr. of the 2005 rev. edn. New Delhi: National Book Trust. 1st edn published in 1980.

Vaudeville, Charlotte (1959): *Etude sur les sources et la composition du Rāmāyaṇa de Tulsī-dās*. 2 vols. Pondicherry: Institut Français d'Indologie.

Vaudeville, Charlotte (1974) : *Kabīr*. Oxford: Clarendon Press.

Vaudeville, Charlotte (2001): *A Weaver Named Kabir: Selected Verses with a Detailed Biographical and Historical Introduction*. 2nd edn. New Delhi: Oxford University Press.

Venkatkrishnan, Anand (2015): "Mīmāṃsā, Vedānta, and the Bhakti Movement". Doctoral dissertation, New York, Columbia University.

Venkatkrishnan, Anand (2018): "*The River of Ambrosia*: An Alternative Commentarial Tradition of the *Bhāgavata Purāṇa*". *Journal of Hindu Studies* 11.1: 53–66.

Venkatkrishnan, Anand (2020): "Philosophy from the Bottom Up: Eknāth's Vernacular Advaita". *Journal of Indian Philosophy* 48.1: 9–21.

Vidyāraṇya (1954): *Panchadasi: A Treatise on Advaita Metaphysics*. Translated by Hari Prasad Shastri. London: Shanti Sadan.

Viethsen, Andreas (2009): "The Reasons for the Descent of Kṛṣṇa in the Harivaṃśa". In: *Parallels and Comparisons: Proceedings of the Fourth Dubrovnik International Conference on the Sanskrit Epics and Puranas, September 2005*. Edited by Petteri Koskikallio and Mislav Jezic. Zagreb: Croatian Academy of Sciences and Arts, 221–234.

Vireswarananda, Swami / Adidevananda, Swami (2003): *Brahma-Sūtras; Śrī-Bhāṣya: With Text, English Rendering, Comments According to Śrī Bhāṣya of Śrī Rāmānuja, and Index*. 5th impr. Kolkata: Advaita Ashrama. First published in 1978.

Warder, A. K. (1992): *Indian Kāvya Literature*. Vol. 6. Delhi: Motilal Banarsidass.

Whicher, Ian (1997): "The Mind (Citta): Its Nature, Structure and Functioning in Classical Yoga". *Nagoya Studies in Indian Culture and Buddhism: Saṃbhāṣa* 18: 35–62.

Williams, Richard D. (2020a): "Dreams, Songs and Letters: Sectarian Networks and Musical Archives in Eighteenth-Century North India". *Indian Economic & Social History Review* 57.4: 583–604.

Williams, Richard D. (2020b): "Sounding Out the Divine: Musical Practice as Theology in Samāj Gāyan". In: *The Oxford History of Hinduism: Hindu Practice*. Edited by Gavin Flood. Oxford: Oxford University Press, 342–361.
Williams, Tyler W. (2019a): "Notes of Exchange: Scribal Practices and Vernacular Religious Scholarship in Early Modern North India". *Manuscript Studies* 3.2: 5–45.
Williams, Tyler W. (2019b): "'Publishing' and Publics in a World Without Print: Vernacular Manuscripts in Early Modern India". *Manuscript Studies* 4.1: 146–168.
Wilson, Horace Hayman (1868): *The Vishnu Purana: A System of Hindu Mythology and Tradition*. Vol. 4. Edited by Fitzedward Hall. London: Trübner.
Wilson, Horace Hayman (1958): *Religious Sects of the Hindus*. 2nd edn. Edited by Reinhold Rost. Calcutta: Susil Gupta. 1st edn published in 1862.
Wulff, Donna Marie (1984): *Drama as a Mode of Religious Realization: The Vidagdhamādhava of Rūpa Gosvāmī*. Chico, CA: Scholars Press.
Yardi, M. R. (1994): *The Rāmāyaṇa, its Origin and Growth: A Statistical Study*. Pune: Bhandarkar Oriental Research Institute.

Web Sources

Gopal Narayan Library. (n.d.): http://bharatpuralibrary.bih.nic.in/index.html (30/05/2023).
Modi, Narendra. (2014): "PM Recalls Swami Vivekananda's Story of a Lion & a Sheep". Narendra Modi YouTube channel. https://www.youtube.com/watch?v=_djE-PpcAmg (28/09/2021).
National Mission for Manuscripts. (n.d.): Ministry of Culture, Government of India. https://www.namami.gov.in/ (28/05/2023).

Index

https://doi.org/10.1515/9783111063836-017